P9-CBR-960

OUR ONCE AND FUTURE PLANET

Our Once and Future Planet

RESTORING THE WORLD IN THE CLIMATE CHANGE CENTURY

PADDY WOODWORTH

The University of Chicago Press
Chicago and London

The University of Chicago Press, Chicago 60637
The University of Chicago Press, Ltd., London

Printed in the United States of America

22 21 20 19 18 17 16 15 14 13   1 2 3 4 5

ISBN-13: 978-0-226-90739-0 (cloth)
ISBN-13: 978-0-226-08146-5 (e-book)

DOI: 10.7208/chicago/9780226081465.001.0001

Library of Congress Cataloging-in-Publication Data
Woodworth, Paddy, 1951–
Our once and future planet : restoring the world in the climate change century / Paddy Woodworth.
pages cm
Includes bibliographical references and index.
ISBN 978-0-226-90739-0 (cloth : alk. paper)—
ISBN 978-0-226-08146-5 (e-book) 1. Restoration ecology.
2. Environmental quality. 3. Climatic changes.
4. Global environmental change. I. Title.
QH541.15.R45W664 2013
577—dc23

2013016605

Supplementary material relating to the book's contents is available online for readers at http://www.press.uchicago.edu/sites/woodworth/.

∞ This paper meets the requirements of ANSI/NISO Z39.48-1992 (Permanence of Paper).

PADDY WOODWORTH was a staff journalist at the *Irish Times* from 1988 to 2002 and is the author of *Dirty War, Clean Hands* and *The Basque Country*.

I would like to acknowledge the generous support of these organizations for aspects of the research for this book:

BORD NA MÓNA

An Roinn
Ealaíon, Oidhreachta agus Gaeltachta
Department of
Arts, Heritage and the Gaeltacht

woodlands
OF IRELAND
Coillearnacha Dúchasacha

iwp International Writing Program

In memory of my brother, David, who left us much too early, with several books unfinished, and with far too many unfinished conversations about practically everything.

And to my beloved nieces and nephews and grand-nephews and grandnieces on all sides of our extended and chosen families, that they and their offspring may live in a restored world.

Contents

Preface *ix*

1 Five Plots, Five Prairies, Reflooding a Delta *1*

2 The Cranes Are Flying—Again *12*

3 From Necedah to St. Louis via Zaragoza: A Restoration Learning Curve *21*

4 Greening the Rainbow Nation: Saving the World on a Single Budget? *49*

5 Awkward Questions from the Windy City: Why Restore? To What? For Whom? *87*

6 Keeping Nature Out? Restoring the Cultural Landscape of the Cinque Terre *133*

7 The Last of the Woods Laid Low? Fragile Green Shoots in Irish Forests *166*

8 Future Shock: "Novel Ecosystems" and Climate Change Shake Restoration's Foundations *196*

9 Dreamtime in Gondwanaland *214*

10 Restoration on a Grand Scale: Finding a Home for 350,000 Species *256*

11 Killing for Conservation: The Grim Precondition for Restoration in New Zealand *287*

12 The Mayan Men (and Women) Who Can (Re)Make the Rain Forest *327*

13 Making the Black Deserts Bloom: Bog Restoration on the Brink of Extinction *351*

14 Walk Like a Chameleon: Three Trends, One Story *384*

15 Conclusions: Why Restore? *434*

Acknowledgments *439* *Glossary* *443*

Notes *447* *Bibliography* *487* *Index* *501*

Preface

We are standing in Rochester Cemetery in Iowa, but we are not being invited to look at its gravestones.

No, we are here to look between them, for new life, or rather, for living traces of an ancient and almost extinct world. You could say that we are looking for the past in the present, and for clues towards possible ecological futures. But I haven't even begun to grasp what any of that means, not yet.

Two local environmentalists are poking around between the graves, trying to find us plants with lyrical names like rattlesnake master, pale purple coneflower, and rough blazing star. Cemeteries, along with railway cuttings, are the last refuges of the once abundant flora of the tallgrass prairie. The intensive agriculture practiced by settlers in Iowa, the "most altered state in America," has come close to erasing this ecosystem in less than two centuries.

It is late autumn, and most of the plants can only show us seed heads at best. Not very lyrical at all, hardly thrilling stuff to the uninitiated or uninterested. Some of our group wander off to decipher inscriptions on worn tombstones or chat about other things.

The International Writing Program at the University of Iowa in 2003 was a melting pot of ingredients from across the globe. We bubbled together happily enough but retained distinctive flavors: poets from Vietnam and Mongolia, a novelist from Chile, a Polish philosopher, a Chinese screenwriter. Not all of us shared the passion for environmental writing that had driven the program director, poet Christopher Merrill, to organize this prairie weekend.

But two at least of us did. Gregory Norminton, a young English novelist, greeted me over coffee most mornings at the Iowa House Hotel with baleful comments about the latest ecological folly, as exposed on the pages of the *New York Times*. He saw very little hope for our overcrowded and degraded planet, and he felt its sickness in his bones, in his gut and in his heart. I tried to resist his pessimism, but did not know enough to offer a more positive reading of the world.

Gregory was up to speed on these things, and I was not. He grasped basic ecological concepts and could discuss debates in the environmental movement with some fluency. He knew his trees, even in America, and I did not, though I probably had the edge on him as a birder—but that does not say much for either of our skills in that department.

Nevertheless, I had been hoping that birds would provide the theme for my next book, though from a human rather than an ecological perspective. I had an idea about exploring the traces left on our cultures by charismatic migrating birds, like cranes, eagles and swallows, from cave paintings to popular music. Even this was a very new field for me. I had been invited into the Iowa program on the back of my first book, an account of terrorism and state terrorism in the Basque Country. Rather weary of writing about people who kill each other, I was looking to natural history for a more congenial subject.

The migration-and-culture project had indeed led me on a wonderful journey, following cranes on their annual journey from the cork oak savannas of Extremadura, Spain's most African landscape, to the melting snows of a Swedish spring. And it would soon lead me to the International Crane Foundation, not far from Iowa in Baraboo, Wisconsin. But by then I would have abandoned that book, and found the seeds of this one.

The night after our visit to the cemetery, we are all invited to participate in a strange little ritual. It seems the antithesis of anything environmentally healthy. There is an acrid smell of gasoline in the air, and soon the pungent stench of damp, scorched vegetation. A rank patch of motley plants, perhaps five hundred meters square and already collapsing before the advance of winter, has been marked out by a firebreak.

Would we like, our hosts Mark and Val Müller ask us, to help *burn their prairie*? We are standing on land that has been farmed for many decades, its natural diversity reduced to a single annual corn crop.

Several years earlier, the Müllers had cleared out the corn, plowed it one last time, and sowed native plant seeds, lovingly gathered locally from remnant prairie patches like Rochester Cemetery. But a prairie community needs the alchemy of periodic fire if it is to flourish: hence the unusual invitation. Gregory is the first to jump in, wielding a drip torch manfully through the flickering shadows. It had rained during the week, however, and the expected conflagration is denied us.

That was my introduction to the counterintuitive world of ecological restoration, where you burn a prairie to make it flourish, slaughter cute and furry mammals to save indigenous birds, and poison healthy trees to bring back native forests.

Working through those difficult issues would come later, however. For that night, despite the anticlimax with the prairie burn, Gregory and I were just fascinated by the notion that a prairie could be "restored" at all—that "natural" status could be returned to land claimed and cleared for human use. Like many people, we had grown up with the idea that, if we wanted to save the natural world, we had to preserve it from any human intervention. The idea that humans could actually participate in nature to our mutual benefit, that we could be the agents of recovery of degraded ecosystems was, we both dared to think, rather inspiring.

"I wonder if this kind of thing is happening in other places?" Gregory reflected out loud. "And if it is, wouldn't it make an interesting book to describe projects like this for general readers?"

My heart leapt and sank in the same instant. I suddenly knew that this was the book I wanted to write, but dammit, my new English friend had come up with it first. One is supposed to be ethical about these things. I kept my frustration to myself for a few days and then one night, over a beer or two, it boiled over. I asked Gregory if he really intended to follow up the restoration book he had talked about. "My dear chap," said Gregory, whose democratic principles have not erased a certain Oxbridge *hauteur*, "That's *nonfiction*. I only write fiction." So would he mind, I asked in some trepidation, if I tried to write it? "I'd be delighted," he said, and then, after a short pause, added mischievously, "as long, of course, as you say it was my idea in the preface."

I am delighted, eight years later, to honor this small promise. Neither of us had the slightest idea at the time whether ecological restoration was a phrase known only to a few nostalgic prairie lovers in

Iowa, or whether it was an idea reshaping new conservation thinking worldwide. Finding some of the answers has been a bigger and more stimulating challenge than I could have then imagined. That quest has taken me to places—and to concepts—I never knew existed, in the company of remarkable women and men. I hope I can convey some of the excitement of that discovery in the chapters that follow. Above all, I hope to offer some grounds, however fragile, for hope itself, in the face of the daily dose of bad news from the environmental front.

A Note on Structure

FOUR STRANDS, ONE STORY

This book is loosely structured and follows the narrative and timeline—more or less—of my personal exploration of restoration. It does not attempt to give a formal academic introduction to the subject. Several books, to which I have had recourse many times, and with which readers will become familiar, have done that very well already in recent years. But I have tried to offer the general reader points of access to all the main topics in the restoration conversation. I hope there is also some useful material here for those who are specialists in one aspect of the subject, or in one part of the world, but have not had much contact with other aspects or other regions. Restoration is place-specific everywhere, in my experience, but each place may also offer general insights not immediately obvious elsewhere. The chapters in this book fall into four main "strands":

Strand One, Traveling toward Restoration, (chapters 1–3) introduces some of the basic questions raised by restoration through an account of my own first encounters with restorationists and restoration practice and theory.

Strand Two, On the Ground, Around the World, visits restoration projects in different ecological, social and geographical contexts. Section 1 (chapters 4–6), "Restoring Natural Capital and Classic and Cultural Landscapes," focuses on restoration of natural capital in South

Africa, the restoration of a "classic" landscape in Chicago and of a cultural landscape in Italy.

Section 2 (chapters 9–12), "Restoration after Radical Changes," focuses on landscapes and cultures that have been drastically changed by human agency, posing radical challenges for restoration that demand radical solutions; in Australia, New Zealand, Costa Rica and Mexico.

Strand Three, Restoration Begins at Home, (chapters 7 and 13) explores the restoration of forests and bogs in my native Ireland. I was not aware of any restoration projects in Ireland when I began researching this book, but I soon found, to my surprise and pleasure, that some very ambitious Irish restorations were underway. This enabled me, or I hope it did, to engage more closely with restoration issues I had encountered internationally on intimately familiar ground

Strand Four, Pause for Thought, Time for Theory, (chapters 8 and 14) outlines the dynamic development of restoration thinking over the past eight years, often in response to climate change, and the promises—and the pitfalls—of proposed new approaches.

Chapter 15 attempts a brief synthesis, in **Four Strands, One Story.**

The chapters in both sections of strand two, and in strand three, can each be read as stand-alone narratives, but the whole book read in sequence has greater resonance around its central question: how successfully can we restore degraded ecosystems, and our own damaged relationship to our environment?

A note on acronyms: The Society for Ecological Restoration (SER) changed its name to the Society for Ecological Restoration International (SERI), and then back again to SER, during the writing of this book. For simplicity, I have used SER throughout. For further information on acronyms, see the index.

1 *Five Plots, Five Prairies, Reflooding a Delta*

In a nutshell, the function of the Arboretum [is] a reconstructed sample of old Wisconsin, to serve as a bench mark . . . in the long and laborious job of building a permanent and mutually beneficial relationship between civilized men [sic] and a civilized landscape. * ALDO LEOPOLD, *a founding father of US environmentalism, on the establishment of Madison Arboretum, which gave birth to several emblematic restoration experiments*

A Rubik's Cube is child's play compared to ecological restoration. So many factors, all interacting with each other. * JEB BARZEN, *chief ecologist at the International Crane Foundation, walking his five prairie plots in Wisconsin*

The relationship between conservation and poverty alleviation is the most important debate in environmentalism right now. * RICH BEILFUSS *of the International Crane Foundation, on a project to restore the Zambezi delta*

* *

This book will follow an eight-year journey into restoration, through a series of encounters with individuals and cultures, with species and ecosystems and landscapes, and with ideas in ferment. I found that the practice of ecological restoration and its related scientific discipline, restoration ecology, are not established fields with agreed or fixed basic principles. They are more like lively adolescents, buzzing with energy—and with vibrant contradictions. Restorationists are, after all, attempting to engage with a global environment whose inherent evolutionary shape-shifting tendencies are being unpredictably accelerated by climate change.

Nevertheless, the spark that the phrase "ecological restoration" ignited for me on that damp night at an unsuccessful prairie burn has become a steadily burning passion that continues to illuminate two hopeful prospects that I glimpsed on that occasion: first, that the natural world is considerably more resilient than I had thought, and that damaged and degraded ecosystems can rebuild a great deal of their complex webs of species, communities and ecological processes, if we give them half a chance to do so; and second, that human beings can assist in beneficially managing this restoration process, and that in so doing we may restore our own relationship to nature.

"Saving nature," then, might not just mean a last-fence stand to preserve shrinking islands of wilderness by keeping people out of them. Through restoration, we might escape from our locked dichotomy between the twin roles of destroyers and preservers and find a more rewarding middle way as facilitators of—and participants in—natural processes.

The stories this book uncovers are a welcome reminder that there is nothing inevitable about humanity's current starring role as the Bad Guy of Planet Earth. The obstacles are daunting, but sufficient scientific knowledge is available to turn us into the caretakers or stewards of the biosphere, of which we ourselves form an integral part. Restorationists say they have found a synthesis which can resolve a familiar dialogue of the deaf between:

- those who see the environment only as a resource to be managed, developed, and consumed, and
- those who see human intervention in the environment only as desecration and damage.

These are false alternatives, say restorationists: we can and should use the Earth, but we must learn to do so without using it up. Public awareness of this new scenario, and the political vision and will to implement it, are both still largely missing. This book seeks to bridge the gap between the promise of ecological restoration and the rapidly growing mainstream concerns about the environment. Of course, I doubt that I would have formulated my thoughts in quite that way on that night, but I think all those elements were there, at least in embryonic form, along with a feel-good factor that, to my own surprise, has survived the often grim developments in the human and natural worlds since then. But I still had no idea whether ecological restoration was only a fad peculiar to Iowa, and perhaps only applicable to prairies in any case. After my discussion with Norminton in Iowa City, my next task was to find out if restoration projects had been attempted elsewhere in the world. I was already due to visit the International Crane Foundation (ICF) in Baraboo, Wisconsin, a couple of weeks later. I had planned to spend most of my time in the library, researching references to cranes in literature and art.

The Feel-Good Factor Takes a Battering

I already knew firsthand from a meeting in Sweden that the ICF's charismatic cofounder, George Archibald, had an exceptionally single-minded dedication to the welfare of cranes worldwide. I had read vivid accounts of his conservation campaigns in Peter Mathiessen's enthralling account of the battle to save the world's fifteen species of crane, *The Birds of Heaven*.[1] But I had no reason to think that ecological restoration might play any role in his cherished cause. So I rather feared he might feel I was breaking the terms of his generous invitation to stay and study at the foundation when I told him on arrival that I was now planning to write about restoration, rather than about the cultural impacts made by his beloved big white birds. But he took my news as the most natural development in the world and led me straight off to meet the foundation's chief ecologist, Jeb Barzen. I found myself out in the field the next morning, at the receiving end of a crash course in some restoration basics. The feel-good factor took a bit of a battering over those few days, in the first sharp breezes over the reality on the ground, but that wind also braced me, and whetted my appetite for more.

Barzen is a stocky, reserved, and thoughtful man. He starts with a question: am I interested in habitat restoration, or ecological restoration? Frankly, I am not very sure. This is the first of many daunting moments—and they recur even now, eight years later—when I realize just how much I still have to learn, will always have to learn. He explains that habitat restoration is directed toward the particular needs of a threatened species—a wetland on the whooping crane's migration route, for example. But ecological restoration is much more ambitious, seeking to put back together all the parts of a damaged sys-

1. **The restoration fire goes around the world, but its purpose changes.** Alexander Kolotiy, production manager of Muraviovka Park in the Russian Far East, uses fire to create a firebreak. The instructor was Jeb Barzen of the International Crane Foundation in Wisconsin. Here in Russia, the purpose is to *limit* excessive wildfires caused by poor management of adjacent lands, not to foster regeneration, so Barzen's experience in prescribed burns for prairie restoration finds another flexible conservation application. (Photograph courtesy of Sergei Smirenski.)

tem, without special regard for any particular component. He takes a Rubik's Cube out of his pocket. "This thing," he says, "is child's play compared to ecological restoration. So many factors, all interacting with each other."

We walk on up through five plots of land he has marked out for a long-term prairie-restoration experiment. They are only a few hundred meters apart. Each has the same orientation to the sun on the gently undulating mid-Wisconsin landscape. Each has similar prairie origins, similar soils, and a similar agricultural history. Each has received the same restoration treatment in different years, between 1990 and 1996: the same prairie seed mixtures planted on bare ground in the fall. Each has subsequently been burned on similar rotations. Control of invasive species has been similarly limited on each site, and there has been no mowing or seed collecting. And yet the result is five radically different prairies, with different plants becoming dominant, and different community interactions, in each of them. He can attribute some of the differences to variations in weather conditions in the years they were planted, and to what ecologists describe as "stochastic" events—unpredictable but not entirely random—like a heavy early frost or the arrival one year of a very large migratory flock of seed-eating birds just after sowing. But there are other differences for which he simply has no explanation, though he will continue experimenting in an attempt to deepen his understanding of how a young prairie behaves. He has tracked unpredictability over seventeen growing seasons and witnessed a bewildering—but one might also say exuberant—turnover of dominant species.

Meanwhile, some of the common factors are disturbing: two key native prairies grasses have peaked on four of the plots and are now in (apparently) ongoing decline, while the alien Kentucky bluegrass is still expanding on all plots, regardless of its initial level of establishment. The five plots may, of course, become more homogenous over time, but he thinks it is just as likely that they may diverge even further. Nevertheless, that very unpredictability fosters biodiversity, because different plant populations become dominant from year to year. The lesson he draws from this experience is that, even though the conditions he is operating under here are enviably stable and controlled compared to many restoration projects, there is still simply no way, with current knowledge levels, that the outcome of a restoration can be precisely predicted. "Research findings," he wrote in 2008, setting out an often-ignored agenda for good restoration, "need replicates among many years to be fully understood or validated."[2]

Wild nature is not a laboratory, there are far too many variables and interrelationships, many of which we are unaware of and many of which we cannot control in any case. This is not gardening, he continues, using an analogy I will hear many times over the next few years. Perhaps the difference with gardening, I speculate, is not only due to our knowledge gaps, but because, by definition, restoring an ecosystem to its natural state must surely mean "letting go" at some point, not attempting to control the outcome. "Yes and no," he responds. "We are not gardening, in the sense that we want to produce a particular configuration of flowers each year. But we are trying to restore prairie here. We do have a general ecosystem in mind, and we would consider this particular restoration a failure if it produced a forest, or a swamp."

Looking back, I can see how on this early outing I was already running up against many of the concepts that would successively puzzle, preoccupy, and stimulate me, and then puzzle me again, over the course of next few years. I did not know the proper terms to describe them yet, though I soon would. What "historical reference system" were we restoring to, in any given case? What "trajectory" might we expect the restoration to follow? Could we nudge the system toward the trajectory the reference system would have followed had no degradation occurred? Was "natural succession"—whereby a series of different plant communities dominate a system, each one then creating new conditions that will bring about its own replacement by its successor—going to determine that trajectory after a particular point in the restoration? And if so, would it lead to a "climax" system, a stable state in which communities of plants and animals would remain in balance with each other for the foreseeable future? Or would the system continue to change perpetually over time, in constant flux, though within a recognizable range of variation? And underlying these technical questions lay deeper conundrums of an existential and philosophical kind that began to take vague form in my mind. What are we talking about when we talk about *nature*? Are we part of it, or do we stand outside it? Can restoration tell us anything about these issues?

Barzen listened to my stumbling first attempts to articulate these questions and pointed me in various directions for answers. He mentioned Aldo Leopold, whose *A Sand County Almanac* I had already found in my room at the ICF guesthouse. Like most Europeans, I suspect, I had never heard of Leopold, but I found he was regarded as one of the founding fathers of the US environmental movement. His famous "shack" where many of the observations he records in the almanac were made is quite close to the ICF.

Keeping Every Cog and Wheel

A quotation from Leopold became one of the central tenets of the American conservation movement, and a very popular maxim among restorationists: "To keep every cog and wheel is the first precaution of intelligent tinkering."[3] The analogy between the dismantled parts of a car engine, neatly laid out on a white sheet prior to a rebuild, and the elements of an ecosystem, fizzing and humming with multidirectional interactions from genetic to population levels, now seems charmingly archaic. Would that nature worked as predictably as a machine! (Though of course it would lose a great deal of its magic if it did). And yet Leopold's phrase is a good metaphor in broad strokes. You must have at your disposal all the biotic elements (living species) and abiotic elements (topography, minerals, climate, etc.) of a system if you are to have any chance of a fully authentic restoration.

It took a little time for my European readers' lenses to see beyond Leopold's rather folksy Midwestern ironies and find in his work a prescient and still painfully relevant critique of American environmental policies—or the lack of them—in the first half of the twentieth century. His writing is rooted in his wide experience with the US Forest Service, where he pioneered the creation of "wilderness areas," before taking the up the first-ever American university chair of game management at the University of Wisconsin–Madison in 1933.[4] In his writings, Leopold flies kites of sophisticated yet easily accessible environmental theory, but their strings are anchored in the sure hands of a man who knows how to dig the soil and how to shoot a deer. He learns an essential lesson of conservation from the dying eyes of a wolf he has just hunted down.[5] At the core of his work is the need for humanity to adopt a new "land ethic," based on the recognition that *Homo sapiens* is just one member of a vast ecological community whose interrelationships we ignore at our material and spiritual peril. At his best—*Marshland Elegy*—he is a writer of great lyrical power and celebrates the abundance of life in every season; he robustly defends and expresses a binocular vision of nature, in which imaginative insights have as sharp and significant a focus as scientific analysis.[6] But he also takes a rather pessimistic view of a central paradox of environmental management in our times: "all conservation of wildness is self-defeating, for to cherish we must see and fondle, and when enough [people] have seen and fondled, there is no wildness left to cherish."[7]

It is another paradox that Leopold's work has been, and continues to be, mined for insights by American restorationists,[8] though his own bias was towards preservation of pristine "wilderness." Although he often uses metaphors that could be borrowed comfortably by contemporary advocates of the restoration of natural capital,[9] he himself believed that "wilderness is a resource which can shrink but not grow . . . the creation of new wilderness in the full sense is impossible."[10] Nevertheless, Leopold was closely associated with the origins of a continuing project that has iconic status in American restoration history. Less than a year after his appointment at Madison, he helped launch the university's arboretum, which he acknowledged in his dedication address would be much more than a tree collection, and have a "new and different" central project: "to reconstruct, primarily for the use of the University, a sample of the original Wisconsin, a sample of what Dane County looked like when our ancestors arrived here in the 1840s."[11]

The argument he used to justify such an almost unprecedented endeavor,[12] at least in public, was not restoration, per se, but primarily was based on the social benefits of research into ecosystem dynamics. He gave his listeners a disturbingly vivid inventory of the damage wrought by bad land management on the state in less than a century; presumably the fact that the Dust Bowl had reached Wisconsin in the form of devastating storms only weeks earlier made them pay rather closer attention than they might otherwise have done. He suggested that studying the "original Wisconsin" would assist in "preserving an environment fit to support citizens. In a nutshell," he concluded, "the function of the Arboretum [is] a reconstructed sample of old Wisconsin, to serve as bench mark . . . in the long and laborious job of building a permanent and mutually beneficial relationship between civilized men [sic] and a civilized landscape."[13]

Two concepts that would become key to restoration thinking were there in seminal form in this address, though Leopold never developed them explicitly as such in his other writings: the idea of a benchmark from the past as a target for the future (the "historical reference system"), and the hope that rebuilding such benchmarks might transform the problematic relationship we humans have with the rest of the natural world.

From Finding a Crane to Restoring a Region

On Barzen's advice, I visit the arboretum shortly afterward, but before I leave the ICF I still have one big question pending: is restoration a quaint practice restricted to prairies and the Midwest, or is it a movement with roots in other ecosystems and other continents? By way of an answer, Barzen brings me to an ICF staff birthday party, which happens to be the last chance to meet the foundation's then Africa director, Rich Beilfuss, before he caught a flight back to Mozambique. And there, in the unlikely context of copious helpings of sponge cake, lemonade, and a little wine, I learn about a restoration enterprise that demonstrated vast ecological (and social) ambition in a setting radically different from the almost extinguished grasslands of Iowa, Illinois, and Wisconsin.

As Beilfuss tells it, the Lower Zambezi Valley and Delta Program started with an initiative to conserve a single charismatic species, the wattled crane, but led to a daring proposal to reflood one of the world's largest deltas. The program aimed at restoring not only thousands of hectares of wetlands but also the livelihoods of the desperately poor people who live in this region, as they emerged from decades of anticolonial and internecine wars. Beilfuss described how the search for a suspected population of this threatened bird led to the rediscovery of a web of precious ecosystems, all in deep trouble. This region had been avoided and then forgotten by conservation organizations during the bloody turmoil of Mozambique's decade of liberation war against Portuguese colonialism, and its subsequent fifteen years of civil conflict. These wars had had disastrous impacts both on people and wildlife in the delta. In obscure hamlets, atrocity and counteratrocity were the order of the day; meanwhile, hungry fighters on all sides, well supplied with automatic weapons if with nothing else, extirpated the black rhinoceros from the region and reduced the world's largest population of Cape buffalo to a few dozen animals.

A ceasefire and remarkably successful peace process in the 1990s revealed a different kind of problem in the region, and its long-term impacts were also devastating. The huge dams built upstream on the Zambezi in the early 1970s, notably the Cahorra Bassa dam, had dried out the delta, reducing its myriad watercourses, ponds, and swamps to a few sluggishly flowing channels. Beilfuss knew what damage this was doing to bird life. But he also understood that conservation per se is rarely a preoccupation of any human society, let alone one with the problems of rural Mozambique. So he set out with colleagues from the Museum of Natural History–Mozambique to find out how this massive hydrological shift had affected local people. They found that livelihoods in shrimping were disappearing fast, and that traditional semiaquatic agriculture was threatened by loss of nutrients once supplied by annual floods, and by invasions of dryland shrub vegetation unfamiliar to delta farmers. There were other impacts, less predictable to an outsider: the mortality rates of women and children in the delta was soaring. Why? Because women, accompanied by their offspring, traditionally spend time at the riverside washing clothes and fetching water. And they were falling victim to the only two large animals that had survived the war well, and were now infesting reducing numbers of channels in every greater numbers—crocodiles, and hippopotami.

Through painstaking investigation and lobbying, Beilfuss and the museum staff have built a coalition in support of a very radical proposal: the restoration of seasonal flooding through periodically reopening the flood gates of the Cahorra Bassa. This coalition includes local tribal leaders; the ruling Frelimo party and the main opposition, made up of the former rebels of the Renamo movement; and a number of Mozambican institutions. It's an impressive achievement, but I could not help wondering whether the broader coalition of ideas—between those who want to restore ecosystems, and those who want to lift people out of poverty, could really prosper.

"The relationship between conservation and poverty alleviation is the most important debate in environmentalism right now," Rich Beilfuss told me when we discussed these issues a year later on a phone line between Ireland and Mozambique. "Some people feel community-based conservation has gone too far in stressing poverty alleviation. They say such projects no longer have any conservation value. Then some people involved with communities say: "You conservationists never really consulted communities from the bottom up anyway, you just went in and told them what to do." People can argue both sides. I think it is clear that very few programs so far have really effectively linked poverty alleviation and conservation interests. Given all that, I think we are seen as making a pretty credible effort at involving local people and defending conservation interests."[14]

This complex initiative is still in process at the time of writing, but the simple existence of such a project in 2003 was enough to convince me that ecological restoration did indeed merit exploration as a global movement with innovative and socially aware responses to environmental crises.[15] Sadly, I never did get to check out this particular project on the ground, but happily there was an immediate opportunity to observe another kind of crane-related restoration much closer to hand.

2 *The Cranes Are Flying—Again*

I stayed steady, the crane elegantly flipped up and rolled on its wingtip, bled off all its speed and settled into third position in the formation behind me. * JOE DUFF, *Operation Migration cofounder, CEO, and ultralight aircraft lead pilot*

* *

We stop talking, as instructed, when we are within two hundred meters of the hide, and advance in silence. We pick our steps carefully in the early dusk and ease our way into an improvised bunker made of hay bales. Through deep, narrow slats in the straw, we can now see the holding pen, another hundred yards beyond us, and the tall, stately birds moving within it; but they cannot see us. The success of this unique restoration project depends, among many other things, on minimum human contact with these young cranes. Whooping cranes are among the very rarest birds in North America, and sixteen of them are foraging in the protected area. They are still feeding steadily after their short but exhausting flight ten hours earlier.

"This is the same number as were left wild in the whole world in 1941," Jeff Huxmann comments in a barely audible whisper.[1] He still sounds a little in awe of what we are seeing, despite his three years' experience as cameraman with the project.

These birds had been hatched in incubators only six months earlier, but they already stand almost as tall as the parents they have never seen: five feet from crown to ground. They still lack the pristine white plumage of adult whoopers, and are a motley lot, splashed and streaked with irregular patches of rusty brown. At dawn on the day in question, fifteen of them had flown twenty-three miles behind three ultralight aircraft. Their upbringing has been carefully planned to make them regard these machines, and their pilots, as surrogate parents. The pilots wear white "crane suits" with only a small visor for their eyes. They look a bit like circus clowns, but this project is a very serious business. If it works, it will greatly enhance the still slim survival prospects of this species, which is almost as emblematic of America's wild places as the bald eagle.

The goal is to use the ultralights to restore a migratory flock of whooping cranes in the eastern half of North America. This would be the first such flock in this region since the end of the nineteenth century, and the second in the whole world today. The ultralights—fondly known as "trikes" because they have three landing wheels—were linked to these birds even before they were born. While they were still in their eggs, the embryonic chicks were played audio tapes of ultralight engines. When they hatched, they were fed by crane puppets. Experience has shown that if young cranes have any direct contact with people, they become too damned friendly with humans for their own good and start dropping in on golf courses and school yards. And so, once they were released into outdoor pens, the first "creatures" they saw were ultralights and their crane-suited pilots. Thus imprinted, they learned to follow the ultralights everywhere, and finally flapped clumsily after them into the air for their first flight.

They are now just two days into a journey in which their human mentors will teach them a migration route of 1,191 miles, from the Midwest to Florida. There are many difficult days ahead, and huge challenges have been overcome just to reach this point.

Aircraft-assisted bird migration was pioneered by Bill Lishman, the sculptor/pilot whose adventures in training geese to migrate was the subject of the Oscar-nominated movie *Fly Away Home*. Lishman was also the first chairman of Operation Migration, Inc., which is in charge of the cranes' journey south. This organization is the most publicly visible link in the Whooping Crane Eastern Partnership, the umbrella group of federal, state, and private bodies, including the International Crane Foundation, which is working toward reintroducing the cranes east of the Mississippi.

Joe Duff leads the team of three pilot-parents, which left Necedah

National Wildlife Refuge in central Wisconsin two days earlier. I arrive there at 6:30 a.m., dispatched by George Archibald in the hope of seeing the young cranes begin the first stage of their journey. I am lucky. Other "craniacs" have been keeping the same early-morning vigil for the previous seven days, but bad wind conditions had made departure too big a gamble. Flying with cranes is a very tricky business, which ideally requires either an absolutely still day or a rock-steady and very moderate tailwind. Anything else greatly increases the chances of a collision between birds and pilots, potentially fatal to both. Operation Migration's representative on Earth at the time of my journey with the birds, Heather Ray, calculates that $100,000 is invested in each young bird, but the financial risk is the least of her anxieties. These cranes have a biological value which cannot be calculated in hard cash. Previous losses—to power lines, illness, and natural predators—have had an impact not unlike bereavement among the tough-minded but softhearted crew.

Timeless Wilderness, Combustion Engines

I climb the wooden tower overlooking the wetlands in Necedah where the young whoopers had spent the summer. Dimly visible in the pale predawn light, a group of cranes is roosting in shallow water off a sand spit. These are sandhill cranes, cousins to the whooping cranes, and as abundant today as the latter are rare. In swathes of mist, tinged the faintest pink by the first rays of sunlight, lovely as a Japanese painting, the sandhills begin taking delicate steps. A trumpeter swan utters its haunting call nearby.

This timeless image of wilderness is abruptly broken by the sound of a combustion engine. Just above the skyline, a tiny flying machine comes into view. It might have been designed by Leonardo da Vinci, being little more than a seat suspended from a primitive wing, powered by a single propeller. There are seven small specks spread out beyond one wingtip, which gradually take the form of whooping cranes. Another trike comes into view, with five cranes in tow, and a third with three. (The sixteenth bird was recovering from an operation and would travel the first stages of the migration in a crate). They pass overhead and quickly out of sight. The long journey south seems well under way at last. But no: a moment later it does not look so good.

Even on the best of days, like this one, things can go wrong, as the sudden drone of a returning trike reminds us. Five of the young birds simply decide they do not want to leave home, and turn back. After making several circles to try to change their minds, the pilot heads back south, but only one crane follows him.

The other four fly back toward their pen, making white and inky-black pen-strokes with their wingtips against the sky. They will have to be captured, boxed, and driven down to the next stop-off in the hope that they will fly with the flock the next day. The remaining birds and their minders have a long slog ahead of them, greatly lengthened by the constraints of the less than ideal partnership between plane and crane.

Under normal conditions, a young crane will learn to migrate by following its biological parents and could cover the distance to Florida in less than a week. Yet the journey with the ultralights was projected to take at least five times that long, and not just because of the many nonflying days dictated by unsuitable weather. In natural migration, the cranes soar on the rising columns of warm air known as thermals and then glide down from a height, covering great distances without having to expend too much energy flapping their wings. But thermals present nightmare flying conditions for aircraft-assisted migration, so these birds have to flap every inch of the way, an exertion which normally limits a day's flight to less than fifty miles.

I catch up with the flock again two stop-offs later, at Sweet Freedom Farm in the Baraboo Hills, where Jeff Huxmann introduces me to the birds in the pen. The farm is owned by Dick and Jane Dana, former Coca-Cola marketing executives who now devote much of their time to restoring prairie habitat on land exhausted by intensive agriculture. Prairie restoration, I am learning, is an increasingly popular trend with Midwestern environmentalists. Not surprisingly, Dick and Jane love being part of a cutting-edge eco-project like the Whooping Crane Eastern Partnership, and they put on a mighty hospitable party for the pilots and the ground crew, including drivers, technicians, biologists, educators, and the aforementioned Heather Ray and Jeff Huxmann. The group is bound together by a camaraderie not unlike that of a touring theater company or rock band. They travel hard, work hard, and sleep (very little) in trailers, or in motels when they're lucky. They are bound together by a shared passion for this extraordinary opportunity to restore a magnificent bird to some small part of its place in American nature. They know how to party, too.

They are joined on Sweet Freedom Farm by a number of staff from one of the Whooping Crane Eastern Partnership's members, the ICF at Baraboo. Among them is George Archibald, who has made it his life's mission to ensure that these "symbols of our untamable past," as Aldo Leopold described them, still have a place in the present and the future.[2]

There is a great deal of drama involved in flying with these vulnerable but independent-minded birds, and an ever-present potential for lethal collisions. At the party at Sweet Freedom Farm, Operation Migration's cofounder and lead pilot, Joe Duff, tells Archibald about a split-second crisis the previous year. One of his charges flew straight at him, having balked at flying over a noisy interstate highway. "I stayed steady, the crane elegantly flipped up and rolled on its wingtip, bled off all its speed and settled into third position in the formation behind me," he recalls laconically. Archibald listens with a mixture of fascination and horror. He manages to combine a boyish love for each individual crane he encounters—he has been known to join captive cranes in courtship dances—with a scientific global strategy for protecting the genus. He admits that it took a lot of hard thinking to persuade him to back a plan where the stakes are as high as aircraft-assisted migration.

Later, Dick Dana invites us to his den on the third floor of a sumptuously converted red barn, that icon of Midwestern rural architecture. He produces a percussion instrument for everyone present, and puts the Moody Blues's *In Search of the Lost Chord* on the sound system. Who said the sixties were dead?

I had to migrate to New York myself on other business the next morning. I did not catch up with the migrating cranes and their minders again until they had gone a long way south. This time, the sociological as well as the geographical environment was very different. Surface-to-air missiles (happily, decommissioned ones) were part of the scenery. Over the intervening six weeks I had followed the migration on the web (http://www.operationmigration.org) through Illinois and Tennessee. In late November, the group had a long hold-up at Hiwassee State Wildlife Refuge near the Tennessee–Georgia state line. The wind just would not blow right, it seemed, but the delay bought me time to escape from commitments in Nebraska and fly to

Atlanta. From there I could drive north and hopefully connect with the birds as they flew south.

Chasing Cranes through Rural Georgia

At seven the next morning, in a motel somewhere on that route north, my phone rings. It is not Heather Ray, as I had expected, but Garry Kenworthy, a man I had never met, inviting me to have breakfast on his farm as the cranes arrive. I find the Kenworthy place easily enough, ignoring the radio exhortations of half a dozen Bible-believing churches to worship with them en route. It is a glorious fall morning, and the sign on the farm gate, "Welcome to Paradise" seems appropriate. A metal cowboy silhouette in black, with a red silk bandana, nonchalantly chewing a straw, suggests a life of productive indolence. Alas, there is to be no time to linger here.

Getting out of the car in the farmyard, I hear an ultralight overhead. There is one of the pilots, flying very high, it seems, so that the cranes behind his wingtip twinkle like small stars as the low sunlight catches their plumage. The timing seems too good to be true—and it is. I am directed to the landing strip, where Heather has just arrived. We introduce ourselves to Garry and his wife. They are ecstatic at the prospect of entertaining the crane migration. They plan on serving beer with breakfast, to the humans at any rate. This sounds just fine in Georgia, where you can't buy booze in a Walmart if it is within three hundred yards of a church.

The cranes, however, have other ideas. It's such a fine morning, with such favorable winds, that they just keep on flying, climbing a steep ridge of hills so that the pilots could lead them to the next farm on the route. With hasty goodbyes to the crestfallen Kenworthys, I chase Heather's yellow Ford Escape back south. That's where the missiles come in. Our next hosts are air force people, with a son in Iraq. The SAMs were a gift from his regimental buddies, who sometimes trained on the farm. American conservation is a broad church, I am learning, and that is no bad thing. And the welcome could not be warmer, with tea specially made in a brown pot for the visiting Irishman and presents of home-produced honey to all of us on departure. I don't mention the war once, honest.

A perfect dawn the next day sees a perfect departure, all sixteen

2. **Restoring the Mississippi/Atlantic Flyway.** Pilot Joe Duff teaches whooping crane juveniles to migrate from Wisconsin to Florida. (Photograph courtesy of Operation Migration.)

birds flying firmly after Joe Duff's ultralight. They are close to adult, all-white body plumage now. As they fly up from the shadowland of the holding pen into the dawn sunlight, that whiteness reflects a gorgeous shift from soft gray to warm pink. Once again, birds and pilots make good progress, and Heather phones back to tell me to skip the next stop-off and head for a town called Buena Vista. One of the crew gives me rough directions to the landing strip. Since there is no way I can make it before the birds, I decide to time my arrival for early evening, taking in some of the strangeness of rural Georgia at leisure en route.

And strange it is. It is not just that clusters of churches, with messages like "God responds to knee-mail," stand sentry in each small town. More telling is the grinding poverty of many of these hamlets. Rows of abandoned houses sometimes sink, quite literally, back into the earth, bushes growing outward through their broken windows. Out in the country, fancy mansions fly the flag of the Confederacy. A philosophy professor in the same state once told me that he still felt that the American South was "an occupied country." At Buena Vista I ask at the gas station for a hotel. "This here is Boo-ina Vee-sta, honey," the lady tells me sadly, "Ain't never been no hotel here."

Finding a hotel is a small problem compared to finding the cranes and their entourage. My instructions—or my misreading of them—leads me down overgrown forest tracks. Heavily armed hunters in camouflage fatigues look understandably baffled, and then increasingly suspicious, when I ask them where I might find a landing strip suitable for cranes nearby. They remain very civil, but I begin to feel like a furtive cocaine smuggler, and head back for the highway. Eventually I make phone contact with Ray to find the cranes have flown on to yet another stop-off further south—more than two hundred miles in one day, a record for this or any previous year. I catch up with the crew just in time for a late celebratory pizza and a restorative shot of vodka.

We all rise five hours later to weather that promises further progress toward Florida, now tantalizingly close. "It's a new day," Joe tells his colleagues at the briefing, "Get the butterflies in your stomach in a row and fly." These guys make it all look so simple in the air. But they take their lives in their hands every time they taxi to the crane pen. We wait in a dark and bitterly cold field for dawn. The first light reveals patches of white at our feet. It is not snow, as my sleep-clogged mind tells me at first, but cotton highlighted with frost. A few minutes later the ultalights are aloft, but it looks bad: the wind has shifted and is actually blowing them backward.

That strange and rather disturbing sight was my last image of Operation Migration, as my own flight that day took me north again, from Atlanta airport, four hours later. But happily the image is not a representative one. Six days later, the trikes led the cranes into the Chassahowitzka National Wildlife Refuge in Citrus County, Florida. There they spent the winter, usually (and voluntarily) roosting in a pen that offered protection from their most successful local predator, bobcats. Otherwise they were now wild creatures, living in the wild.

Bugle Calls on the Mississippi Flyway

That was a great achievement, but the real miracle is what followed. In the spring, each of these yearlings migrated back to the Midwest unaided and of their own accord, some of them all the way back to the wildlife reserve in Necedah where they had been reared. Without any instruction, the young adults latched onto the thermals, and made the journey north a whole lot faster—and easier—than their maiden migration.

Out of the 147 cranes that have made human-assisted first migrations to Florida since 2001, and the 56 juvenile birds that have been released directly among the adult cranes in the new flock in a more recent strategy, 108 survived as of March 2013.[3] This means that the total number of these magical birds in the Mississippi/Atlantic Flyway now constitutes more than one-third the number in the previously sole surviving wild flock that migrates from Canada to Texas.[4] There have been some tragedies along the way, especially in 2007, when 17 of the 18 juveniles that had reached Florida that year were killed in the one storm. A second wintering site for juveniles has now been established to reduce the potential impact of another such disaster.

The biggest challenge, however, comes from the failure of the new eastern flock to regenerate itself. Cranes do not normally breed for their first four years, so the consolidation of the flock was always bound to be a slow process. But while 34 mating pairs have formed, and many of them have laid eggs, only 5 chicks had fledged successfully by the breeding season of 2012.

Scientists are still trying to determine why there is such poor breeding success in the restored flock. It would be a tragic irony if, after these birds learned to migrate in such difficult circumstances, a mysterious incapacity to raise their own young should prevent this flock from ever become self-sustaining. But Joe Duff remains optimistic: "Black flies are a factor and we are working to mitigate that influence. We also have a new reintroduction site outside the range of black flies at Necedah. I am confident we can overcome that hurdle."[5]

Meanwhile, the little planes of Operation Migration must continue to lend a hand, or rather a wing, until parent birds are guiding their own offspring over the flyway, with the evocative bugle calls that our civilization had silenced for more than a century, and that had very nearly fallen silent forever.[6]

3 *From Necedah to St. Louis via Zaragoza: A Restoration Learning Curve*

Ecological demands are often presented as a sacrifice. The truth is the reverse. The environment is not a luxury. It produces absolutely essential services to our well-being, which are not currently factored into market economics. People will be better off, not worse off, on a restored Earth. * ROBERT COSTANZA, *ecological economist*

Ecological restoration is the reframed environmental movement, restoring the future. * KEITH BOWERS, *former Chairperson of the Society for Ecological Restoration and restoration entrepreneur*

Restoration of natural capital—RNC—is the next big thing. * JAMES ARONSON, *restoration ecologist*

RNC is a moral failure that threatens to tear at the fabric of restoration ecology. * MICHAEL J STEVENSON, *Yale School of Forestry and Environmental Studies*

We need to treat economics as if nature mattered, and ecology as if people mattered. * JAMES BLIGNAUT, *ecological economist*

* *

You may be wondering whether the restoration of a migratory flock of whooping cranes to a flyway from which they had long vanished really is *ecological* restoration, strictly speaking.

And the answer must be no, it is not. The target of this project is the reintroduction or restoration of a single species, not the restoration of the vastly more complex web of topographic, climatic, mineral, animal, and vegetable elements and ecological processes that constitute an ecosystem. Nevertheless, such a reintroduction does form an essential part of any ecological restoration that aims to reassemble the entire suite of species that are appropriate to a system. Always assuming, of course, that extinction has not made the reconstitution of such a suite impossible, short of the faint possibility of recreating extinct species from genetic material at some stage in the future.

Moreover, a species reintroduction project may act as a catalyst for full-scale ecological restoration, since the project cannot succeed unless the target environment has the suite of species and processes the reintroduced species needs to prosper. And with a migratory species like the crane, species reintroduction may spur a number of such projects along the migration route. Indeed, large and conspicuous animals like the whooping crane (or the panda or the wolf) are known in the conservation field as "charismatic megafauna" because of their capacity to act as funding magnets for less immediately popular projects. So wetlands may be restored to assist the reintroduction of cranes, or bamboo forests for pandas. In some cases, the reintroduction of a single species may also assist the restoration of a whole system in purely ecological terms, especially if a top predator is involved. The return of wolves to Yosemite National Park, for example, has reduced excessive deer populations. This has in turn facilitated the revegetation of many areas degraded by overbrowsing.[1]

Where a broad-vision conservation group like the International Crane Foundation is involved, the concern for a single species, like the wattled crane, may create awareness of the need for a major regional restoration like the Lower Zambezi Valley and Delta Program, as we saw in chapter 1. Internationally, a crane's combination of charismatic status in many cultures, plus its generally very long migration routes, can stimulate joined-up thinking across numerous environmental—and political—frontiers. I learned from George Archibald that the progress made by the Whooping Crane Eastern Partnership had persuaded the ICF to negotiate a similar program for the Siberian crane, which requires the cooperation of countries as disparate as Afghanistan, Azerbaijan, China, India, Iran, Kazakhstan, Mongolia, Pakistan, Russia, Turkmenistan, and Uzbekistan.[2] The partnership also highlights a typically counterintuitive aspect of ecological restoration. It aims to return a species to its natural setting but is forced to use the most "unnatural" means—surrogate mechanical parents fused with humans in fancy dress, plus a battery of technical ground support—to achieve this end.

Before I followed the cranes and their trikes south to Sweet Freedom Farm, I spent some time looking around Necedah and found that this refuge had long experience with restoration and reintroduction before the Whooping Crane Eastern Partnership ever set up its pens and runways. The refuge manager, Larry Wargowsky, was the first of many public service conservation professionals across the world who generously introduced me to restoration projects. I particularly recall his view that many conservation breakthroughs, big and small, come about through accidental processes. You start off clearing prairie sedge for nesting duck, he said, and find that you have a precious population of Karner blue butterflies. Ecosystem restoration projects at Necedah ranged from upland pine savannas to sedge meadows. Species as varied as the red-headed woodpecker and the aforementioned Karner blue benefitted from associated prescribed burns. The fact that so delicate an insect as this lovely butterfly could benefit indirectly from a management technique as volatile as burning its habitat seemed particularly striking. The list of species reintroduced at Necedah is also impressive, and serves as a salutary reminder of how some species and populations have flourished, not diminished, in recent decades. It seems hard to credit today, but it was deemed necessary to reintroduce the Canada goose here in 1939, the wild turkey in the 1950s, and the mallard in the 1960s. One hopes that the 1990s reintroduction of trumpeter swans may enjoy similar success—though, on reflection, perhaps not quite such an exponential increase as that of the now ubiquitous and very problematic Canada goose.

While scanning a marsh for these swans in the still, early morning, my eye was caught by the presence of several bald eagles roosting on dead trees. As the sun rose higher, I saw them several times in flight, great wingspans soaring until they were but specks against the blue, but somehow making more visible the vastness of the sky. The return of these magnificent birds from the brink of extinction due to an intense protection and public education program by the US Fish and Wildlife Service over several decades, prompted me to remember that the perception of an unbroken bad-news narrative from the conservation world is a very partial one. I wondered what part restoration might play in broadcasting a different kind of story.

But a trip to the University of Wisconsin–Madison Arboretum, conveniently situated for my Midwestern travels that fall, would cure me of me any impression that the news from the restoration front would be uniformly sunny. And this was not just a question of appearances. Early winter is, of course, a less than ideal moment for the layperson to appreciate the arboretum's remarkable achievements in prairie restoration. As at the crane foundation, the untidy tangles of brown seed heads atop collapsing plants recalled Rilke's lines on autumn in Venice:

> The summer like a bunch of puppets dangles
> Headforemost, weary, done away.

Appreciating the aesthetics of prairies when they are not in flower is an acquired taste. Nevertheless, I was lucky in that I had as my guide Dave Egan, then editor of *Ecological Restoration,* the influential journal for restoration practitioners published by the university.[3] But what remained with me, after we walked dank paths under a threatening gray sky, was less the historic importance of the Curtis and Greene prairies as long-term restoration experiments, and more the persistence of a number of severe problems after seventy years of meticulous work. I learned that, while a rich variety of prairie plants were present, and some were increasing, it had proved impossible to eradicate invasive alien plants (IAPs). This was an issue that was to recur so frequently on restoration projects I visited that "IAP" will become the most common acronym in this book. And some IAPs at the arboretum, like reed canary grass, were becoming more dominant, not less, over time. Nor had the re-creation of a native plant community produced the hoped-for rate of return of native animals.

Even in the relatively secure context of a long-term university-owned site, with the benefit of consistent monitoring and adaptive management in recent years, restoration success was elusive and problematic. Part of the problem, of course, is that the arboretum is very small relative to classic prairie landscapes and does not exist in splendid isolation. Instead, it sits in a matrix of landscapes that bear no resemblance to the pre–white-settlement surroundings of the original prairie site. Both scale and the degree of connectivity of sites to similar ecosystems repeatedly emerge as key factors in restoration projects. So, for example, urban storm water from off-site locations repeatedly floods through parts of the arboretum's prairies, bringing in new supplies of the seeds of IAPs, as well as cargos of toxins. Egan also stressed that while part of the Curtis prairie had never been plowed—and that sector, probably significantly, still held the greatest richness in prairie plants—much of it had been cultivated for almost a century. All of it had been subject to agricultural grazing. The changes in soil and seed bank that that history had created could not be easily reversed. This was another theme that would dominate later restoration discussions.

Excited as I was about the restoration narratives I was encountering, it was becoming increasingly clear to me that this was a very complicated story, and that neither ecosystems nor human communities always reacted well, or at the very least quickly, to restoration projects. Around the same time, I saw a report in the *Washington Post* about the damage allegedly done to the social fabric of Montana by a scheme—the Conservation Reserve Program—in which the federal government paid farmers to put croplands into fallow.[4] The purpose of the scheme was to halt erosion—and stabilize grain prices—but there was a by-product of spontaneous restoration, and native plants and wildlife were certainly benefitting. But according to the report, small towns were being drained of people, because stores were closing down, as farmers were buying much fewer chemicals and seeds, much less fuel and equipment. And young people with families were blocked from buying farms, as older farmers stayed longer on land that was yielding an income with minimal labor. Clearly, restoration projects of all kinds would have to be designed with the human as well as the ecological context in mind if they were going to work.

Because the arboretum has such an iconic place in the history, the science—and the folklore—of Midwestern restoration, one imagines that it might be prone to propagate historical romanticism: the "if we burn, it will come back" approach. A brief introduction to Joy Zedler, research director at the arboretum, put me straight on that score. She stressed the importance of learning from mistakes in restoration and insisted that failure to achieve project targets was much more common than success, though restoration literature has often given the opposite impression. At the time I imagined that the tough and unsettling skepticism she expressed in that casual conversation was largely a response to my own gushing and uninformed enthusiasm as a total newcomer to the field. I have learned since that Zedler had long been a cautionary voice raised against optimism-rich, data-poor advocates for restoration. She was one of the first to warn that any full restoration of ecological processes and communities on degraded sites would take much, much longer than most restorationists were suggesting or, even more rashly, promising. She went further, and said out loud—almost a heresy—that in many cases achieving this ambitious target would be impossible. As if that were not enough, she was starting to argue that accelerating human impacts on the biosphere, especially climate change, would make achieving restoration targets significantly more difficult in the future. In short, she was my first introduction to the current fierce debate about whether the emergence of "novel ecosystems" is making the aims of the restoration movement obsolete before they are fully formed.[5]

Hurricanes, Tsunamis, and Restoration Basics in Zaragoza

But it was Egan who directed me toward the organization where I would witness such issues being thrashed out in numerous conferences and papers over the next few years. "You have to attend a conference of the Society for Ecological Restoration," he said. And it so happened that the first SER conference ever to take place outside North America would be located very near my old journalistic beat in the Basque Country, in Zaragoza on the fringes of the Spanish Pyrenees, though it would not take place for another two years.

I spent most of that period focused on finishing a second Basque book, but two catastrophic environmental events over that period inevitably made a deep impression: the 2004 Indian Ocean tsunami, and, just weeks before the Zaragoza meeting in September 2005, Hurricane Katrina. I could not help wondering what, if anything, restoration could offer in response to such massive natural disasters. I was agreeably surprised to find out, in one of the first plenary sessions, that restorationists, alongside ecological engineers (a profession I had not heard of previously), had been thinking long and deep about hurricanes on the Gulf coast, long before Katrina had been christened. "We were predicting for years that New Orleans would take a direct hit from a hurricane unless the natural barriers once provided by Louisi-

3. **What happens when you don't restore.** The damage done by Hurricane Katrina was increased exponentially by failure to implement a plan to restore the Louisiana barrier islands and coastal wetlands, which would have buffered its impact. (Photograph courtesy of Sally Pesavento.)

ana's lost wetlands were restored," said keynote speaker Bill Mitsch of Ohio State University. "We told the authorities there would be an unmitigated disaster. But they were totally unprepared."

Mitsch is the doyen of ecological engineering as applied to wetlands. He and others had advocated a highly ambitious and very complex project to restore these coastal ecosystems, estimated to cost $14 billion. And someone in a position of authority had been listening, because it had received federal approval. But the first tranche of funding only came through a few months before Katrina made landfall. Much too late for the people of New Orleans. As we know, thousands of people have died, and the US government was faced with a bill many times higher than the cost of restoring the wetlands. Mitsch railed against the folly of rebuilding the city below sea level without restoring the region's natural defenses against storms. "It's vital that they do not put the city back the way it was. Building bigger and bigger dikes just sets us up for bigger and bigger disasters," he told me after the session. "As sure as I'm standing here, new dikes will be breached again." The Mississippi must be given time to recover at least some of its estuary structure, he continued. The natural sandbanks that would develop from that process could absorb much of the shock of an incoming storm before it reached the city.

His ideas resonated well with the thinking behind "The Values of a Restored Earth," an keynote presentation by Robert Costanza, leading light of another growing but little-known field, environmental economics.[6] Costanza is a silver-maned and charismatic star of the environmental conference circuit. He is most famous—or notorious—for a 1997 article cheekily blurbed (apparently without his prior approval) by *Nature* as "Pricing the Planet." He and his coauthors attempted to put cash values on the entire world's natural capital and ecosystem services.[7] "Ecological demands are often presented as a sacrifice," he told me in an interview at the conference. "The truth is the reverse. The environment is not a luxury. It produces absolutely essential services to our well-being, which are not currently factored into market economics. Nor does the market tell us the hidden costs of our current production system in terms of ecological damage. People will be better off, not worse off, on a restored Earth."

This was heady stuff. The idea that we should factor ecological costs and benefits into mainstream economics was completely new to me. Costanza's arguments sounded so cogent I wondered why I had never heard of them before. But I also had to wonder why most other people had not heard of them either. I was disappointed to find that Costanza seemed decidedly vague about how the world's markets, or the world's voters, could be persuaded to listen to this message. He cited psychological surveys that showed that increased economic growth and consumption do not make people happier. Ergo, people would somehow suddenly embrace sustainable economic strategies. The fact that almost every society in the world was then (and still is) moving—and fast—in the opposite direction, lured by the mirage of eternal consumer-led growth, appeared to the great man as an almost irrelevant and rather irritating detail.

It is not that Costanza was entirely unaware of the issue. In the conclusions to his presentation he told us: "A sustainable and desirable future is both possible and practical, but we first have to create and communicate the vision of that world in compelling terms." And

PANEL 1 • The SER Restoration Primer: The Basic Concepts Behind Ecological Restoration

The conference also introduced me to the *SER International Primer on Ecological Restoration*, a kind of manifesto on the basics of restoration that had been published, after much robust and occasionally acrimonious debate, only the previous year (Clewell, Aronson, and Winterhalder 2004; see also Higgs 2003, 93–130). The principal authors are Andre Clewell, James Aronson, and the late Keith Winterhalder, though other SER heavyweights like Carolina Murcia, Denis Martinez, John Rieger, Richard Hobbs, James Harris and Eric Higgs (the latter three now among its chief critics) also contributed. Its first two paragraphs clarify the aims of ecological restoration and make clear that it is usually a very complex exercise, which will often be executed imperfectly:

> **Ecological restoration** is an intentional activity that initiates or accelerates the recovery of an ecosystem with respect to its **health, integrity and sustainability**. Frequently, the ecosystem that requires restoration has been **degraded, damaged, transformed or entirely destroyed** as the direct or indirect result of human activities. In some cases, these impacts to ecosystems have been caused or aggravated by natural agencies such as wildfire, floods, storms, or volcanic eruption, to the point at which the ecosystem cannot recover its **predisturbance state** or its **historic developmental trajectory**.
>
> Restoration attempts to return an ecosystem to its historic trajectory. **Historic conditions** are therefore the ideal starting point for restoration design. The restored ecosystem will not necessarily recover its former state, since contemporary constraints and conditions may cause it to develop along an altered trajectory. The historic trajectory of a severely impacted ecosystem may be difficult or impossible to determine with accuracy. Nevertheless, the general direction and boundaries of that trajectory can be established through a combination of knowledge of the damaged ecosystem's pre-existing structure, composition and functioning, studies on comparable intact ecosystems, information about regional environmental conditions, and analysis of other ecological, cultural and historical reference information. These combined sources allow the historic trajectory or reference conditions to be charted from baseline ecological data and predictive models, and its emulation in the restoration process should aid in piloting the ecosystem towards improved health and integrity. (Clewell, Aronson, and Winterhalder 2004, 1; emphasis added)

PANEL 1 *(continued)*

The desired outcome of most restoration projects, then, is not a carbon copy of some past state, a caricature that is still attached to restoration, sometimes by very surprising sources (see chapter 14, p. 423). Recognizing that ecosystems are dynamic, not static, the aim is restoration towards its "historic trajectory"—that is, the path the system would have developed along, had degradation not occurred. It is also clearly recognized in the primer that it will rarely if ever be possible achieve this very ambitious aim in its entirety. The bottom line is the creation of a system that enjoys better "health, integrity, and sustainability" than its current degraded state. These adjectives are closer to metaphors than to scientifically verifiable conditions, an issue that has caused some skeptical comment, but their use as communications tools with a broad public is self-evident (for a lucid critique of the problems that arise when such metaphors are taken too literally, see Hilderbrand, Watts, and Randle 2005, 19). In general, health and integrity are translated into more concrete criteria in restoration practice. Three of the core criteria are an ecosystem's **biota**, or **native biodiversity** (expressed through "communities" or assemblies of native species), its **abiotic structure** (topography, minerals, climatic range, etc.) and its **functions** (such as soil formation or nutrient cycling). Regular **monitoring** of biota and functions is vital both during and after restoration, to enable ongoing **adaptive management** deal appropriately with unexpected ecosystem responses to restoration interventions, and learn lessons that may be applicable to other projects.

There is much more detailed information to be found in the primer, including a section that distinguishes restoration from (more—or less) similar environmental interventions with which it is often confused: rehabilitation, reclamation, revegetation, mitigation, habitat creation, and ecological engineering (Clewell, Aronson, and Winterhalder 2004, 12). These activities can overlap with restoration, but only to the degree to which they conform to its distinctive principle: "restoration initiates ecosystem development along a preferred trajectory [based on an historical reference system], and thereafter allows autogenic processes to guide subsequent development with little or no human interference" (12). The text makes it clear, however, that, due to the dynamism of ecosystems, there will usually be not one, but rather several, historical reference systems from which an historical trajectory may be mapped. That is, over the centuries, the system will have experienced a **historical range of variation** that offers a variety of potential targets for its restoration.

Likewise, looking to the future, a restoration may, depending on current and future conditions, move in several different directions, leading to any

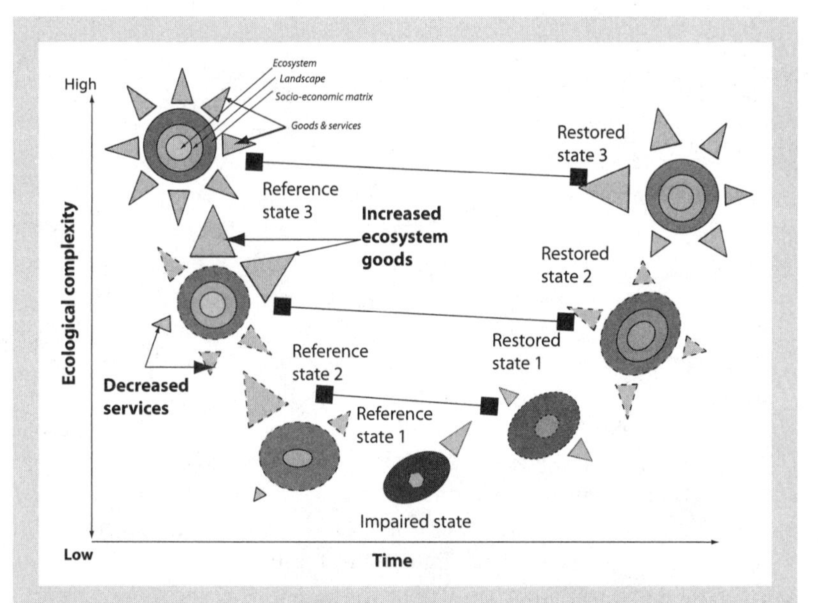

4. **Ecosystems, evolution, impairment, restoration.** All ecosystems evolve dynamically over time, but impairment through human agency will decrease ecological complexity and biodiversity, and the ecosystem goods and services (EG&S) that flow from them in varying measures. Restoration will never produce a carbon copy of a past point on its historical range of variation, but it can recover much of its biodiversity—and EG&S—as it evolves in the future. (Courtesy of Andre Clewell, James Aronson, and Island Press.)

of several possible **multiple stable states.** This phrase is not used in the primer, but has been commonplace in ecology since classic natural succession theory began to be challenged in the middle of the last century. It reminds us that succession is not a fixed linear sequence but a dynamic process that can lead to a variety of outcomes. The outcome of any given restoration project, therefore, is not strictly predictable, and narrow targets can be misleading.

Another important concept in restoration, again not dealt with in the primer, is that of the **thresholds** that ecosystem trajectories may cross due to varying degrees of degradation and damage. These are "tipping points" that radically change the dynamics and trajectory of a system. Each threshold crossed will call for a different level of restoration action to bring the system back within its historical range of variation. Broadly speaking, a system that has crossed a biotic threshold will usually only need manipulation of its flora and fauna to return within this range. A system that has crossed an abiotic threshold will need modification of the overall environmental context—topography, hydrology, and so on—perhaps through

PANEL 1 *(continued)*

some form of engineering, to do so (see Whisenant et al. 1999, 101–7). And some thresholds may prove to be like one-way gates through which it will prove exceptionally difficult, or even impossible, to bring back the new trajectory within its historical range.

It is becoming increasingly common (see, e.g., the cover image of the *Economist*, May 27, 2011) to describe our geological period as the **anthropocene**—one in which all ecosystems are, to a greater or lesser extent, modified by human impacts. Much recent—and controversial—thinking about restoration has focused on the implications of this unprecedented stage in the development of our planet. This has led to the description of "no-analog," or **novel**, ecosystems, where one-way thresholds pose daunting challenges for restoration and conservation, not foreseen in the primer. There has been much passionate debate in the restoration movement in the last five years about how prevalent such systems really are, and whether they require a radical rethink of restoration fundamentals.

Restoration, then, is not based on historical determinism, but ultimately on human choices about which ecosystem state—within a range of given conditions—is most desirable, and possible. The concepts outlined above will underlie much of the debate and discussion about restoration, both practical and theoretical, in the chapters that follow. (For further information on restoration ecology theory, see Van Andel and Aronson 2006; Falk et al. 2006; and Perrow and Davy 2002.)

he certainly had created and communicated that vision to many SER members, though one veteran commented acidly to me that his tendency to use computer-generated projections to leap from local realities to grandiose global scenarios was "modeling gone mad." In any case, he seemed to have no idea how to mainstream that vision beyond the conference walls. This blind spot about communication, beyond the self-selecting circles of restoration scientists and restoration practitioners attending the conference, proved characteristic of SER in general, with honorable exceptions. To be sure, those circles were much larger than I had originally expected—there were 850 delegates from 77 countries present at the event. But despite the state-of-the-art facilities at the daily press conference, I was the only journalist in the audience, apart from—occasionally—members of the regional media.

In contrast to the upbeat rhetoric of Costanza, Richard Hobbs's smaller but standing-room-only session at Zaragoza sounded a strong note of caution, though he was speaking about much more modest projects than the restoration of the entire planet.[8] It echoed the admonitions I had heard from Joy Zedler, and it raised, though I did not fully realize it at the time, some questions about the thinking in the SER primer (see panel 1). He warned us that the simplest restoration project involves more complex factors than most scientific experiments. The only certainty in this field, he said, is that the restoration of a degraded environment will not achieve the precise result the restorers seek. This gap between aspirations and achievement sets up a potentially dangerous tension between the expectations that a project might arouse in the community that funds it, and the actual results that they have to live with.

Hobbs is an infectiously charming man who presents complex ideas in a witty and accessible style and a resonant Scots accent. "Prediction," he reminded us, quoting the dry aphorism of the Danish physicist Niels Bohr, "is very difficult, especially about the future. Under these conditions," he continued, "it's amazing we do as well as we do." He insisted again and again that restoration projects cannot guarantee a return to a "desired state," some supposedly ideal point in the past of a degraded or disturbed ecosystem. He piled on challenging questions relating to climate change and reference ecosystems: to what extent should we value the past when the past is no longer a reliable indicator for the future? But he also pointed out that unpredictability means that that restoration projects produce benefits that are, well, unpredictable: the outcome of a project may be very different from its target ecosystem, but it may at least be richer in biodiversity than the degraded state had been, and perhaps even—heresy again—richer than the target state itself.

Hobbs described our era as one of "post-normal science," where "facts are uncertain, and values are in dispute." He recognized the difficulties such uncertainties create in trying to attract funding and public opinion to restoration, which brought us back to the problem of communication again. "Simple messages are good and effective in terms of PR," he told me after his session. "But science keeps telling us that the message is more and more complex." The media reduces public discourse to sound bites; a public overloaded with raw data craves clarity; science offers only increased uncertainty. Communicating restoration, I realized, was never going to be a black-and-white news story. At bedrock, though, despite the number of variables his school of thought was introducing into the comforting stability of earlier visions of restoration, Hobbs still shares—or at least he did then—some of their infectious optimism. Why are you involved in this field, I asked him. "Why? Because it is not all about negativity. Most of the news we get in the world is so bad. This stuff is good news."

Over the next six years, as I absorbed a series of new Hobbs papers and presentations on climate change, "novel ecosystems," and restoration, I felt that some of that optimism was seeping away. His increasingly finely honed analysis of human impacts and ecosystem dynamics led him to question what appear to be core principles in the field, without which the word "restoration" seems to be drained of any commonsense significance, and thus of the hope that is inherent in any restoration project. The questions that Hobbs and some of his colleagues have been raising in recent years are very uncomfortable, but cannot be shied away from. It is unfortunate, however, that they have been seminal in the growth of popular science articles that champion a gung-ho, bring-it-on approach to futuristic "no-analog" ecological scenarios. These tend to dismiss restoration's high valuation of historical ecosystems of reference as hopelessly nostalgic, in favor of a confident (or blind?) embrace of whatever new and exciting ecological assemblies the future may bring. Other colleagues in SER are deeply disturbed by this development, and feel that Hobbs's recent work crosses a line between legitimate—indeed, essential—attention to the historically unprecedented rates of ecological change occurring in our period, and a troubling tendency to disregard what can still, in many cases, be salvaged for the future through restoration. This is a remarkable evolution for the editor of the leading journal in the field, and one of the brightest stars in its firmament.

But that is to get beyond ourselves at this stage. Aspects of this debate will recur as we examine restoration projects around the globe, and we will return to this key debate in later chapters.[9] For the moment, however, the overall message I took away from the conference was that restoration was indeed a very lively force in conservation, already engaged with a very wide range of ecosystems and engaging a great variety of human beings. It was clearly a global force, too. I had learned about applications of restoration ranging from reforestation to combat climate change in China to the Marsh Arabs's remarkable recovery of the wetlands so cruelly drained by Saddam Hussein. And I had begun to learn something about restoration's potential social and even political benefits. I had heard an Israeli describe how the restoration of the Alexander River was bridging divides between his community and their Palestinian neighbors. Yet his Palestinian colleague had not been allowed to travel to the conference by the Israeli authorities. Hopes raised, hopes dashed.

And as I scrambled to up the first steep slopes of my own learning curve toward grasping the basics of restoration science, I found that fine minds who had devoted many years to mastering it talked of knowledge gaps, contradictions, and paradoxes. I had learned that soil was the basis of all restoration, but that very few restorationists paid it much attention. And that those who did freely confessed that they understood very little about it how the microfauna and microflora under our feet actually interact. If restoration was part of the solution to the global environmental crisis, it was clearly no silver bullet. But I had also heard a rousing address by Keith Bowers, then chairperson of SER, telling delegates that "ecological restoration is the reframed environmental movement" and that that reframing was focused on "restoring the future."[10]

As soon as I heard that phrase, I knew I had a guiding theme—long the provisional title—of my book. And I had a growing sense of excitement about the breadth and potential of restoration, and fair warning that it was an extremely complex enterprise. Just how complex would become much clearer over the next few months. Out of that excitement I wrote a newspaper article about the conference[11] and sent copies to some of the people I had met there, including James Aronson. I had encountered this intense, edgy, and dynamic American scientist at the Zaragoza sessions on the media and publishing. He had listened with apparent interest to my concern that the story of restoration was not being communicated well to a wider world.

A Baptism of Fire at the Gateway to the West

To my amazement, my phone rang only hours after I had sent him the article. Would I like to come to a symposium in St. Louis to help workshop a book on restoration, specifically on the restoration of natural capital? He said it as though he were inviting me across the street for a coffee. I could write an article about it if I liked, he continued. And as I had been talking about writing a book, and one of his publishing colleagues would be attending—Aronson edits a series on the science and practice of restoration for Island Press—why didn't I bring an outline proposal of its contents and structure? I wasn't at all sure what "natural capital" was, though I recalled that Costanza had mentioned it in Zaragoza. In any case, Aronson's enthusiasm was so infectious that I think I said "yes" to both suggestions before I asked him for any further background—and without a word of my own outline yet written. Aronson, as I would learn again and again over the next few years, is a man who is always in a hurry, a ball of energy with a half-dozen articles and sometimes as many books on the go at the same time. Passionately dedicated to the cause of restoration, he is as extraordinarily generous as he is fiercely impatient.

He leaps into polemical confrontations with colleagues, sometimes without much forethought or sensitivity, whenever he feels the cause of restoration is misrepresented in any way. He could charm a flower off a tree, and yet he can, quite unintentionally, alienate an audience when his patience snaps or when he misjudges the mood of a room. He yields no quarter on principles he is passionate about, yet is exceptionally open to accept tough criticism himself, as I have found in six years of commenting on his works in progress. I am happy to count myself among the many people he has mentored generously, and to have learned once again, adapting the aphorism of William Blake, that opposition can be an essential part of true friendship.

The "Restoring Natural Capital" symposium, in the marvelous setting of Missouri Botanical Garden, was a baptism of fire. Aronson and his two South African coeditors on the project, fellow ecologist Sue Milton and economist James Blignaut, had brought together about forty colleagues from around the world to finalize the structure for an ambitious book. The aim was to express a radical new convergence of economics and ecology worldwide, via the restoration of natural capital.[12] "Restoration of natural capital—RNC—is the next big thing," Aronson told us in his introductory presentation. The range of concepts and projects discussed—and the people discussing them, from mathematicians to foresters to philosophers—certainly suggested something new was afoot. From the poverty-relief aspects of the Working for Water program in South Africa to the profitability of restored forests in Hawaii, from restoring endemic birds on Tiritiri Matangi Island in New Zealand to strategies to improve human and environmental well-being through the creation of ecological corridors in Madagascar, the book was obviously going to be rich in themes and comprehensive in its coverage. They were talking the language I had begun to learn from Costanza and applying it to vastly varying conditions on the ground.

One now often-cited example of creative—and very practical—RNC thinking comes from the rediscovery of the linkages between a quintessentially urban environment and its rural hinterland. In the 1980s, New York City's water supply was found to be far below federal purity standards. This was due to the degradation of the Catskill–Delaware watershed from which it flowed. The city could have built very expensive new filtration plants to get cleaner water. Instead, it restored the watershed, repairing septic tanks, planting trees to restore riverbanks, and so on. The city used restored natural capital rather than built infrastructure to filter and purify its water supply. It is reckoned to have saved at least $1 billion—estimates vary—in capital costs, as well as major annual savings in current operating costs. And these figures do not factor in the less immediately tangible benefits in health and well-being to both residents and tourists in the restored areas.[13]

Phrases that were new, stimulating and sometimes disturbing to me swirled around the symposium discussions: "restoration of natural capital, but also of social capital"; "concern for the livelihood impacts of restoration projects"; "classic ecosystems are often economically obsolete." There were plenty of questions: "How do we factor in species that have no obvious economic value?" "Beauty, how are you going to talk about that?" "Can we reconcile sustainability, consumerism, and equity, or do we have to drop one of them, and, if so, which?" I began to develop a sense that ecological restoration is a catalyst concept: one of those ideas that, when it meshes with other disciplines, sets off creative thinking in many directions. I learned that concepts like natural capital and ecosystem goods and services had already been taken on board by the United Nations in the comprehensive and groundbreaking Millennium Ecosystem Assessment that had just been completed that year [14] (see panel 2). However, this kind of thinking was still not finding its way into the mainstream media, nor into the discourse of policymakers and politicians—not even, for the most part, into the thinking of those who defined themselves as "green."

But what exactly is natural capital? And what exactly are ecosystem goods and services? At one level, the answers are fairly obvious, and most laypersons would guess them intuitively, even though they may not have thought about the world in these terms before. This in itself suggests that they are apt and insightful metaphors for our relationship with the natural world. Natural capital is made up of the "assets" with which the environment endows the planet. We might say that it extends from the soil under our feet, across the landscape before our eyes, to the weather system above our heads. And from these assets flow the "dividends," or "interest," of goods and services supplied to us by ecosystems. Many of these goods and services are essential to our survival and most of them enhance our well-being, whether physically, economically, aesthetically or spiritually. A single ecosystem may have capital assets that produce a whole range of goods and services. A wetland may offer fish stocks for businesses, waste filtration for communities, inspiration for artists, and recreational opportunities and uplifting landscapes for all of us. It also supplies a natural habitat for biodiversity, and may act as a flood barrier. And then some. (Of course, one must not forget that nature also does us some grim *disservices* in the form of tsunamis, hurricanes, volcanic, eruptions and so on. But the evidence suggests that restoring natural capital minimizes the impact of such natural disasters.)

In a little more detail, the symposium followed a definition proposed by the Millennium Assessment, that natural capital comprises renewable resources (living species and ecosystems), replenishable resources (clean air, clean water, fertile soils), cultivated lands (crops and forest plantations), and nonrenewable resources (subsoil assets, such as coal, minerals, oil, and diamonds).[15] But this is a new field, and the debate about definitions is still in play, with central figures like Aronson fine-tuning their own understanding with every fresh discussion.

This is hardly surprising, given the enormously complicated and dynamic interactions and processes we are trying to describe here.

And many knowledge gaps remain to be filled in restoration science, as I discovered again and again in researching this book. This is particularly true in regard to the ability of ecologists to predict thresholds or tipping points, those moments when degradation of ecoystem functions and overexploitation of natural capital leads an ecosystem to flip over into a radically altered state, which may be irreversible by current restoration technologies (see panel 2).

The RNC approach uses broad strokes for its descriptions, borrowing metaphors from economics; these metaphors can certainly illuminate many matters, but they may also cause some confusion.[16] Slippage between the applications of these concepts remains a problem. Some of the components of natural capital listed above, for example, might more logically be described as ecosystem goods (crops) or services (the filtration by a wetland that provides clean water). The Millennium Assessment in turn divided ecosystem goods and services into four categories: supporting (nutrient cycling, soil formation, etc.); provisioning (food, fresh water, wood and fiber, fuel); regulatory (of climate, flooding, disease, water filtration); and cultural (aesthetic, spiritual, educational, recreational). But again, these categories are porous, even slippery. Some of them are obviously mainly goods (provisioning) and some services (regulatory), but some, like water, may be both.[17] Some specialists now argue that the supporting services are not services at all, but ecosystem functions, because they do not impact directly on people but rather underpin the other three sets of goods and services.[18]

You may be wondering by now, as I sometimes do still, whether any of this really matters, or whether it is all an exercise in semantics for the pleasure of pedants. On reflection, these definitions matter a great deal, though they need not detain us much longer here. Having a name for everything under discussion, and having each name mean just one thing, is vital to coherent scientific discourse. This is especially true for a project that attempts to translate ecological information into economic data. And that translation is vitally important if natural capital and ecosystem goods and services are ever to be recognized in the national and international balance sheets that influence policymakers, or in the local accounts that may persuade a local council, community or private landowner that a restoration project is worthwhile. The ongoing failure to agree a set of clearly

PANEL 2 • The Millennium Ecosystem Assessment

The Millennium Ecosystem Assessment (MA) was initiated in 2001 by the UN to "assess the consequences of ecosystem change for human well-being." It involved 1,360 experts worldwide over 5 years, and produced a number of reports. Their key messages included:

- Everyone in the world depends on nature and ecosystem services to provide the conditions for a decent, healthy, and secure life.
- Humans have made unprecedented changes to ecosystems in recent decades to meet growing demands for food, fresh water, fiber, and energy.
- These changes have helped to improve the lives of billions, but at the same time they have weakened nature's ability to deliver other key services such as purification of air and water, protection from disasters, and the provision of medicines.
- Among the outstanding problems identified by this assessment are the dire state of many of the world's fish stocks; the intense vulnerability of the two billion people living in dry regions to the loss of ecosystem services, including water supply; and the growing threat to ecosystems from climate change and nutrient pollution.
- The pressures on ecosystems will increase globally in coming decades unless human attitudes and actions change.
- Measures to conserve natural resources are more likely to succeed if local communities are given ownership of them, share the benefits, and are involved in decisions.
- Even today's technology and knowledge can reduce considerably the human impact on ecosystems. **They are unlikely to be deployed fully, however, until ecosystem services cease to be perceived as free and limitless, and their full value is taken into account. (Millennium Ecosystem Assessment 2005; emphasis added. All quotations in this panel are from this statement.)**

The MA's Bottom Lines

"In the midst of this unprecedented period of spending Earth's natural bounty . . . it is time to check the accounts. That is what this assessment has done, and it is a sobering statement with much more red than black on the balance sheet.

"Nearly two thirds of the services provided by nature to humankind are found to be in decline worldwide. In effect, the benefits reaped from our engineering of the planet have been achieved by running down natural capital assets."

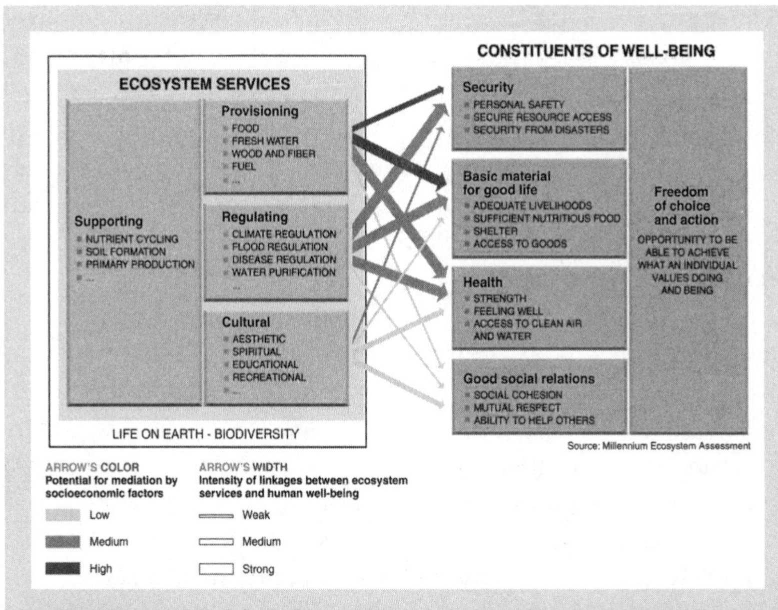

5. **The "interest" that flows from natural capital.** Healthy ecosystems provide us with a vast range of goods and services, ranging from food to flood protection to the sense of aesthetic or spiritual well-being generated by a beautiful landscape. (Figure by Philippe Rekacewicz and Emmanuelle Bournay of UNEP/Grid-Arendal, courtesy of the Millennium Ecosystem Assessment.)

"Many . . . services of nature do not appear on conventional balance sheets, but they are . . . essential for the survival of modern economies. Their true worth is often appreciated only when they are lost."

Unknown Unknowns: Tipping Points and the Need to Invest Now

"If natural systems were well understood and behaved in a predictable way, it might be possible to calculate what would be a 'safe' amount of pressure to inflict on them without endangering the basic services they provide to humankind.

"Unfortunately, however, the living machinery of Earth has a tendency to move from gradual to catastrophic change with little warning. Such is the complexity of the relationships between plants, animals, and micro-organisms that these 'tipping points' cannot be forecast by existing science.

"Investing in the health of natural assets could therefore be seen as a form of prudent insurance against abrupt changes and the risk to human well-being that they pose."

defined categories for these terms can indeed layer confusion over necessary complexity in these debates. The struggle to make comparative ecological and economic analysis accessible, to policymakers and the public, is both difficult and vital. Ecological economists have not yet won over many of their colleagues in other branches of their own discipline. Despite advances, RNC thinking has still not been adequately mainstreamed into public discourse and policy. Much work remains to be done.[19]

Aronson and his colleagues made it clear at the symposium that they conceive of RNC as a human-oriented strategy, a movement that attempts to reconcile humanity and nature, conservationists and advocates of development, especially but not only in the desperately poor countries still emerging from colonialism. If hungry people see conservation as a luxury, they argued, the answer is show them that the restoration of natural capital serves their needs—indeed, that without such restoration their plight will only worsen. Social capital, in other words, cannot be restored and augmented without the restoration of natural capital. Despite the difficulties in reaching fully satisfactory definitions for all its elements, RNC thinking is undoubtedly based on a compelling and fruitful metaphor: ecosystems are natural capital, and the ecosystem goods and services essential to our existence and our well-being flow from restored natural capital in a manner analogous to the flow of dividends or interest from well-invested financial capital. A core principle of economics is that capital should be conserved and that only a portion of the interest it produces should be consumed, with the balance reinvested to produce more wealth in the future. Yet our approach to natural capital has been almost entirely heedless of this principle. We are consuming renewable resources faster than they can renew themselves. We have degraded replenishable resources and cultivated lands to a critical degree. We are exhausting stocks of nonrenewable resources without developing adequate substitutes. In short, we are depleting or damaging these vital stocks like spendthrifts on a drunken spree.

But can natural capital, once depleted, be restored? Restoration ecology offers positive evidence that the depletion of stocks of renewable natural capital can in many cases be halted and reversed. The "productivity" of degraded ecosystems can be augmented through restoration. It must be added, of course, that nonrenewable natural capital, by definition, *cannot* be restored, and therefore must be husbanded with great prudence. Once I began to grasp this way of thinking, it seemed shocking—and it still does—that so few public figures are paying attention to the lessons it offers us. There is a glaring omission in our international and national accounting practices, because they do not quantify or evaluate natural capital. If we do not find ways to factor in the losses from our reckless exploitation of environmental assets, indicators like gross domestic product are as misleading as a bank account that fails to register a series of substantial and recurring withdrawals. As the Millennium Assessment puts it, part of the problem is precisely that we perceive ecosystem services as if they were "free and limitless."[20]

The dialogue of the deaf between mainstream economists and ecologists does not help. Conventional "neoclassical" economists refuse to recognize that renewable as well as nonrenewable natural capital is finite and limited. They do not accept that if we consume more natural assets than we restore, we are cutting from beneath our feet the ground that sustains us. They argue that humanity has always found substitutes for forms of natural capital that we have exhausted, and that our technologies will always be capable of inventing new products that can substitute for diminishing stocks of natural materials.[21] Remember, they say, that great seafaring countries like France and Britain long feared that their forests would one day no longer be able to meet the demand for wood to manufacture ships. But from the mid-nineteenth century the invention of the ironclad hull spelt the eventual obsolescence of wooden ships, and propelled international maritime commerce (and, less fortunately, warfare) to unprecedented capacities.

Invisible Nature, the Witchcraft of the Markets

Impressive as such examples are, the argument that we will always find ingenious substitutes for exhausted natural capital ignores several key issues. First, it puts no value on the cost of that exhaustion to nature in itself—entire bioregions are devastated by our rapacious appetite for consumption and growth, with a cultural, aesthetic, and spiritual deficit for humans that only poets can calculate, and poets don't do accounts. Second, the reckless exploitation of natural capital in the developing countries, with its accompaniment of resource depletion, increased flooding, erosion and infertility, destroys the well-being, livelihoods—and often the lives—of many millions of people. Having squandered much of their own natural capital in past centuries, the developed countries now extract it elsewhere. You need to see the world through grossly distorting spectacles to believe that our globalized market economy really works to the benefit of humanity in general, and that imperialism is a thing of the past. Third, this century's exponential expansion of both population and consumer demand puts such unprecedented pressure on natural capital stocks that it would test the faith of the truest believers in technological substitutability, if they allowed themselves to think about our real situation at all.[22]

But they don't, and they don't have to, because neoclassical economists airbrush the natural capital crisis out of our global economic indicators. In the memorable words of Pavan Sukhdev, himself a leading figure in Deutsche Bank's global markets strategy in Asia in the 1990s, "the witchcraft of the markets has made nature invisible."[23] Sukhdev is now the leading public face of a ground-breaking study, *The Economics of Ecosystems and Biodiversity* (TEEB), which puts substantial flesh on the bones of the Millennium Assessment.[24] The five TEEB volumes set out the fundamentals of ecological valuation for five target audiences—scientists, business people, national and international policymakers, regional and local administrators, and citizens in general. The volume for scientists fine-tunes and advances the thinking first developed comprehensively by the assessment.[25] The citizens' volume online sets out the case for valuing natural capital and ecosystem services with exemplary clarity, using a variety of media.[26] Each volume leads us back down myriad ecological pathways, again and again, to that bank account in which massive and rapidly increasing losses are not registered, so that we think we are in credit even when our overdraft vastly exceeds our ability to pay. Sukhdev never loses an opportunity to stress that it is the poor who suffer most from the depletion of natural capital, and who stand to gain most from its restoration. One of the difficulties, he says, in raising public awareness of the value of natural capital is that the developed world gains (in the short term) from this depletion, and the poor who bear the brunt of its immediate impact have hardly any voice:

> There has always been a biased thinking that every time you try to attend to conservation it impoverishes people, but the broad economic analysis shows that this is not the case. The GDP of poor households are heavily dependent on free flows from nature and if you end up depleting fresh water reserves or filling up wetlands you hurt the livelihoods of the poor and cause them a huge amount of insecurity.[27]

Conventional economic wisdom tells us that the limiting factor on economic growth is the availability of capital, financial or manufactured. But it is becoming increasingly clear that the limiting factor today is very often the availability of natural capital, not of man-made capital, according to ecological economists. For example, it is the stock of fish, not the stock of fishing boats, that now limits the growth of the fishing industry, as the economist Herman E. Daly pointed out in a letter to the St. Louis symposium. Humanity is effectively liquidating its core assets, consuming the proceeds, and undermining its ability to create future income. We are living off sales of the family silver, and it is becoming increasingly obvious that this inheritance—just think oil—is running out.

"All we are asking," James Aronson told me after the conference, "is for market economics to obey its own rules." The current rates of depletion of natural capital, he continued tartly, "represent massive market failure."

But the fault does not only lie with mainstream economists and unbridled market forces. If conventional economists leave natural capital off balance sheets, conventional ecologists, environmentalists, and conservationists all too often leave human nature, and human needs, out of their account of the Earth's biosphere. But they are both talking about the same place. They need to listen to each other if we are to advance matters. It will not be simple to enable them to hear each other. There will be, inevitably, huge problems (and self-interested disagreements) in reaching global or even local agreement on how to calculate the value of natural capital. And, as we have seen, ecologists can rarely tell us with any certainty when natural capital depletion is approaching a tipping point beyond which restoration would be impossible.

But it does seem prudent to apply a precautionary principle here. "We should not be hovering on the brink of catastrophe," ecological economist Joshua Farley told the workshop. "When our survival is at stake we need to be standing well back from the edge." Persuading policymakers and legislators to adopt a precautionary principle, however, is never easy. Public opinion tends to be very reluctant to forego today's pleasure because it will bring pain tomorrow, and is much more reluctant still if the pain is a half-century down the road. Our short and shallow political cycles also tend to cripple any truly long-term policies.

Two other thorny issues cast a shadow on the general optimism about RNC's potential at the symposium and were not resolved there, nor are they likely to be resolved anywhere in the near future, though they are both of the utmost urgency. These are the interrelated problems of rising consumer demand and rising population growth. We may be able to restore natural capital globally at sufficient levels to meet basic needs, but if those needs keep being revised upward to include, for example, a car for every family, and then two cars, and then three or more, there is simply not enough natural capital on our planet to produce ecosystem goods and services to satisfy our ravenous appetites. And if, at the same time, we continue to breed more and more millions of people, each one culturally programmed to be addicted to ever-rising consumer expectations, the future remains bleak indeed. These were the questions I had raised with Costanza, and convincing answers were no more forthcoming at St. Louis than they had been at Zaragoza, though Sue Milton, one of the book's co-editors, acknowledged them with a bleak scenario of her own.

Writing to me after the conference, she described the two problems as they impact on her native Karoo region of South Africa, but extrapolated from them to a global scenario:

> Each one of us born makes the world a little poorer for every other person (and most animal and plant species) on earth. The recognition of the need for ecological restoration represents a recognition that the human species is at carrying capacity. There is not much left to exploit. Soon the exploitation effects will become evident, as, in common with amoebas in a test tube or overcrowded rats we compete, fall ill, starve and rapidly thin our type out in most unpleasant ways. Restoration is a part of the solution. Frugality is another (even less popular) option.[28]

Obviously, such huge and contentious issues cannot be resolved by restorationists alone. If RNC is indeed even part of the solution, then it is a positive. But it should be added that RNC also faces a challenge from much closer to home, as it were, though it seems to be diminishing. The RNC advocates reject the tendency, present in many schools of environmentalism, to see the world through a man-versus-nature lens. For Aronson and his colleagues, human needs are themselves an essential part of what he calls "socio-ecosystems."

But one sector of the restoration movement remains deeply suspicious of attempting to put a dollar value on the environment, fearing that the intrinsic value of nature for its own sake will be sacrificed to the commodification of ecosystem functions, goods, and services that are beneficial to humanity. One of them has described RNC as "a moral failure that threatens to tear at the fabric of restoration ecology," as Michael J Stevenson put it in a September 2000 editorial in *Restoration Ecology*.[29] "I don't take for granted that the bottom line is the prime motivator for people on ecological issues," says Bill Jordan, who briefly attended the symposium but left after a prickly public encounter with Aronson. Jordan was clearly out of sympathy with what he understood the RNC agenda to be at the time. Jordan speaks with considerable authority as one of the founding fathers of the ecological restoration movement in the United States. He recognizes that the RNC argument could be an effective weapon in mobilizing support for conservation, but he thinks it is dangerous to identify it with ecological restoration in the fullest sense of the phrase: "The distinction between restoration of whole ecosystems, including all their features (species, processes, etc.) for their own sake, and restoration of elements or features selected for "ulterior" reasons, whether these are economic, aesthetic or whatever, is of immense importance. Nothing less than the future of classic landscapes and the non-economic species that make them up (that is, the large majority of species) depends on it."[30]

We will see similar arguments cropping up in subsequent chapters, and we will return to Jordan's points in greater detail in our conclusions, because they represent a valid and important pole of thought within restoration. Overall, however, I found myself reasonably comfortable with the synthesis the South African economist (and the third coeditor) James Blignaut offered at the end of the conference:

"We need to treat economics as if nature mattered, and ecology as if people mattered."[31]

In any case, I left St. Louis with my head ringing with new, exciting, if often conflicting and sometimes confusing ideas. Above all, I needed now to get out of the lecture hall and find out how these ideas were actually working on the ground. One project frequently mentioned at the conference struck me as uniquely significant in attempting to combine ecological restoration, the restoration of natural capital, and the relief of human poverty—or "restoration of social capital,"—in a single program. I pitched sixty media outlets to do an article on Working for Water in South Africa. *The Scientist* magazine said yes, and I was on the way to my first full-scale restoration field trip.

4 *Greening the Rainbow Nation: Saving the World on a Single Budget?*

This is one ballsy program! * KADER ASMAL, *former South African minister for Water and Forestry Affairs*

It all started in such a rush that there was no strategic ecological plan. They had to spend money immediately, and this was ecologically disastrous. Instead of focusing on the areas worst-affected by invasive aliens, and working systematically, they spread the program far too wide across the country for political reasons. * PATRICIA HOLMES, *ecologist*

How can the clearing have been "ecologically disastrous"? It may not have been suitably prioritized, but what would be ecologically disastrous would have been not to do anything. * GUY PRESTON, *chairperson and national Program leader, Working for Water*

Working for Water is a roaring success. With all its flaws, it's magnificent. * RICHARD COWLING, *ecologist and advisor to Working for Water*

My comments seem rather blunt and suggest that I do not support WfW's work—which is not true! My main point has always been (since the 2003 review) that WfW should consider itself an ecological restoration programme and NOT simply an alien clearing programme, because there is a fundamental difference between those two stances that can lead to very different outcomes. * *Patricia Holmes*

* *

An image of a coffin made of cheap pinewood fills Guy Preston's cell phone screen. He is sitting in gridlocked traffic in Pretoria, South Africa's administrative capital, conducting an interview with me that transfers without pause for breath from the airport to his car, then to a café, and back to his car again. Preston is in his fifties, tall and restless, with graying hair, tired sad eyes that somehow still sparkle, and the energy of five ten-year-olds. He is a man with a passionate mission who does not seem to stop working for a moment, day or night. Some people say he has not stopped working since he launched the Working for Water (WfW) program, which he still heads, in 1995.

"This is from our eco-friendly coffins pilot project," he says, passing over the phone. "It gives the poor a chance to bury their dead with dignity, without paying extortionate prices."[1] The production of cheap coffins seems a bizarre line of business for WfW to be promoting. This program was set up, as its title suggests, mainly to help restore and conserve South Africa's national water supply. It attempts to do this through clearing invasive alien vegetation and restoring indigenous biodiversity. The theory behind this is that native South African plants consume and displace significantly less water than generally larger and—it is assumed—thirstier foreign imports, like eucalyptus and Australian acacias. "Ecocoffins: Reducing the Burden of Bereavement" is just one of many unpredictable offshoots of WfW. And, like almost every aspect of this polymorphous program, coffin production is itself a multifaceted project. As one of the WfW's secondary industries, it adds value to some of the huge stocks of timber produced by the WfW's massive clearances of trees and shrubs. This in turn reduces the fire risk represented by leaving the wood stacked up in the bush. Finally, it provides direct assistance to impoverished families. South Africa's raging AIDS epidemic has, of course, made coffins an increasingly frequent necessity on the shopping lists of the poor.

WfW has been jokingly accused, even by friendly critics, of "trying to save the world on a single budget." It is certainly one of the most ambitious programs of its kind. Indeed, there is nothing in the world quite like it, and it is widely seen, within South Africa and elsewhere, as an innovative model for combining ecological, economic, and social goals in the developing world. WfW has brought ecological restoration right into the mainstream of state administration, and the program has been supported—a rather mixed blessing at times—by no fewer than fifteen government departments. It is not surprising, then, that WfW, with triple targets—ecological, economic, and social—is one of the projects most often cited as exemplary by the RNC strand

6. **Endgame for invasive species?** Guy Preston, Working for Water National Programme Leader, playing chess with one of his colleagues, Bulelwa Dikana, on a school desk incorporating a chess board. Preston can cite multiple benefits from this single artifact: the desk is made from a cleared invasive alien eucalyptus, river red gum; previously unemployed people crafted it; and it promotes cognitive thinking in schools in poor areas. (Photograph courtesy of Guy Preston.)

within the ecological restoration movement. It now employs about 30,000 people part-time each year, equivalent to 10,500 full-time jobs, on an annual budget of close on $100 million.[2] It claims to be restoring a variety of ecosystem goods and services worth much more than that. WfW, in the very tricky context of postapartheid South Africa, appears to tick all the right boxes for right-thinking people. Its saving-the-world activities range from organizing AIDS education classes for its workers to setting up crèches for their children.

But the very scale of this program's ambitions, and the rather easy ride it has received from a well-disposed national and international media, should make us cautious. Is it really possible to do so many things at once without doing at least some of them very badly? As we shall see, its multiple goals often come into conflict with each other. Inspirational as WfW undoubtedly is, it is a model whose serious flaws are as instructive as its remarkable achievements. Before we examine those flaws and achievements more closely, however, it is necessary to understand how the redoubtable Guy Preston and his colleagues got such an unprecedented program off the ground in the first place, and grasp the ecological and social context in which it operates. The origins of WfW lie in a problem that has undermined South African development since the colonial period. For a country famously well endowed with many natural resources, it suffers surprising shortages of two key forms of natural capital: wood and water. Worse, attempts to increase stocks of wood have almost certainly made water scarcer still. South Africa's natural woodlands are few and very far between, not because of massive clearances but because of its climate and ecology. Forests cover less than one-quarter of 1 percent of the land surface, and never covered a great deal more in recent centuries. Patches bigger than one square kilometer can only be found along parts of the Cape's Garden Route—especially the temperate rain forests around the holiday town of Knysna—and scattered through the Lowveld Escarpment further east. These forests were and are sources of fine timber with evocative names, like Outeniqua yellowwood and black stinkwood. But they could not even begin to supply the general needs of a fast-growing economy. So, by the mid-nineteenth century, South Africa had become a global pioneer in establishing plantations of exotic trees, like eucalypts, pines, and Australian acacias, on both public and private land.

Around the same time, it was becoming evident that the country's rainfall and aquifers could hardly yield enough water to meet the demands of a rapidly expanding agricultural sector, much less the massive urbanization that was to follow. And water appeared to be getting scarcer every year. Farmers suspected from a very early stage that the massed ranks of trees, occupying already vast and ever-increasing tracts of the landscape, were to blame for streams running sluggishly and wells drying up. This was hard to prove, of course, but their argument was taken seriously by the South African administration. So the government set up the Jonkershoek Forestry Research Station as early as 1935, in the hills above the vineyards of the Western Cape. Despite its title, its brief was and is "restricted mainly to the effects of afforestation on water conservation," in the words of its first director, C. L. Wicht.[3] That research soon indicated that the farmers' instincts were probably right on the money. Experiments showed that IAPs, especially tall trees, were usually a lot thirstier than indigenous vegetation—at least under certain specific conditions. But it would take more than research findings to stop the foresters, and the plantations grew bigger and bigger.

Invasive Aliens, Transformers—and Apartheid

Many of these alien plants were proving themselves to be both "invasives" and "transformers." That is, they escaped their plantations, and displaced native plants, and then radically changed everything from the soil composition to the hydrology in the ecosystems they colonized. Many of them favored the banks of rivers where their effect on the water supply was likely to be maximized.

Fast-forward to the heady transition from apartheid to democracy led by Nelson Mandela in 1994: most South Africans were, quite naturally, focused on one of the most notable and hopeful political dramas of the last century. South Africa's diverse races of *Homo sapiens* were, after a long and bitter struggle, finally winning equal rights and some measure of mutual respect. Meanwhile, however, a small panel of scientists was meeting in the Fynbos Forum to consider a crisis affecting other South African species, but with big implications for humanity also. They reckoned that, however the political battles panned out, the country's rich *bio*diversity was on a steep slope to catastrophe. The displacement of South Africa's plant communities by invasive alien plants (IAPs) threatened the survival of many of the endemic species that make up South Africa's vegetation, especially its unique fynbos flora. From a botanical point of view, the concentration of biodiversity found in fynbos can hardly be overstated. Look at Ronald Good's designation of the world's six floral kingdoms: you will find that, while five of them stretch over great swathes of the globe, the relatively small Cape region of South Africa makes up an entire kingdom all by itself.[4] This in turn is largely thanks to fynbos, a shrub system named after the Afrikaans word for "fine-leaved bush." This plant community covers only 6 percent of the country but holds one-third of its species and exhibits extraordinary variety within each of its families.

For example, there are six hundred species of Erica, or heaths, in the Cape; the rest of the world has just twenty-six.[5] Other familiar plant families with many unfamiliar members to be found in fynbos include proteas, pelargoniums, freesias, gladioli, irises, and daisies. Its concentration of species makes this the richest biome per square kilometer in the world, more than three times richer than South American rain forest. Eight sectors of the Cape, most of them fynbos, have been designated a single UNESCO World Heritage Serial site.[6] It is recognized as one of only eight biodiversity hotspots[7] in the whole of Africa by Conservation International, because 70 percent of its plants are found nowhere else in the world—and because one-third of them face extinction. The major extinction threat, the scientists at the Fynbos Forum agreed, came from not from urbanization, as might be expected—it is well known that Cape Town sprawls over two particularly valuable fynbos sites—but from IAPs. These plants had now spread into remote regions with little human presence, but with knock-on impacts on the whole country. The Fynbos Forum members committed themselves to a campaign to roll back the invasion and restore native biodiversity. Realistically, however, they could not have had much hope of success. Such a campaign was going to be very expensive and could only be effectively funded on a national scale out of the public purse. But between the exultation surrounding the birth of Mandela's "Rainbow Nation," and the colossal and pressing social challenges that his inexperienced African National Congress (ANC) government faced, it seemed most unlikely that anyone in the new cabinet was going to get too excited about rare plants.

To be sure, Mandela and his colleagues were more environmentally aware than many more experienced governments, and they were generally keen to show that the "new" South Africa was going to be a world leader in solving the problems of the twenty-first century. But, in the late twentieth century, millions of South Africa's citizens were still living in makeshift shacks, euphemistically called "informal housing." The outrageous gap between the (mainly white) rich and the (mainly black) poor was still widening. Political equality was buttering some choice economic parsnips for the new black middle class, but the masses remained as dirt poor as ever. Moreover, conservation in South Africa had been seen, with some justice, as a hobby for a white elite. There was a feeling that animals had often been treated considerably better than the indigenous human population, as blacks had been evicted from their homes in order to establish the country's landmark national parks. These reserves were often regarded simply as lavish holiday resorts for whites by the majority of the population. So you would imagine that biodiversity was not exactly top of the ANC's short-term agenda. And yet, within a year of Mandela's inauguration, a budget of $4 million had been allocated to the clearance of IAPs. Hundreds of millions of dollars have been spent since then, and tens of thousands of people employed each year.

What persuaded the new government to dedicate such a substantial slice of its overstretched resources to a project like this? The answer is a lot of lateral thinking, both from the Fynbos Forum scientists and from the first postapartheid Minister for Water Affairs and Forestry. The scientists, who included future WfW executives and advisers like Guy Preston, Christo Marais, and Brian van Wilgen, realized that their concerns would receive little more than lip service unless they could give their cause an economic dimension. At the most obvious level, a number of cash benefits are associated with fynbos. Its botanical glories attract many tourists, especially but not only in the spectacular flowering season, when the subtle greens, grays, and browns of the landscape bloom into brilliant and kaleidoscopic colors. The proteas and many other plants are harvested for the dried flower industry, which is worth about $4 million annually,[8] and the fynbos is also a valuable source of tea (rooibos), thatch, and pollen for honey. But these elements did not add up to a large enough package to convince the new government to commit serious resources to confronting the biodiversity crisis.

The scientists needed something much bigger, and they found what they needed in the role that fynbos plays in protecting soil and water resources. Its tightly woven net of small plants prevents erosion and maintains the region's sandy soil as a filter, guaranteeing a flow of clean water to the rivers and aquifers. The onward march of the "enemy aliens" was having the opposite effect. The big plants degraded or damaged the soil, and now infested many of the major watersheds. According to the scientific consensus, and a lot of anecdotal observation, this was causing rivers to shrink and even vanish, and the water table to drop. One much-cited paper researched in 1995 predicted that, as a result of these IAP infestations, Cape Town faced a loss of 30 percent of its water supply.[9] This was very bad news at the very moment when millions of its impoverished citizens expected the new political system to deliver running water to their homes for the first time, and luxury use by more fortunate residents was also rising fast.

So the Fynbos Forum was able to argue that the protection of native biodiversity was crucial to restore a critical stock in South Africa's bank of natural capital—the supply of fresh water. Its members added that IAP clearance on agricultural land would also produce significant economic benefits. These were not new insights, as we have seen. In fact, the Department of Forestry had controlled the invasive aliens quite effectively in the 1960s, but the situation had deteriorated a lot since. "We used to be up there among the best in the world for integrated catchment management then," says botanist Richard Cowling, one of the world's foremost experts on fynbos and an advisor to WfW.[10] But control slackened with decentralizing policies under the presidency of P. W. Botha in the 1980s, and the IAPs had achieved critical mass by the 1990s. Despite tightened border vigilance under democracy, new species still arrive every year. This has created an unprecedented level of threat to the country's most vital ecosystems and to the natural goods and services that flowed from them.

Bouncing Biodiversity onto the Cabinet Agenda

Guy Preston found a listening ear for these arguments in the new minister at the Department of Water Affairs and Forestry (DWAF). Kader Asmal, who died in 2011, was a distinguished human rights and antiapartheid activist, who had spent decades in exile teaching law at Trinity College, Dublin. His new brief was completely unfamiliar to him when Mandela invited him into his first cabinet in 1994, but he seized on it with energy and vision. He quickly appointed Preston as an adviser to his department, and together they brainstormed the ideas from the Fynbos Forum, which would give birth to WfW. However, Asmal felt that despite the project's goal of augmenting the national water supply, it still lacked full-on appeal at South Africa's first democratic cabinet table; poverty alleviation was the first item that every minister in this new government wanted to be seen to address. Asmal was quick to grasp a basic link between water and poverty. The most acute problem faced by the poor was, in fact, lack of access to clean water, so anything that increased the water supply should benefit them. But he still needed another element to convince his skeptical colleagues, and he found it in the form of a public works program. If the IAPs were to be cleared manually by "the poorest of the poor," and they could be paid for their work, it would give the project huge political clout. As Asmal told me, he saw that he would be uniting three powerful strands "in one ballsy program."[11] At a stroke, South Africa would be protecting its indigenous biodiversity, restoring natural capital in the form of water and putting money in the pockets of thousands who had never earned a day's wages before in their lives, thus also restoring social capital. The newly christened Working for Water was thus imbued by an innovative "win-win-win" trinity of values which represented the new South Africa in the best possible light.

Guy Preston remembers the moment he knew it was going to happen: Asmal confirmed that he had requisitioned a budget of 25 million rand (US$4 million). This was paltry by current WfW standards but more than Preston was sure he could handle in the first time frame they were given. "I went to Christo Marais, and asked him if he could spend 25 million on our new project by the end of the year. Christo said 'yes' immediately, and WfW was born." As we will see, however, the remarkable velocity of the program's takeoff would create many problems down the line. Asmal recalls with glee "the horrified response" of senior officials, all from the old regime, "all conservatives, of course," to the plan he, Preston, and others presented. They proposed to offer very poor people, mostly the permanently unemployed, five-year part-time contracts to clear botanical aliens for 30 rand (US$5) per day. That was radical enough and risked the ire not only of conservatives but of the newly enfranchised trade unions, because the remuneration was well below the minimum wage. But they were going further: they proposed to put the poor partly in charge of this operation themselves and ultimately create business and employment opportunities for their own communities.

Indeed, reflecting the ANC's flight from socialism and embrace of free enterprise, this was to be a public works program with a most unusual twist: it would attempt to grow entrepreneurs at ground level. WfW would seek out "contractors" from among the chronically unemployed. They would entrust each one with the budget and the tools to hire ten of her or his peers as a team to clear invasive alien vegetation, mostly on public land, for two years in total over a five-year cycle. Meanwhile, carrot-and-stick laws were in the pipeline to motivate private landowners to root out IAPs on their own properties. It was envisaged that a significant number of these contractors would, on exiting the program after their five years were up, find new work for their teams selling their services to landowners. The program founders envisaged the creation of hundreds, if not thousands, of micro-enterprises in ongoing IAP clearances across the country for the foreseeable future. From the outset, no one at the top of WfW ever thought small.

Battalions of the Dispossessed

Asmal also felt that, by pitching battalions of the dispossessed into the front line of a huge environmental project, he would be able to radically alter the image of conservation in South Africa. So, again from the start, WfW involved a strong element of public education, rooting environmental concerns among the people who had been alienated from all aspects of civil society for so long. "Conservation is a middle-class affair in most countries," said Asmal, and this is especially true in South Africa. In fact, a kind of precedent for WfW can be found in the middle-class voluntary sector. There is a long and honorable history of locally organized "hacks" that target invasive alien vegetation at weekends, but their activities have been limited to areas like Betty's Bay and Blouberg in the Western Cape, which are affluent and still mainly white. A number of "hackers" would come to feel rather alienated themselves, as WfW teams came into their areas with little awareness of, or respect for, their experience. But that is a small matter, compared to WfW's very positive impact on the life of many black South Africans.

Soon, the distinctive yellow-and-green T-shirt of WfW could be seen in most parts of the country. The program has employed and trained many thousands of impoverished blacks, giving most of them their first ever job. It had cleared some two million hectares of IAPs by 2011. Follow-up clearances—returning to initial clearance sites to halt reinvasions—covered more than five million hectares.[12] These clearances were estimated to have restored some fifty million cubic meters of water each year, though this figure—and its cash value—are problematic, as we shall see.

Even today, there is something about WfW's win-win-win optimism that recalls the heady spirit of South Africa's first Rainbow Nation days. That relatively peaceful final transition from apartheid to democracy seemed like a joyful miracle to the world, and most of all to South Africans themselves. But then South Africa's first postapartheid decade witnessed harsh realities: a crassly mishandled AIDS epidemic, corruption scandals at cabinet level, and limited delivery on promises to the poor. More than a little of the shine has worn off that early optimism. But it is still reflected in WfW, which, along with a genius for good publicity, helps explain the program's enviable status in South African public opinion. It became a flagship program—probably *the* flagship program—of the "new" South Africa. WfW was instrumental in making Kader Asmal one of the most popular members of the government, within his own party and with the general public.

Asmal was certainly a most dynamic minister, and his legacy includes the National Water Act of 1998, which he said "liberat[ed] water from the tyranny of the landowner."[13] The image of big swimming pools and gushing lawn sprinklers in white areas, while millions of blacks had to walk long distances to a pump to draw drinking water, often from contaminated sources, had indeed been a classic illustration of apartheid injustice. Among many radical provisions, Asmal's act set much higher values for luxury use of water than for necessary consumption. Big farmers had also had almost unlimited and often heavily subsidized access to this scarce resource. Asmal served a second term as DWAF minister, then moved on to the Ministry for Education. He had retired from the cabinet for health reasons at the time I interviewed him in 2006, though he remained a member of Parliament. His successors at DWAF, Ronnie Kasrils, Buyelwa Sonjica, and Lindiwe Hendricks, have continued to give enthusiastic support to WfW, as have other ministers from partner departments.

Why has the program been so popular? "Because we delivered," Asmal says pithily. Curiously, though, he responds with Olympian disdain to questions about the many points where WfW's delivery falls far short of the program's own targets. These issues are "just matters for local government," he told me.[14]

We will have to look elsewhere for answers to the serious charges about WfW's shortcomings made by South African ecologists and economists, many of them advisers or executives on the program.

Among these criticisms, to give just one example, is a chronic failure to monitor and control the regrowth of IAPs in cleared areas. There is general consensus that the program's worst flaw has been to allow some of these areas to revert to conditions as bad, or worse, than pre-clearance levels of degradation, due to inadequate follow-up operations. Before we consider those questions, however, it is worth spending a little time with WfW on the ground, starting, just to keep things complicated, in the air.

Planting Spekboom in Baboon Canyon

"Farmers hang themselves here," says Mike Powell, looking down from the helicopter sweeping us above the muddy Groot River, as it cuts its tortuous way through the stark sandstone cliffs of the Baviaanskloof (Baboon Canyon) in the Eastern Cape Province.[15] We have just flown over a single prosperous-looking farm. Its citrus trees march smartly in deep green checker-board formations across a little plain that makes for a brief pause between the harsh mountain ridges. But almost immediately we plunge again into steep valleys between twisting cliffs, and most of the human habitations look abandoned, though large flocks of sheep indicate that some farmers are still struggling on. But dense stands of invasive Australian Black Wattle seize the once (relatively) fertile ground along the river margins. It is easy to imagine that they are visibly sapping its stream flow, as the Groot River is only moving in fits and starts, but in high summer it probably never ran very fast anyway. Higher up, however, there is no doubt about the visibility of a different problem. Overgrazing has stripped the slopes of ground cover. At one point a good fence draws a textbook line between the fuzzy greens of the region's characteristic subtropical thicket vegetation and the bald grays and browns scoured by herds of goats.

But things change quickly when you're flying, and we are suddenly cresting a broad grassy hilltop and scattering a family of kudu. The muscular bull stands his ground for a few instants before tossing his fine corkscrew horns and galloping off. Baviaanskloof is one of the eight areas in the Cape with UNESCO World Heritage Site status. It is rich in game and vegetation—except in the many sectors where IAPs and consequent degradation have taken hold. It is a huge wilderness area, popular with adventurous hikers, who follow the river for some two hundred kilometers between two mountain systems. Another abrupt change in perspective reveals hundreds of blackened stacks of wattle trunks laid out in neat rows for several kilometers along a broad valley floor. Like many restoration projects in their early stages, it looks like devastation, but it represents renewal. Elite WfW teams have been flown in here—some valleys are impossible to reach by road—and they have burned and cut broad clusters of the invasive plants. But vast stands of wattles further down the valley shows that it will take Herculean endeavors to drive all the invasive aliens out of this mountain stronghold.

Today we are not visiting these clearance sites close up, but looking at another enterprise, rather unusual in the WfW context. If successful, it will become a prototype for a whole new program, Working for Woodlands. (Working for Water has already given birth to two other programs—Working for Wetlands and Working on Fire). The Subtropical Thicket Restoration Project is unusual because IAPs are not the main problem it deals with. Furthermore, the project involves active restoration rather than relying on the spontaneous regeneration of indigenous plants. Powell is the manager of this experiment, which is testing both the ecological feasibility and the economic value of undoing the damage done by overgrazing. On one optimistic reading, this project could ultimately "restore biodiversity to the transformed landscape, earn carbon credits on international markets, reduce soil erosion, increase wildlife carrying capacity, improve water filtration and retention, and provide employment to rural communities."[16] That is the theory, and it will take a number of years to test it. The practice lies in the valley now swooping up below us. The helicopter drops into a clearing, among what Powell tells me was once pristine subtropical thicket, but has now lost much of its vegetation to goats and sheep.

Tall, urbane, and witty, with cropped hair, stylish floppy bush hat and immaculate pale khaki shorts and shirt, Powell might at first sight be taken for a tourist on his first day at a game farm. A pair of Clark Kent–type glasses adds an even more incongruous and rather bookish touch to the appearance of a man who works outdoors daily in such remote valleys. But perhaps he needs to be a Superman to take on the job of attempting restoration here. He combines a sharp scientific curiosity with savvy practical knowledge of his environment and the human beings who live in it. Completely at home in the bush, he strides off toward a group of figures, barely dots in the dry and shimmering midday heat.

The sides of the Baviaanskloof range from very steep to vertical. Halfway up a 45-degree slope, under the supervision of Abbey-gail Lukas, a spirited and attractive nineteen-year-old, ten men and women are carrying what looked like police batons, each sprouting a few tentative leaves. These thick cuttings are from a succulent shrub locally known as spekboom. The workers begin to plant them in circles three meters apart. The spacing of the plants varies from one zone to another, Powell explains, to test which configuration permits the fastest regeneration. Nearly two years into the project, he is moderately pleased with the results. Most of the cuttings are taking, and surviving. Spekboom is the characteristic ground-cover plant of subtropical thicket in this area. There is a healthy patch of it five hundred meters away, but where Lukas's crew is working the earth is bare, as dry and rough as sandpaper. Pick up a stone and it will burn your hand in an instant. A century of overgrazing by goats and sheep has left only a scattering of canopy trees standing to give some shade. Their shadows do not extend far enough to shelter their own root systems; without restoration, the days of these jacket plums are themselves numbered. Where there is no spekboom cover, their roots are baking in soil that has becoming a slow-cooking solar-powered oven. Some have already shriveled to gray skeletons and keeled over. Desertification is just one short step away, and tipping the scales in favor of restoration will be a close-run thing.

Abbey-gail and her team have to carry the cuttings, ten kilos per person at a time, from the nearest dirt road, almost a kilometer distant. The workers are paid 40 rand ($6.60) a day. (The rate roughly doubled over the next five years, but remains very low). If they don't plant the spekboom at the appropriate intervals and depth for their zone, they don't get paid at all. They would like more money, but in a province where unemployment runs at 50 percent, they say they feel lucky to be working. "I am learning so much about our native plants, for me it is a pleasure to work here," says Cedric "Sappie" Kleinbooi in halting but precise English. Several of his colleagues look decidedly skeptical. "Having a job is nice, but I don't like going up and down the mountain," says Tersia Noubous.[17]

The team works willingly enough, but Lukas does not have an easy job. It is hard in any culture for a nineteen-year-old girl to give instructions to thirty-five-year-old men and to women old enough to be her granny. In traditional rural South Africa, her age and gender are major handicaps, though her tough but humorous manner is an asset. She tells me shyly that she has a young child. "The father is a *Zulu*," she says with a curious emphasis, smiling through teeth that a Californian model would die for. It is clear from her tone that her partner's race is an occasion for both amusement and pride.[18] WfW's crèche facilities do not reach such an isolated area, and she has to pay a babysitter. "This is hard work and we need more money. But there isn't any other work."[19]

As manager, Powell has his own problems. On the day of my visit, he challenges Lukas over a recent absence from work, of which she had not informed him. She left the team without a leader for several days. He tells her he may have to fire her, or at least demote her as supervisor. They have a long, tough, but mutually respectful discussion. She convinces him that she had to go to Port Elizabeth because of a query from the tax authorities, and he decides to let her stay on. "Bureaucracy makes confusing demands on people here, and travel without a car can take forever. Black people have been exploited for so long that it is natural they have problems with time-keeping now that they have some freedom," he reflects later. But clearly, training this team has been an arduous task, and he rather dreads having to train another one when its contract expires. This is an issue that repeatedly undermines continuity in WfW projects. The limited employment contract is justified on the grounds that it allows a maximum number of the very poor to access work on the program. However, the failure to create suitable job opportunities in plant clearance in the private sector means that this rotational policy often simply plunges people who have briefly had the benefit of training and a minimal income straight back into unemployment and acute poverty.

WfW operates in many contexts, and my next field trip takes me to the Ninth South African Infantry Battalion army base near Cape Town. The perimeter of the base is heavily infested with two of the worst invasive aliens in the locality: Port Jackson willow and, to a lesser extent, rooikrans. My guide on this occasion is Christo Marais, the ecologist

7. **High noon.** Mike Powell, manager of the Subtropical Thicket Restoration Project, has a robust discussion with Abbey-gail Lukas about her role as team supervisor, in the midday heat of Baviaanskloof. (Photograph by Paddy Woodworth.)

from the Fynbos Forum who picked up Preston's challenge to help put the program in motion in 1995. He is a white Afrikaans speaker and frankly admits that, while he never thought apartheid was a great political system, he grew up without ever doing anything in particular to oppose it. But now he is a passionate believer in his country's ability to build a genuinely multiracial and fair society, though he has no illusions about the huge difficulties involved in achieving that goal. When I met him in 2006, he was WfW's executive manager for partnerships with government departments and programs.

Perhaps the most unlikely of all these many and varied partnerships is the one we are visiting, Operation Vuselela (Renewal). At its launch in 2003, DWAF Minister Ronnie Kasrils was moved to recall the biblical injunction to turn swords into plowshares. Vuselela, he said, would require "the active deployment of military veterans, armed with 'ploughs,' to further serve their nation."[20] In short, Vuselela offered old soldiers work combating IAPs on land managed by the Department of Defense. In the new South African context, of course, the old soldiers came from several different armies. They included former members of Umkhonto we Sizwe (Spear of the Nation), the armed wing of Mandela's ANC, and corresponding groups related to the more radical Pan-Africanist Congress and the Zulu nationalist Inkatha Freedom Party. These rival resistance groups clashed, sometimes violently, during the apartheid period. And Vuselela went further in pursuit of reconciliation, recruiting former members of apartheid South Africa's security forces, once pillars of the old regime.

At the army base, a group of veterans is doing follow-up work after clearance, spraying herbicide on the cut stems of acacias. "Did you think that fighting in the liberation war would lead you to this kind of work?" I ask the men. "We did not think we would live to do any kind of work at all," one responds laconically. "The work is not bad, it is better than staying at home," says another, rather noncommittally. Even the poorest of the poor will hardly get very excited about working for such low rates. "We have learned about these plants, our natural enemies, we can advise our neighbors not to plant them," says a third. "But we need proper management, we need to get paid on time."[21] This is a common and well-grounded complaint among WfW workers, which I hear again from a team working on Table Mountain, and another near the Kogelberg Reserve on the Garden Route. From the bottom up, WfW does not always look quite as glossy as its annual reports. Marais, who has been observing me at a discreet distance, is talking to the Vuselela project manager Mncedisi Mciteka. The two discuss difficult management issues with a refreshing frankness that I find characterizes WfW staff. They support the program with passion, but they do so with their eyes wide open. There is obviously a strong bond of trust between these two men, yet Mciteka's life experience has been worlds apart from that of Marais. He was a senior ANC military commander, a role that cost him years in prison, and he was a significant figure in the very first stages of the talks that ended apartheid. When the last apartheid state president, F.W. de Klerk, was meeting Mandela secretly in his prison cell, there came a point when the ANC president insisted that he be able to consult with senior ANC figures, including the military leadership. Mciteka was one of the Umkhonto we Sizwe delegates released temporarily to participate in those talks. With credentials like that, Mciteka could have become comfortable in an ANC sinecure after liberation, like many of his comrades.

Instead, he chose to serve in a demanding position in WfW, not very well paid and working alongside men whom he once might have regarded as enemies, like Marais, and like Bandile Joyi, another WfW manager who is on site today.

Joyi also spent time in prison, but as a leader of the military wing of the ANC's more radical rival, the Pan-Africanist Congress, famous for its chilling slogan, bordering on black racism: "One Settler, One Bullet." Listening to the ease with which these three men animatedly debate the creation of value-added industries based on the cleared wood stacked around us, there is a palpable sense that Mandela's vision of a Rainbow Nation could be a reality, albeit a distinctly delicate one.

"There is an extraordinary quality of patriotism about men like Mciteka," Marais comments afterward. Some months later, Mciteka died suddenly. Marais felt his death was a terrible loss to the organization. He was deeply moved when he found he was the only white man invited to speak at his funeral, an unlikely place, as he put it, "to find an Afrikaner farm boy like me."[22]

WfW: A World Leader with a Negative Scorecard?

WfW does good things under difficult circumstances at many levels. The program has much good news to tell to a world battered by catastrophes. So it can seem churlish to go behind the high aspirations, the headline rhetoric, and the genuine feel-good moments and assess its contribution to restoration in a cool and objective manner. It has captured imaginations worldwide because of the huge ambition of its linkage of ecological, economic, and social aims, and has won a number of prestigious international awards. "The program is recognized as a world leader in its field," says Asmal.[23] He claims that President Bill Clinton backed the Comprehensive Everglades Restoration Program in Florida, in some respects the most ambitious restoration project in North America, because he was impressed by WfW. And in South Africa its national profile is undoubtedly impressive. "There has been a huge return in awareness, in mainstreaming nature to the general population," says WfW adviser Richard Cowling.[24] In short, it is the kind of program most people of good will would just love to see succeeding. Moreover, there has been enough tangible success on various levels to justify at least a part of the program's enviably high reputation.

Nevertheless, a closer examination does confirm that WfW has many significant deficiencies. This should hardly be either a surprise or a motive for hand wringing. As a pioneer project it is natural that there is as much to learn from its failures as from its achievements. Nor is it surprising that many of these failures derive precisely from its bold attempt to combine ecological, economic, and social goals. And such an examination is made much easier by the great frankness with which its executives and ordinary workers are willing to discuss its problems—an openness that is a welcome feature of South Africa's young democracy. WfW identifies its central goal as a South Africa where "invasive alien species are controlled or eradicated, and introductions of potentially invasive species are prevented, in order to contribute to economic empowerment, social equity and ecological integrity." Its far from modest list of ancillary targets includes "improvements in the fields of ecology, hydrology, and agriculture, socioeconomic empowerment, and economic and institutional development."[25] The improvements in agriculture relate to restoration of soil quality and the reduction of fire risk by IAP removal.

Even on the essential question of IAP eradication and control, however, WfW's scorecard remains, on the face of it, distinctly negative. Over the past twelve years, despite the hundreds of millions of dollars spent and the many thousands of people employed, the stark fact is that IAPs occupy significantly *more* territory than they did when the program started. They are advancing faster than the program can clear them. That's not all: there is a consensus that climate change has already made the war on invasive plants more difficult than ever and will introduce new and unexpected obstacles in the future. Most disturbing of all, even where the program has cleared invasive aliens effectively, big questions remain about the extent and effectiveness of follow-up clearances to halt reinvasions, despite very significant improvements in this area since 2006. Continuing failures to monitor and follow up still undermine WfW's ability to deliver on its hydrological and ecological targets. And despite WfW's energy, imagination, and commitment on diverse social issues, it has also had severe problems in meeting its socioeconomic objectives. Its senior executives and advisors acknowledge many of these difficulties, often differing only in the degree of shortfall that they accept has occurred.

"The invasive alien plant population is increasing exponentially; WfW is still only at the margins," says Cowling bluntly. Preston says that Cowling is exaggerating—"the growth is nowhere near exponential"—but then concedes, "We are losing the battle so far."[26] The defeats inflicted have forced some hard reassessment. Christo Marais and Brian W. van Wilgen have estimated that "at current rates of clearing . . . many species will not be brought under control in the next half century."[27] WfW's original target date for overall control was 2020. However, Preston contends that losing the battle to date does not mean that WfW has made a bad start on the war. On the contrary, he believes the war would now be pretty well unwinnable if the program had not taken off when it did. The spread would have been much faster, he says, and it would now be almost impossible to get any kind of a grip on the problem if clearing had not started in 1995.

This is a valid point, yet some of WfW's problems today are due to the very haste with which it was set up and to its own very rapid growth as a public works program. The best intentions of WfW have been victims of the program's own extraordinary political success; its strengths have contributed to its weaknesses.

Good Politics, Bad Ecology?

"It all started in such a rush that there was no strategic ecological plan," says Patricia Holmes, an ecologist who participated in an external audit commissioned by WfW in 2003.[28] She is a precise-minded woman who expresses her views bluntly: "They had to spend money immediately, and this was ecologically disastrous. Instead of focusing on the areas worst-affected by invasive aliens, and working systematically, they spread the program far too wide across the country for political reasons."

Similar criticisms are also leveled against planning on a local scale. There are certain basic practical rules for clearing IAPs from a watershed. You should start at the highest level, because otherwise the plants at the top of the catchment will continuously reseed the lower levels. You should clear the lightest areas of infestation first, because otherwise these areas will continue to rapidly spread the infestation while your team is tied down clearing the well-established and denser thickets. "I do think we have made a few very serious mistakes in some areas when it comes to approaching the work systematically," admits Christo Marais, "but definitely not in all areas."[29] It is important to note that there is no significant disagreement about methods between Marais and Holmes. Indeed, they coauthored a book chapter in 2007 where they laid out a clear and comprehensive roadmap of priorities for "restoration by removal of invasive alien plants" in a fynbos context.[30] But Marais has to operate in the practical setting of an unwieldy public works program, while Holmes's role has been to assess that program from the standpoint of theoretical best practice.

Holmes acknowledged even in 2006 that there had already been significant improvements on both the national and local scales, and shares the consensus that things have got better since then. But her point that the sociopolitical priorities of the program clashed, at least initially, with its ecological priorities is telling. It is quite natural that a public works program, financed by the state's poverty alleviation fund, should direct its efforts to areas of greatest economic need. Understandably, too, there is political pressure in each locality for a slice of such a popular initiative. But if WfW's primary target is ecological—the eradication and control of invasive aliens—it would make much more sense to focus on those areas where infestation is most acute. Many of the worst IAP-infested areas are in the Western Cape, a region with a large and generally affluent white population. Many of the worst black poverty zones are in the east and north, where infestation is not yet a major problem. But, as Holmes says, political pressures dictate that WfW teams should be created in more or less equal numbers right across the state. So the Western Cape does not have nearly enough people working on clearance, and KwaZulu-Natal may have more than it really needs. This is one of a number of instances where the politics of the program runs counter to its science. To adapt the language of RNC, it is evident that restoring social capital may conflict with restoring natural capital, at least in the short term. "The bottom line for us is getting poor people into work," was how one regional manager put it.

The pragmatic counterargument is that there was no option but to kick-start WfW and extend it to every province, if it was to get anything approaching the resources it required in any province. "This program would not have taken off at all," said Asmal, "if it had not had the élan, the spirit of that extraordinary time."[31] However, that élan created a hasty dynamic that did not always resonate with the best principles of scientific practice. No clear assessment of ecological priorities was drawn up before the minister famously cut the first invasive alien tree with one blow of a slash-hook in 1995, earning himself the sobriquet of "one-slash Asmal," which he rather enjoyed. And no such full assessment and plan have been plotted out since, though again there have been significant improvements in this area.

"There is inevitably a clash between our social and ecological goals," says Preston. "We have worked where there is acute poverty; we have worked where there is unrest and violence, because we are a public works program dedicated to providing opportunities to the most disadvantaged people." But, referring back to Holmes's assertion, he asks, "How can the clearing have been 'ecologically disastrous'? It may not have been suitably prioritized, but what would be ecologically disastrous would have been not to do anything."[32]

Christo Marais "absolutely agrees" that strategic ecological planning was lacking at the outset. But he asks, "Doesn't the restoration of social capital form part of the restoration of natural capital? . . . We were in a very significant window of opportunity, and if we had waited five years to get all the plans and science in place we would not have got the political buy-in and budget."[33] Marais also believes that the pressing political and social realities of South Africa dictate that poverty relief cannot take second place on the program to IAP control, despite a categorical recommendation from the program's first external auditors—in a report known as "Common Ground"—to the contrary[34]: "We have to remember we are controlling invasive alien plants on poverty relief funds." It will only be possible to fully prioritize ecological goals, he says, "when we've sorted out the unemployment in the country."

And everyone knows that may take a long time. Mike Powell concurs: "Kader Asmal and Guy Preston had the vision to start WfW up in tough times; it would have been much easier for that government to build more houses, for example. For all its problems, we have to keep at it; it would be suicide to let it go now."[35]

WfW has come under close research scrutiny from planners and scientists on two occasions, both prompted, to its credit, by the organization itself. One was the Common Ground external audit; the other was a research symposium held at Kirstenbosch, Cape Town, in August 2003, which generated a series of very illuminating published papers.[36] What is perhaps most surprising, looking at these papers and the external audit, is that the key hypothesis which gave birth to WfW—that invasive aliens consume significantly more water than indigenous vegetation—has not been clearly substantiated by the program's own clearances.

Estimates of the amount of water saved by WfW, says a key overview paper by the highly respected South African ecological consultant Ian Macdonald, "are based in the main on the extrapolation of results from a few paired catchment studies" predating the program. These experiments at Jonkershoek took place in conditions different from those in most catchments where WfW has been operating, and the biomass/water-use model employed may not be appropriate. But, on balance, Macdonald accepts that "the water benefits alone . . . are enough to justify the programme."[37] So, while few doubt that the program has in most cases increased water supply, no one can say with any precision how great that increase has been. It was claimed at the symposium that the project so far had yielded between forty-eight and fifty-six million cubic meters of additional water annually, roughly the capacity of a medium-sized dam. But this remains closer to a guesstimate than a fully scientific calculation.

Indeed, Brian W. van Wilgen, the senior South African ecologist who was instrumental in setting up WfW through the Fynbos Forum and who remains one of its key advisors, concedes that a "pressing question" remains as to whether "the predictions of significant benefits arising from the control of invasive alien plants can be substantiated by good science."[38] This is a remarkable admission, to put it mildly, from such a well-informed figure. One of van Wilgen's colleagues at the Council for Scientific and Industrial Research, the hydrologist David Le Maitre, is also cautious. "There generally is more water in rivers and aquifers as a result of clearing [of invasive aliens], but not all species are equal in the amount of water they use," he told me.[39] "We do not really know how much water the natural vegetation which regrows in cleared riparian areas will use compared with the invaders; this will almost certainly be less than that of tall tree invaders—but for medium trees and tall shrubs, we are less sure."

"What we need is hard data, paired catchments, experiments over ten years." A great opportunity to do such research in the context of the program's clearances has been missed, he believes. Echoing van Wilgen, he continues, "The program is having a hydrological effect, but we don't know how much."

While Preston accepts that more hydrological research is needed, he argues that the scale of clearance so far must mean, in itself, that water gains are substantial. And he adds that WfW does "more research than any other conservation program, and certainly more than any other public works program."[40] He also points to progress toward the program's other targets, including the recovery of soil quality on agricultural land and the reduction of fire risk. Mike Powell agrees that more scientific evidence is needed to confirm how much water is being restored, but he says the anecdotal evidence he and his colleagues hear from regularly local residents of cleared catchments cannot be dismissed: "It's not just a myth, we are told repeatedly 'Thank God you guys came along, the streams are running again.'"[41]

Economist Stephen Hosking says he remains unconvinced about water, soil, and fire benefits.[42] But then, he identifies a completely different strand of argument in the program's defense, using the contingency valuation method (CVM). This is an attempt to put a price on nontradable assets, for example, the "existence value" of an aesthetically pleasing landscape, by running opinion polls to establish how much the public would be prepared to pay to preserve them or restore them. Some economists regard these methods as far too subjective to be reliable. Others say the CVM is a very useful way of guiding public policy on goods that would otherwise be regarded, rather vaguely, as invaluable and therefore, somewhat paradoxically, as of no economic value at all. CVM research that Hosking has supervised at the Nelson Mandela Metropolitan University in Port Elizabeth suggests strongly that most South Africans are willing to pay for the restoration of their native vegetation as a public good. He argues that this means that WfW could justifiably continue to spend tax revenue even if it were failing to meet other goals.[43]

The visual impact of IAP clearance and the restoration of fynbos can certainly be truly spectacular and very pleasing to the eye, bringing not only aesthetic benefits to residents but also direct economic benefits in terms of tourism. However, as we will see in later chapters, this kind of valuation can be a two-edged weapon for restorationists. Some invasive alien vegetation is aesthetically unpleasing to a majority of the population—especially where it causes the obvious degradation of a landscape. However, other invasive aliens may be very popular with the general public: eucalyptus in California, silver poplar and Norway maple and even European buckthorn in Chicago, beech in Ireland, and mimosa in Provence. In any case, in a country with South Africa's problems, the acid test of a program like WfW will undoubtedly be whether it is meeting its stated economic, social and ecological goals. CVM-related benefits certainly add value to its overall worth, but they can hardly tip the scales in WfW's favor if the balance sheet on its priority targets is not positive.

What Is Water Worth?

The problem in establishing how much water WfW has saved is compounded by differing economic assessments of the cash value of this resource. Beatrice Conradie of Cape Town University, who dismisses CVM as "crazy," claims bluntly that "as an engineering argument for water, the program is stupid," though she concedes that WfW is "a very clever marketing ploy, making a public works program sexy by linking it with ecology. But it costs a hundred-rand investment in public funds for every twenty rand of water restored."[44]

Christo Marais contests this figure. "Our research work on the watershed supplying the town of George [in Western Cape Province] and elsewhere has shown that clearing invasive aliens from riverbanks is about the cheapest water augmentation option you'll find, except stopping leaking pipes and getting people to use less water." In the case of George, his research shows that clearance resulted in significant net savings in water cost. "Conradie's arguments are based on very, very low values of water," Marais continues. "There are serious debates about the value of water worldwide. The value of water in water-scarce countries such as South Africa is very often seriously underestimated. And she also does not take the broader ecological and agricultural benefits into account."[45]

However, the reality remains that WfW has simply not generated enough data to settle such arguments conclusively, despite the fact that the program represents "the best opportunity anywhere in the world for research on landscape-level modifications of the environment over a wide range of ecosystems," according to Ian Macdonald.[46] And this absence of information can again be traced to the frenetic haste with which the program started up. The resultant "undermanagement" (Preston's word) led to what is perhaps WfW's greatest flaw: the program has very often failed to monitor even its own basic clearance work. All too frequently, neither the successes it claims, nor the failures alleged by its critics, can be fully substantiated.

"Record-keeping at the beginning was simply abominable," says Barbara Schreiner, formerly a top civil servant at the Department of Water and Forestry Affairs and now an independent consultant. "Monitoring has improved significantly, but we're not there yet. But please remember that WfW was born at a time of great turmoil in this country. And while we still face very complex challenges to get the management right, they are not impossible to resolve."[47] Richard Cowling laments the "fragmentation" of research and implementation in WfW, saying that it is "tragic that there was not more put into monitoring these field-scale experiments" from the very beginning of the program.[48]

Nothing remains static in these debates, and the WfW leadership responds energetically to challenges—adaptive management in action. Christo Marais has more recently collaborated on several papers attempting to establish the amount of water saved by IAP removal. In one of them, he supports the proposal that payments to WfW, or to independent contractors clearing IAPs, should be "strongly linked to water supply delivery targets."[49] If it is implemented effectively, this proposed measure could become a landmark model for payment for ecosystem services. Moreover, because taxpayers and their political representatives would no doubt closely scrutinize it, this measure might perhaps finally remove much of the doubt over both the water impact of IAPs and over the economic value of water. In a parallel paper, Marais and other colleagues attempt to estimate "conservative" national figures for that water loss (streamflow reduction) due to IAPs, and conclude that 4 percent is being lost under current conditions, a figure that would quadruple in a "fully invaded" scenario where control was abandoned.[50] Valuable though such studies are, they still fall short of the "hard data" and long-term experiments that

Le Maitre argues are needed for a higher degree of scientific certainty on this issue.

Marais returned to the fray again a year later with a paper that teased out the costs and benefits of IAP clearance in terms of water yields with more precision than any previous work, separating out the impacts of different aliens in different ecosystems.[51] The paper concludes with what he calls a conservative estimate that 7 percent of IAPs have been cleared by WfW, yielding 34.4 million cubic meters per year at a cost of 116 million rand. This contrasts very favorably with the new Berg River dam scheme, which has yielded 81 billion cubic meters of water per year at a cost of 116 *billion* rand. While much remains to be done in terms of long-term data collection and analysis, it is hard to argue with their conclusion that WfW's costs and benefits already represent "a very good investment."[52] In a further article with other colleagues, Marais advanced the view that WfW was moving toward becoming the vanguard of a nationwide system of explicit payments for ecosystem services.[53] Along with analog services in carbon sequestration, water purification, and fire protection from its sister organizations, it could become an umbrella for an emerging network that marries social needs with conservation goals.[54]

Monitoring streamflows and, especially, aquifer levels is undoubtedly a demanding task. But WfW has often failed at the much simpler but crucial work of follow-up—monitoring the impact of each clearance, and returning for further clearances where the IAPs are resprouting, as often happens. Obviously, just cutting down tenaciously invasive trees and shrubs is not enough—indeed, it may simply amount to "pruning," so that the offensive vegetation springs back, even denser than before within months. There are catchments where the alien infestation is significantly worse after a WfW clearance than before.

"This is a cardinal sin," concedes Preston. Then he candidly adds, "I don't even know how often this has happened, and it's shocking that I don't know, because of poor reporting and bad management."[55]

Preston's honesty is admirable, but his admission raises doubts as to whether WfW can ever get a firm handle on the fundamentals of the huge task it has set itself. Some well-informed critics question whether it really comes close to fulfilling its brief to "improve the ecological integrity" of the land it clears. The standard WfW prac-

8. **Degradation and restoration.** A fence line keeps overgrazing livestock out of the thicket on Baviaanskloof, where the manual planting of spekboom is assisting recovery. (Photograph courtesy of Mike Powell.)

tice is to clear invasive alien vegetation manually, using slash-hooks and, where necessary, chainsaws. There is usually some subsequent treatment of stumps with herbicide to prevent resprouting. But there is very little active restoration of indigenous vegetation. It has been generally assumed that fynbos will come back of its own accord from seedbeds established over many previous decades. While this is true in some contexts, it is not in many others, especially where long-term IAP infestation has formed a closed canopy forest, denying light to any indigenous vegetation that remained beneath. Despite this, WfW usually practices passive rather than active restoration in all circumstances, once the basic clearing has been done (the Sub-Tropical Thicket Project is clearly an exception). And, quite often, reliance on passive restoration simply gives the invasive aliens an opportunity to surge back with renewed vigor, so that the last state of things really is worse than the first.

Patricia Holmes says WfW teams are indeed sometimes responsible for several kinds of ecological deterioration. "There is no ecological monitoring whatsoever. How do you train the poorest of the poor to distinguish native and invasive alien species? There are instances where indigenous plant populations are actually reduced after clearance."[56] This is an argument I have also heard from frustrated "hack" volunteers, who claim that precious patches of native vegetation, which they have protected for decades, have been destroyed by WfW intervention. One cannot discount the possibility that such arguments by volunteers are sometimes motivated by resentment at the intrusion of WfW in areas the hackers have worked lovingly for many years. And the suspicion lingers that they may also be tainted by racist attitudes, conscious or otherwise, to poor black workers doing a skilled ecological job. But neither can one dismiss these arguments on such suspicions alone.

Holmes points out that soil systems developed over millennia to support fynbos are often badly degraded by just a few years of alien plant infestation. Clearing the invasive aliens without any follow-up work may simply make matters worse. In some contexts, the disturbance involved in their removal will loosen and expose this impoverished soil to rapid erosion unless great care is taken. Clearance, she says, should be coupled with active ecological restoration where necessary, by planting native species, which will rebind and rebuild the soil. Relying solely on spontaneous restoration of the indigenous species works in some circumstances but not others. She would like to see key fynbos species replanted or reseeded in a number of the cleared areas, followed by supervision of the overall recovery of the ecosystem. Initially, soil stabilization by mechanical means may also be necessary. Unless these kinds of measures are taken, she says, "you may just be removing one alien only to see it replaced by another," because many invasive nonnative species are more successful at colonizing disturbed sites than indigenous plants.[57]

On these issues, Preston mostly disagrees. "We should not be starry-eyed about restoration. In a perfect world, yes, it would be best. But I've yet to see figures that would justify it. We would have to be sure that the net benefit [of active restoration over spontaneous regeneration] would be greater. I'm not convinced of that."[58] Marais points out that WfW has in fact invested substantial funds in researching active restoration options, but when I first spoke to him he doubted that there were many circumstances where it can be economically justified. However, in a 2008 article he cautiously endorsed a much more active restoration policy in some instances, especially following eucalyptus clearance.[59] And he has a keen awareness of the ecological degradation some WfW activities cause. Leaving cut trees and brush stacked after clearance can foster fires at far greater heat than a "natural" fynbos fire or even a forest fire, and result in sterilization of the soil. He showed me dramatic illustrations of this damage when I last interviewed him, in the same year.

A Problematic Marriage of Two Key Ideas

The roots of many of the problems in this field lie, once again, in South Africa's social context: how can you justify spending money, which is largely derived from a poverty alleviation fund, on apparently abstruse ecological practices? Can cash drawn down for basic wages be diverted to pay fees to the middle-class specialists needed to do the kind of research that might demonstrate how well the program is working? Yet, inside and outside WfW, cogent voices argue that skimping resources on active restoration, monitoring, and research risks wasting more public money in the long term. As one WfW regional manager puts it, "the higher initial cost of restoration may be compensated by lower future management costs," because a properly restored watershed will be much less prone to re-invasion in future. But balancing priorities in a program with so many goals is clearly a bit of a nightmare.

Jennifer Gouza, a young assistant manager in the Eastern Cape, sums up the situation like this: "We can't go forward unless we sort out these issues. But trying to raise the poor one step up, and the environment one step up, and marrying those two ideas, is very difficult."[60] Just how difficult that marriage can be is also revealed by the problems WfW has had in achieving its ambitious social goals. As we have seen, the program not only aspires to provide short-term poverty relief but has sought to empower its workers to find continuing employment for themselves when their two-year contracts expire. This process was to have been facilitated by team leaders, who operate as independent contractors within the program.

The hope was that, on exiting the program, they would form their own plant-clearing businesses, employing former WfW colleagues. The theory was that there would be a big demand for their skills from private landowners, who would be firmly nudged by new laws to clear their own land of invasive aliens, and keep it clear. Guy Preston still hopes that regulations drawn up under South Africa's 2004 Biodiversity Act will tighten enforcement on landowners to initiate and maintain IAP clearances after WfW has moved on. This development is essential both to maintain the ecological impact of the program, and to provide ongoing work opportunities for its former employees. While not shirking the program's share of the blame, he believes that WfW's lack of a strong autonomous mandate has made it very difficult to move ahead in this area. WfW's dependence, not only on the DWAF, but on the fourteen other government departments that have a partnership role in the program, can be a source of strength—but clearly it can also create bureaucratic paralysis. The failure of the Department of Agriculture to make landowners obey the new laws and its reluctance to devolve enforcement powers to WfW lie, he believes, at the root of the problem:

> This failure has led to avoidable inefficiencies. Perhaps the most poignant example is the failure to create an "enabling environment" for workers and contractors exiting the program. Were disincentives [like fining landowners who fail to clear invasive plants] to be used intelligently, it would be possible to create work opportunities for those who have completed an adequate training and empowerment program within Working for Water. But that has not happened. In essence, we over-subsidize landowners, instead of using our laws and incentives to secure sustainable solutions. Linked to this lack of a mandate and decision-making authority have been administrative weaknesses, resulting in delays in procurement.[61]

He argues that dependence on other agencies has led to WfW's failure to always pay contractors and their workers within the stipulated time frame, as the workers on the military base and elsewhere had told me. Such late payments, he has acknowledged, are "totally unacceptable in a poverty relief programme."[62]

There is a further key complication in the post apartheid context. This is the very understandable demand for what is called "transformation," a euphemism for redressing the disproportionate race and gender balance in senior positions. This balance has hugely favored white males in the past. WfW's record in this area is very strong, and, according to Preston, unprecedented. In 1995, the management committee was made up of twenty-one people, twenty of them white men. By 2003, there were twenty blacks and fourteen women on a committee of twenty-six. This seems laudable in broad political terms. But then, one surely has to ask, does such "transformation" cause downskilling, if people are being promoted on the basis of race and gender quotas rather than talent and experience? Preston is well aware of this danger. He is adamant that "whoever has a job must be able to do it. Where this goes wrong, the victims are the poor, who suffer from bad administration. So there is a lot of mentoring involved. Perhaps we did not do this as well as we might have, but is essential that we made the transformation."

Marais believes my question is back-to-front: "You should be asking, 'Has transformation contributed to the *development* of skills?' My answer is yes; a number of youngsters who came through my projects can walk into senior jobs on merit and not based on the color of their skins. So yes, maybe we did lose skills but we've also gained skills."[63]

The jury must remain out on WfW's overall performance for years to come, partly because of the time required to judge results—an almost universal issue with restoration projects—and partly because of the extraordinary scale and diversity of its ambitions. As we have seen, it has yet to show that it can meet any of its key objectives in ecological restoration and the restoration of natural capital on a national scale, though it has significant achievements to its credit on some local and regional scales. As a poverty alleviation program, it has certainly brought employment opportunities, training, dignity, and increased incomes to thousands of people who desperately needed them. And the additional services it has provided, from AIDS education to ecocoffins to crèches, provide vivid evidence of the commitment of its staff to a better South Africa. Overall, the costs of running the program are reckoned to exceed the direct economic benefits by about 20 percent annually, but if one takes into account the indirect benefits, including tourism and CVM benefits, and especially the medium term negative cost of allowing IAP infestations to spread unchecked, it probably more than pays it way.[64] However, one has to remember that WfW wages are well below the market rate. And, most critically of all, the program's undoubted achievements still fall far short of the targets that it has set for itself.

But at this stage it would be a mistake to judge the program too harshly for its shortcomings, especially since many of its most perceptive critics either work within the program, or act as advisors to it, and their feedback is taken very seriously at the highest levels of management.

"The New South Africa Is Making History Here"

One of these advisors, Richard Cowling, a passionate ecologist, has serious concerns about WfW's performance, as we have seen, but nevertheless considers it a "roaring success. With all its flaws, it's magnificent. We should not be too fixated on direct economic outcomes," he says, given the program's achievement in bringing ecosystem health to the forefront of public opinion.[65] And that mainstreaming of ecological issues into political, social and economic life may indeed be WfW's greatest legacy. Says Preston:

> For all the many mistakes, I would again go the same route, rather than a more cautious one. Where I would have changed things, in retrospect, would have been to have used my political position to secure far greater autonomy for the program. I think we have shown this can work with the comparative successes of recent sibling programs like Working on Fire and Working for Wetlands. The fundamentals of WfW are good, and its problems are primarily ones of management, including the decision-making powers of those engaged in operations. These are factors that Working on Fire (surely the best program running in South Africa at present) and the very impressive Working for Wetlands have gotten right. In terms of the WfW "model," they are showing that it can work.[66]

Christo Marais, in an e-mail after a long discussion of the program, concludes: "This year's [clearance and follow-up] figures are looking much better than last year's. It is a matter of sorting out some challenges, and we will truly be one of the best conservation programmes in the world."[67] From most other executives, in most other companies, this would sound like empty rhetoric. From Marais, who acknowledges just how big those challenges are, it is an ironic understatement linked to an assertion of unflagging commitment.

Nosipho Jezile, now a senior civil servant with the Department of Environmental Affairs and Tourism who has extensive previous experience in WfW, says that the program's development has been "a very tricky process." But, she adds, "In its defense, it has been very innovative. It takes time and energy to build an institutional framework to support sustainable development. The new South Africa is making history here."[68]

Looking to the future, there is a general consensus that WfW must not only overcome the shortcomings already identified, but will need strong support from departmental and private partners in several key areas if it is to stand a real chance of achieving its goals.

The first such area is the prevention of new introductions of IAPs, and the rapid elimination of "emerging invasive aliens," which have just begun to pose a threat. With the steep increase in travel and commerce across South Africa's borders, this poses a major challenge for the country's customs service. Prevention may also require new forms of inter-provincial control, as in Australia, where cars can be searched and even sprayed if they are suspected to be carrying IAPs or their seeds, deliberately or otherwise, across the country. Climate change is also likely to make it easier for some aliens, hitherto innocuous, to turn into significant invaders and transformers, which will make prevention more difficult still.

Second, WfW should be able to create far more added value through secondary industries than it has done to date. Great potential wealth, from biomass-based industries through chipboard and fine furniture to traditional sculpture, is locked in all those stacks of wood now piled up across the country. Profits from secondary industries could significantly subsidize the costs of the program. Furthermore, processing this wood will also remove the added fire risk that WfW operations have created in some areas.

Third, WfW needs to make much more use of what Mike Powell describes as "the sleeping giant in our arsenal"—biological control, or, biocontrol. One of the reasons IAPs spread so quickly is that they are usually free of the creatures that feed on them or otherwise control their populations in their native environment. Biocontrol consists in importing these predators and parasites and using them to eliminate or control invasive alien populations. South Africa has been a world leader in this field since the 1930s. This method is particularly useful where the alien has a positive economic impact—some species of eucalyptus are useful for honey production, for example, and poor black communities use rooikrans for charcoal. With biocontrol, especially through the introduction of seed-eating insects, the spread of the alien is contained, but some of the existing alien plant stock can still be maintained and harvested.

There is, of course, a risky paradox in introducing an alien to eliminate an alien. The downside of biocontrol occurs when the imported parasite has a negative effect on native species in its new home, as in the notorious case of the introduction of the cane toad to Australia. In South Africa, biocontrol parasites are restricted to insects and microbes, and are subject to rigorous seven-year tests to ensure that they are "host-specific"—that is, they will only eat the plant they have been brought in to control. Benefit-cost ratios within WFW for biocontrol of invasive alien species are excellent, ranging from 8:1 to as high as 709:1. But the program needs to use this weapon on a much wider basis if it is to be fully effective.[69]

* * *

Most of this chapter was drafted in 2007, based on articles I had written in 2006 for the *Scientist* and the *World Policy Journal*,[70] and then updated after another visit to South Africa in 2008. Returning to the story in 2011, I could see no reason to change its structure, because the issues raised in WfW's first decade, as it strove to advance its bold triple restoration agenda—ecological, economic and social—remain very relevant.

However, it is only fair to the program to record here briefly some details of its often-remarkable progress over the intervening three years. When I recontacted Guy Preston in May 2011, I found him as indefatigable as ever, despite a recent encounter with cancer about which he was both extraordinarily brave and excruciatingly funny. He was still in the midst of treatment when we spoke by Skype, but was more than willing to talk late into the night about a major organizational revamp that he believes has liberated WfW from a number of disabling constraints. "We have turned the corner," he says. "We had one hand tied behind our backs, now we can fight with both hands in front of us."[71]

The program had just moved from its original home in DWAF to semiautonomous status within the Department of Environmental Affairs. This should mean that it will be much less subject to the vagaries of civil service bureaucracy. Meanwhile, WfW's head office staff is both doubling in numbers (to one hundred) and upskilling. "We were outrageously undermanaged," says Preston, adding that "he has never felt more positive" about the program.[72] For once, it is possible to be reasonably confident that this not a case of a quango feathering its own nest. At least as long as Preston remains in charge, waste will be cut to the bone, and hard work and commitment will be the norm for senior staff. Clearly, then, WfW remains a flagship project in a South Africa that is no longer "new" and though often turbulent and troubled is still vigorous; its rainbow more than a little tattered and faded, but still inspiring. And Preston says that the sister projects that WfW has spawned, like Working on Fire, Working for Wetlands, Working for Woodlands, and Working on Land are doing even better.[73]

There has also been significant progress on all three key targets outlined for WfW in our conclusions above. A specialist early-detection and rapid-response unit has been set up to deal with new invasive aliens and has had significant successes. WfW is becoming the key national agency dealing with biological security.[74] The value-added secondary industries associated with the program are thriving, and the ecocoffin project has expanded into the rather happier field of school desk manufacture. Indeed, there are far more ambitious schemes afoot: Working on Energy proposes to use the biomass produced by WfW to generate electricity, though government approval is still pending. If the projections are accurate, this idea could significantly augment the purely economic arguments for WfW's state funding and eventually cover much of its costs.[75] And biocontrol has moved into the forefront of the program, as can be seen from even a brief visit to its website.[76]

That's the good news. The bad news, or the worst of it, is that the threat from IAPs is even greater than anyone imagined when the program was established. Preston now says that the 1997 statistics for land critically invaded were underestimated by 50 percent.[77] This means that the program, which in any case continues to fall behind its original targets, has to run even faster just to stand still. And Preston volunteers that "data management remains a frustrating weakness in WfW."[78] This makes it hard to assess his claim that, based on just four criteria, WfW has saved South Africa 45 billion rand since its inception.[79]

It is still harder to assess whether he is right when he asserts that "there is no question" that clearing of IAPs on new sites (and follow-up clearances on reinfested sites) should always be WfW's priority over active restoration after initial clearance.[80] Some observers still think that this failure to engage with active ecological restoration remains a worrying blind spot in Preston's remarkable vision, and could yet undermine what the program has achieved. However, I found that criticism is less harsh than it was in 2006. But Preston still tends to refer to restoration as "rehabilitation" and discuss it as though it were an aesthetic rather than an ecological issue.

Patricia Holmes, who was one of the toughest critics of WfW's failure to restore in 2006, stresses that she has always supported WfW. But she continues to insist that it "should consider itself an ecological restoration program and NOT simply an alien clearing program, because there is a fundamental difference between those two stances that can lead to very different outcomes." And she adds, "I agree with Guy that at the national scale active restoration (i.e., planting natives) is a much lower priority than ensuring follow-up clearance or clearing of recently dense alien stands that still have good natural recovery potential from the seed bank. However, there are a small proportion of sites that seem to never recover—even after the 8th follow-up—and clearly an earlier intervention to re-introduce native seed or plants could have interrupted the degradation cycle and possibly been as cost-effective as continuing follow-up control."[81]

Significant research has been done in this field over the last five years, and much of it bears out Preston's contention that, where monitoring and repeated follow-up clearance is not neglected, fynbos vegetation will regenerate spontaneously and hold off alien reinvasion in most cases. But not in all of them. Karen Esler, who coedited a special issue of the *South African Journal of Botany* with Holmes and others on riparian vegetation management after alien invasion,[82] argues that riverside and some other sites are especially in need of active restoration: "Riparian zones are at the receiving end of everything that goes on in the catchment, and disturbance is the norm. This makes management of these zones particularly challenging, as biotic and abiotic thresholds may have been passed—more so than in many fynbos habitats." She continues, "There are some examples where active restoration in fynbos may also be necessary—for example in sand plain fynbos, where nutrient stocks may have been changed to an extent where remnant native species are simply unable to recover without active intervention."[83]

The best guarantee of long-term success for WfW is that these debates should continue, backed up by research, so that its development will always be guided by its exceptionally adaptive management. WfW is undoubtedly an inspiration like few other programs for the new environmental movement which, as natural capital restoration advocate James Blignaut put it at the St. Louis symposium, "treats economics as if nature mattered and treats ecology as if people mattered."[84]

And whatever its final outcome, WfW will remain a strikingly useful model for developing countries, not least because it has made serious mistakes, has acknowledged them, and has attempted to correct them. Its experience encapsulates many of the challenges of sustainable development on a planet where the depletion of both natural and social capital has reached a critical level. There really is nothing quite like it anywhere else in the world, for its scale, its multiple ambitions and, above all else, the heroic commitment of its leading figures.

5 *Awkward Questions from the Windy City: Why Restore? To What? For Whom?*

Stormy, husky, brawling
City of the Big Shoulders
CARL SANDBERG, "Chicago"

So numerous indeed and so powerful are the causes which serve to give a false bias to the judgment, that we, upon many occasions, see wise and good men on the wrong as well as on the right side of questions, of the first magnitude to society. This circumstance, if duly attended to, would always furnish a lesson of moderation to those, who are engaged in any controversy, however well persuaded of being in the right. * ALEXANDER HAMILTON, *Federalist Papers*

These "restorations" look like the deforestation of the Amazon. * BATHSHEBA BIRMAN, *antirestoration environmental activist, Urban Wildlife Coalition*

These are not people who love nature, these are not people you see at natural history meetings. * STEVE PACKARD, *cofounder of the North Branch Restoration Project, on critics like Birman*

People are learning each other's disciplines, each other's language. I hope we can bring together people who have been mutually suspicious. * LIAM HENEGHAN, *ecologist and cochair of the Chicago Wilderness Science Team*

* *

My initial encounters with ecological restoration left me, I have to admit, just a little bit starry-eyed. I had become rather world-weary after many years of reporting on the bitter and intractable Basque conflict, and indeed, in an earlier journalistic incarnation, of writing about the sometimes equally vindictive Irish arts scene. So it was enormously refreshing to engage with people who had a vision of restoring the beauty, health, and productivity we associate with natural environments. Sure, nobody pretended that restoration was easy, and my first field trip researching Working for Water had shown that it could often be damn hard—and fraught with social as well as scientific tensions. But it was a story from Chicago that made me realize what painful questions can be raised by the urge to restore, that the road to restoration hell can indeed be paved with the best of intentions.

The city's North Branch Restoration Project (NBRP) is—or used to be—legendary among US restorationists of a certain vintage. I was urged to visit it at a very early stage in my research by a senior SER member, who had herself been deeply involved in it as a volunteer. The rediscovery and restoration of a lost ecosystem right in the heart of an industrial megapolis certainly sounded like a significant and dramatic environmental narrative. I wrote to Bill Jordan about it, since he lived in the region and had given the project very honorable mention in *The Sunflower Forest*.[1] To my surprise, he suggested that my first port of call should be in Dublin, a meeting with Irish soil ecologist Liam Heneghan, who was coming home on holiday from his position in DePaul University, where the two men had established the Institute for Nature and Culture.

I met Heneghan, appropriately enough, in a restored distillery. He is exhilarating company, a blend of wiry energy, discriminating intelligence, and mischievous humor. He told me many good things about the North Branch and related projects in the city. But he also told me a cautionary tale about the strains between the citizen volunteers who made those projects happen and the ecologists who wanted to research—and hopefully improve—the restoration process. The great problem on the sites where they worked was infestation by an invasive alien, European buckthorn. He found that volunteers were frustrated by repeated and very vigorous reinvasion after laborious clearances. He proposed setting up a series of experimental plots to test soil treatments—various mulches, cover crops and so on—that might prevent this from happening. But the steward (the lead volunteer on the site) refused to allow him to set up any plots where native plants were present, which would have made the experiment useless. Her rationale was stark: "I feel like I would be sanctioning a thalidomide experiment," she told him.

Heneghan was disturbed that someone in her position would use such an analogy. He pulled out within twenty-four hours and found a more accommodating site elsewhere. He stressed to me that the work the steward was doing on the first site, which he preferred to not to name "because people would be hurt," was and is superb in many ways. He added that she genuinely believed the experiment would do more harm than good. But her unwillingness to sacrifice some cherished plants in the short term in order to find ways to protect their species better in the long run indicated an attitude toward science that he found was quite widespread among volunteers and boded badly for maintaining restoration successes into the future.[2] He (and many others) have spent the intervening years attempting, with increasing success, to build a "respectful companionship" between scientists, volunteers, and planners. But it has been a roller coaster ride, through territory that has been fiercely contested, heavily conditioned by the context of a much broader controversy about restoration that has created deep and lasting fissures in the Chicago environmentalist community.

Indeed, the story of ecological restoration in the Chicago metropolitan area has more than a little of the nervy human energy that fuels the popular TV dramas based in the city, like *ER* and *The Good Wife*. It reflects Carl Sandberg's robust characterisation of his beloved "City of the Big Shoulders"—"stormy, husky, brawling." It is a narrative packed with big human characters, dramatic advances and severe reversals, bitter and damaging conflicts, and the kind of larger-than-life outcomes that Sandberg might have relished. In the very year—1916—in which Sandberg's epic collection *Chicago Poems* was published, something else happened that would express the city and its hinterland in, at first at least, a quite different and much quieter tone than the poet's. Deer Grove, five hundred acres of upland forest punctuated with wetlands and scored with ravines, became the first land acquired by the Forest Preserve District of Cook County, newly established after many years' debate.

"Preserving Country Naturally Beautiful"

The setting up of a forest preserve district (FPD) in Cook County—and another simultaneously in neighboring DuPage—was a remarkable initiative. It was very out of character, not only with the city's rambunctious stereotype, but also with the dominant zeitgeist of the late nineteenth century in the whole country. The expanding and urbanizing United States was busily chewing up the vast Midwestern prairies and the region's smaller but still significant forests. Yet at this very moment, when images of felled trees and plowed grasslands were deemed unequivocal icons of land improvement, a group of influential Chicago citizens dared to articulate a different vision. Landscape architect Jens Jensen, along with Dwight Perkins, one of the founders of the Prairie School of architecture, were among them. They were members of a science committee that had studied the surviving natural areas surrounding Chicago, and they argued that the city needed not only more of the elegant and urbane parks of which it was already proud but also something much more radical.

"Instead of acquiring space only, the opportunity exists for preserving country naturally beautiful," they declared in 1904. They went on to namecheck a dozen localities that are today familiar as (often contested) sites of restoration, including the North Branch of the Chicago river, Palos Heights, and the Calumet River. They also stated, rather prophetically, that "all of these should be preserved for the benefit of the public in both the city and its suburbs, and *for their own sake and scientific value, which, if ever lost, cannot be restored for generations*."[3]

How right they were, yet they could not have imagined the complex dilemmas and fierce debates that, ninety years later, would attend the legacy their apparently innocent words bequeathed the city. Initially, however, the preserves settled into a period of remarkable stability and growth. As the city and its suburbs surged outward again after the second World War, FPDs were established in the neighboring Lake, Kane, and Will counties, as well as in Lake County in Indiana. As recently as 1971, another of the "collar counties" around the city, McHenry, set up a conservation district to acquire and manage public preserves. Meanwhile, the Indiana Dunes on the southern rim of Lake Michigan, where Henry Cowles had researched his seminal theory of natural succession, had finally achieved the status of National Lakeshore in 1966. Add in the Midewin National Tallgrass Prairie, converted from military use in 1996, and some other reserves and refuges, and we find that the Chicago metropolitan area has no fewer than 370,000 acres of "natural preserved" land in its midst.[4]

Few, if any, modern urban bodies can claim so many green veins. Chicagoland environmentalists have much to celebrate, and the region may still become a model for new kinds of urban ecological relationships. There is something inspiring and hopeful in the fact that this is happening in a place whose name has been the epitome for dynamic (or destructive, take your pick) industrialization, and whose iconic architecture is the signature of twentieth-century humanity's aspiration to dominate the Earth and stretch up so potently into the sky.

Cook County FPD's mission statement, which dates from 1913 and reads like an environmentalist's dream, has a reference to restoration that seems remarkably prescient: "To acquire . . . and hold lands . . . containing one or more natural forests or lands connecting such forests or parts thereof, for the purpose of protecting and preserving the flora, fauna and scenic beauties within such district[s], and to *restore*, restock, protect, and preserve the natural forests and said lands together with their flora and fauna, as nearly as may be, in their natural state and condition, for the purpose of the education, pleasure, and recreation of the public."[5] How, one might well wonder, could such a positive rubric lead to the mother of all restoration rows and result in the banning of all restoration activities in some of the preserves at the end of the twentieth century? A more sober and skeptical parsing of the mission statement offers some clues. What exactly, for example, is a "natural" forest? And why should forests be the privileged ecosystems for preservation in Illinois, which is officially designated the *Prairie* State? Above all, is there real agreement as to what, if any, single "natural state" is the target for the restoration, restocking, protection, and preservation of FPD lands? What is your *reference system*, an ecological restorationist, if such a beast had existed at the time, might have asked the authors.

In fairness to the undoubtedly visionary founders, not all of these questions would have made much sense in 1913. While the existence of groups like Jensen's Prairie Club of Chicago—and indeed of the Prairie School of architecture itself—suggested a sentimental attachment to prairie landscapes, almost all of the prairies had already disappeared under the plow. Little was understood about how they had functioned ecologically, and it would be a long time before park staff anywhere accepted that fire, for example, might be a useful or even essential tool for land management in many ecosystems. The writers of the mission statement had taken a bold step away from the manicured "Nature" of European-inspired urban parks in creating the forest preserves. But they were still influenced by European thinking, more appropriate perhaps to New England than to the Midwest. They saw forests and natural places as more or less synonymous. Indeed, they could also claim, had they been so inclined, to be expressing very advanced scientific ideas in choosing such an arboreal definition of wilderness. Did not the pioneering research by the local giant of contemporary botany, Henry Cowles, indicate that forests are the inevitable culmination of natural succession, the process that he had just demonstrated in the Indiana Dunes?

Whatever the founders' intentions were, the mission statement was long interpreted as literal and unproblematic by the FPD managers. Since they were supposed to be running forest preserves, they took the injunction to "restore" (which at that time had acquired none or very few of the nuances we are exploring in this book) to mean "restore to woodland." Most of the land they acquired had been cleared for farming, more or less recently, and no effort seems to have been made to determine what plant communities had been dominant on any given patch of land prior to those clearances. In terms of contemporary ecological restoration, they were working without any historical reference system to guide them, though of course that concept was not yet current. Crucially, they understood "protection" to include protection from fire. Without fire, of course, these fragments soon went through a process of apparently "natural" succession that led to mostly closed-canopy woodlands. Finally, since nature was understood to mean "unmanaged by people," the job of preservation was taken to mean something very straightforward—leaving the new woods almost entirely alone apart from maintaining some areas for recreational activities ranging from hiking to picnics.

This concept of preservation, very common among conservationists for most of the twentieth century, almost seems to have derived from the culinary analogy of "preserves" and is really not "natural" at all. To create a well-stocked kitchen pantry, you put the cherries, onions, or whatever in with the right sweet or sour mixture. Then you seal the jar, and you can be confident that the fruit or vegetable will be fit to eat in the future, months and even years after they would have decomposed in a state of nature. In states of nature, however, there are no long-term preserve mixtures, and certainly no sealed jars: everything is connected, and nothing stays isolated for very long. The forest preserves each carried varying traces of their previous, usually agricultural, use. They were fragmented, certainly, from each other, but they were intimately linked with their urban matrix: birds, animals, and insects, as well as wind and water, carried seeds, microbes, and mycorrhizal threads across their porous boundaries for decades. Regardless of the intentions of the founders, the preserves were changing all the time.

Wonderful Stuff *Here,* Bad Stuff *There*

And so, by the time Steve Packard sought out the forest preserve sites along the North Branch of the Chicago River in the late 1970s, much of what he found was very different from the "natural forests" that the founders had hoped future generations would enjoy. According to his own account, Packard was at this time a rebel without a cause and without a job, a full-time antiwar activist made redundant by America's recent withdrawal from Vietnam. The new environmentalism attracted him. He says an eighty-six-page book, *The Prairie: Swell and Swale* by Torkel Korling, changed his life. His growing fascination with tallgrass prairies and then with tallgrass savannas led to his becoming one of the most inspirational—and problematic—figures in the contemporary US restoration movement. Packard was particularly struck by an introductory note to Korling's book by Robert Betz, a leading prairie ecologist. He learned that a mere handful of intact remnants of tallgrass prairie remained in Illinois. Insofar as restoration was on the agenda at this stage, it was conceived of as reconstruction, like the pioneering work Betz was doing, literally from the ground up, on cleared land at Fermilab.

"I drove around with my girlfriend looking for the preserves, they were not even marked. We would walk in, and find this wonderful stuff *here*, but with all this bad stuff growing *there* that he [Betz] talked about. And I thought, "There is something missing here, something should be done, and I know how to do it, I know about community organization and so on. The problem was that no one cared about them [the preserves]. Those who did traveled around, looking for best ones, but from day to day, even month to month, there was nobody there looking after them."[6]

The "wonderful stuff" consisted of the patches of native prairie, forest, and—he later learned—savanna plants that could still be found on the North Branch preserves. The "bad stuff" was invasive vegetation, and it looked like it had already won the battle. He described the scene in his first major—and still his best—ecological article, "Chronicles of Restoration: Just a Few Oddball Species: Restoration and the Rediscovery of Tallgrass Savanna":

> The most obvious symptoms of this deterioration are infestations of European buckthorn [*Rhamnus cathartica*], Tartarian honeysuckle (*Lonicera tatarica*) and garlic mustard (*Alliaria officinalis*). These aliens create thickets so dense, green up so early in spring, and hang on so late in fall, that they often drive out everything else. An especially sad (and common) landscape features forlorn, aristocratic old oaks in an unbroken sea of buckthorn—the understory kept so dark by the dense, alien shrubbery that not one young oak, not one spring trillium, not one native grass can be found. Except for the relic[t] oaks, whose decades are numbered, the community is dead. Early publications . . . in the 1920s show gracious open groves with table clothes spread for picnics. To traverse some of the same ground today would require an armored vehicle, or dynamite . . . in some places you can only explore the preserves by crawling along for long stretches on bare dirt under the dead, thorny lower branches of buckthorn.[7]

In recognizing this dramatic scenario, which was then barely acknowledged as a problem by the Cook County FPD management, and hardly at all by the general public, Packard saw the new challenge he had been looking for. It became, and remains, his mission to organize the liberation of the biological communities evicted or suppressed by invasive plants in the preserves.

"So I started restoration," he told me, initially using just a few simple hints in the Torkel book as his manual. He would later write his own handbook.[8] "You cut brush, you needed to burn once in a while. I asked for an OK from the FPD to do some work. Not on the best sites—I presumed they were for scientists only—but I started working on the North Branch, and it soon became obvious that we were a whole lot better equipped to do this than anyone else on the scene."[9]

Packard's ability to size up a crisis and tackle it with tireless energy and great imagination would prove to be the strength of the North Branch restorations. His organizing skills transferred easily from street demonstrations to restoration days, and he and his close associates soon had dozens, then hundreds, and eventually thousands of citizen volunteers clearing, burning, planting, and managing North Branch sites. In the meantime, he became a respected professional conservationist, working full-time for the National Audubon Society. His ongoing and rather hubristic conviction that he is the right person in the right place would become something close to a tragic flaw. Yet he is not an obviously arrogant man and, both in conversation today and in his 1988 account of his literally groundbreaking work at Somme Prairie Grove, he displays doubts, a commitment to testing hypotheses—up to a certain point—and a self-deprecating sense of humor.

"Would you say," I asked him, "that you stumbled upon ecological restoration?"

"That would be far too graceful a way of putting it," he responds dryly.

Much of Packard's story has been very well told elsewhere,[10] but some of it needs to be recapitulated here, because the North Branch narrative highlights in especially sharp and often painful perspectives questions that are fundamental to ecological restoration everywhere:

- Should we restore at all?
- If so, where, and to what target ecosystem, based on what reference system, and according to which criteria?
- To what extent are the answers we give to these questions colored by cultural ideologies and personal aesthetic preferences rather than by science?[11]
- How much does that matter?
- Who should determine management policy for public land designated as "natural areas" or "for conservation": bureaucrats, scientists, citizen-activists, or some combination of the above; or

should the final word lie with politicians, subject to all the vagaries of democracy?

- What issues need to be factored into the answers when the context of restoration is urban, as it is likely to be for most human beings in the twenty-first century?

So what follows is not in any sense an attempt to establish the full facts of the "Chicago controversy," in the sense of who did what, when, and where. It is much more an effort to unpack the arguments that underlie it, and hopefully illuminate useful answers to these questions.

Rediscovering an Ecosystem Lost to Science

To many local restoration folk in Chicago, and to some further afield, Packard is an environmental hero who has the extraordinary distinction of not only restoring, but *rediscovering*, a precious historical ecosystem, one previously unknown to science. He has certainly pursued, like few others, the ideal of restoring a "classic landscape," in Bill Jordan's sense of that phrase,[12] though Jordan himself has not had an easy relationship with him. And yet to others, especially to some residents in neighborhoods close to his restoration sites, Packard is a villain. They accuse him of being driven by a God complex to insist that he and his besotted followers know what is best for the forest preserves. The North Branch Restoration Project, its critics say, is hell-bent on imposing a uniform ecological template on each and every one of the preserves, regardless of the local preferences of the citizens who use them, and, some add, ignoring even the best advice of scientists.

He leaves no one who meets him neutral, and even his close associates admit he can be divisive and manipulative. I found him to be charming and quietly charismatic, and very well informed in ecology though formally untrained in the science. I also found him to be dismissively resistant to information and theories that might challenge the cherished conclusions he reached about the ecology of Cook County in his "oddball species" article, or to efforts by philosophers and sociologists to situate his project in a broader context, even when they are well disposed towards his enterprise. He portrays himself as utterly weary of the conflict that has surged around his restoration

9. **The seeds of restoration.** Erin Faulkner, an NBRP volunteer and high school science teacher, displays the seed of woodland puccoon collected in Harm's Wood Forest Preserve. It will be broadcast in appropriate habitats on other restoration sites on the North Branch of the Chicago River. (Photograph courtesy of John and Jane Balaban, Master Stewards, Forest Preserve District of Cook County.)

practices periodically over two decades. Above all, however, I found him to be a man utterly engaged with the piece of earth he has been restoring since 1980, Somme Prairie Grove. Even his critics grant that this work is superb.

"Enjoy Prairie Grove," Packard's *bête noire* Petra Blix enjoined me as I left her home for his after listening to her well-informed and sometimes searing critique of NBRP practices. "It really is a beautiful site. It is as good as it gets."

Somme Prairie Grove and Somme Woods, another forest preserve site, lie between the west and middle forks of the North Branch of the Chicago River, in the comfortable and leafy Northbrook Village area, about twenty-five miles north of the city's downtown district. Both are within a five-minute walk of the home Packard shares with fellow restorationist Linda Masters. In line with NBRP's long-established but (to some) controversial practice of appointing individuals as "stewards" to manage restoration in specific areas, Masters is steward of Somme Woods, and Packard of Somme Prairie Grove. To walk through the snarled and snarling traffic on Waukegan Road and into the ninety-acre grove is to experience a total shift of mental and sensory gears. Beyond the soundproofing shelter of a belt of European buckthorn—and native ash and elms—Packard leads me onto a magic carpet of flowers in which he seems to know every stitch, to appreciate every nuance of ecological weave and warp. The mosaic of oak woodland, shrub patches, savanna, prairie, ponds and swales creates a bewitching variety of vistas. Nevertheless, repeated encounters with invasive vegetation—still on more than 50 percent of the site—is a reminder of how much remains to be done after thirty years' very hard work.

Packard's standard trail through the site leaves the best to last, but because it is a very hot May afternoon he reverses the route, starting with the relatively shady Vestal Grove, his *pièce de résistance*. Here the profusion of scarce species is richest, flourishing under an open savanna canopy of burr oaks, some of them already two hundred years old when the first white settlers arrived in the late eighteenth century. He points out white blue-eyed grass, woodland sunflower, carrion flower, Solomon's seal, Joe Pye weed, thicket parsley, touch-me-not, star campion, columbine.

"Almost everything you see growing here is rare, almost everything is here because we threw handfuls [of seed] in the air. . . . We did not even clear the buckthorn, we just burned it," he says, a succinct summary of knowledge gained out of back-breaking labor and heartbreaking disappointments in the 1980s. As Packard tells it in his 1988 article, the restoration of Somme Prairie Grove and other North Branch sites in that period was "a trial-and-error process using hundreds of varying uncontrolled restoration experiments."[13] He makes no apology for not doing the kind of meticulous replication on comparative plots that would have provided "good, solid data." There was simply neither the time nor resources to meet those kinds of criteria. He and his volunteers had one priority: get rid of the "brush"—a term that would become very contentious—so the prairie and woodland flowers would come back.

Easier said than done. There were blocks of brush between forty and four hundred meters wide separating the oaks from what remained of the prairie; indeed, the brush crowded in right under the big trees. At first reluctant to burn too near the ancient relicts of the system, volunteers tried burning from the prairie side, but found the brush, having shaded out the fuel previously provided by a grassy understory, very resistant to fire. So they cleared it by hand with chainsaws and scythes, and broadcast and raked carefully collected prairie plant seed onto the exposed earth. Hopes were raised when some prairie species took quite well, but within three years it became evident that the brush was winning again in the open areas; meanwhile, under the trees, the prairie seeds never did well at all. The volunteers had expected reinforcements to rise from long-dormant seedbanks, but the new recruits never materialized. Perhaps they had been dormant too long; perhaps they required hotter fires to germinate. Packard and his colleagues simply did not know.

Several years into this process, Packard was near despair, especially when he found three unfamiliar species of grass growing where the oak canopy was fairly open. He suspected that they were aliens taking advantage of the clearances. But over the following winter he identified and researched them. He discovered they were in fact natives, though not particularly associated with either woodland or prairie. A new intuition began to dawn on him: "We were thinking too much about prairie and were not picking up what this other community—the savanna—was trying to tell us."[14] Up until then, Packard had subscribed to the conventional notion that savanna was an ecotone—an intermediate system—between closed oak woodland and open prairie, and did not have characteristic species or communities in its own right. But now he began to argue that this definition of a savanna as "a prairie with trees" was "profoundly misleading."

"In many ways, the savanna is as different from prairie and forest as these communities are from each other" he argued.[15] If he was right, savanna would have its own distinctive species. He was, he concluded, trying to restore the wrong plants.

He now saw himself in the line of Bob Betz, who had found relict prairies in the 1960s where no one had expected them, in old cemeteries. The authoritative Illinois Natural Areas Inventory had finally overcome its skepticism about Betz's research to list many of his relicts as the real thing in the 1970s. But it found hardly any remaining savanna. "What they sought—and did not find—were undisturbed stands of oak with prairie flora underneath," Packard wrote. "Maybe they were looking for the wrong thing." The old savannas, "protected" from fire by the settlers, and then by FPD managers, had shifted into woodland exceptionally quickly, according to several accounts he tracked down. For all intents and purposes they had simply disappeared before the science of ecology developed. Against the deep grain of received scientific wisdom on the question, Packard undertook a remarkable piece of detective work in search of clues about this missing ecosystem. He brooded over old plant lists made by early settlers that might be relevant to savannas, also variously called "oak openings" or "barrens." He drew up a hypothetical master list of 122 species. Poring over botanical textbooks as if they were "ecological mystery novels," he "sensed the possibility that a lost community was emerging from the nether edge of oblivion."[16] Now he had found a truly remarkable mission: restoring an ecosystem that was unknown to contemporary science.

He began to find many of the often unfamiliar plants on his list in odd corners of the county, with conditions roughly analogous to savanna: roughs on golf courses, railway rights of way, and forest path edges. And he began to collect their seed and broadcast it on the hitherto unproductive ground he now identified as likely former savanna on the preserves. Some academics he consulted warned that what he was doing was "dangerous" and could lead to the creation of artificial plant communities in the name of restoration. Packard's response was a daring practical exploration of the concept crystallized in a seminal restoration essay by A. D. Bradshaw that argued that ecological restoration was the best method for proving an ecological theory: "The acid test of our understanding is not whether we can take ecosystems to bits on pieces of paper, however scientifically, but whether we can put them back together in practice and make them work."[17]

Things looked up for Packard's thesis when a colleague recalled that Middle Fork Forest Preserve in Lake County had been regarded as being good for neither woodland plants nor prairie plants, but was in fact rich in the species on Packard's new list. And then he found another old list, made by a country doctor in 1846, which seemed to clinch his case. S. B. Mead's "The Plants of Illinois" organized species by habitat.

One hundred and eight of these species were designated "B," for barrens, the common nineteenth-century term for savanna in Illinois. A great many of these species overlapped with Packard's master list. He felt that he had encountered "a Rosetta Stone for the savanna."

In the spring of 1986, the seeds the NBRP had broadcast the previous autumn, based on his new master list, began to sprout in profusion. Blue-stemmed goldenrod, silky wild rye, starry campion, bottlebrush grass, and sweet black-eyed Susan, among many others, began to form a new community under the dispersed oaks. Packard brought his article to a rousing conclusion: the North Branch restorations, he wrote, potentially represented "the resurrection of a complex, dynamic, splendid ecosystem that no ecologist has ever seen."

This is a very big claim, and the scientific jury is still out on it—and may be for decades to come. "'Oddball Species' had a case to make and made it powerfully," says Liam Heneghan. "But it is more like a philosophical piece than a scientific piece. The work he is doing at Somme Prairie Grove is of great conservation significance; it is absolutely encouraging, when I visit the site I believe it, but it is not a replicated experiment. That is *not* to say," he continues, carefully treading through the double negatives with the nimbleness of a disciple of Heidegger, "that it is *not* a phenomenon in the world that's true."[18] What he would like to see is replicated experiments across a range of sites to test the truth of Packard's thesis and see how it could be best and most widely applied elsewhere. In some ways, the question of whether tallgrass savanna is an ecosystem in its own right or an ecotone or intergrade between systems as obviously distinct as forest and prairie, is semantic. Clearly an area with 80 percent canopy cover is very close to forest, and one with 20 percent canopy cover is very close to prairie. But what Packard does seem to have proved, in practice if not in peer-reviewed and replicated experiments, is that the zone between those levels of cover fosters many species and associations of species not generally seen in either ecosystem. Heneghan says, "When you read all his stuff, you can be confident, as a shorthand for describing a certain set of environmental conditions, that savannas exist and that certain species require that combination of conditions. In the absence of the work that Packard inaugurated we simply would not have those species [here today]." If we value biodiversity, then, Packard's call to restore this system is compelling, however we categorize it.

Certainly, many Chicagoans felt compelled, even inspired, to join in the task of savanna restoration in the late 1980s and early 1990s, as the volunteer movement mushroomed statewide and won the powerful patronage of The Nature Conservancy. Laurel Ross, a veteran "North Brancher" and site steward, became organizer of the Volunteer Stewardship Network for The Nature Conservancy and wrote an informative (and revealing) article about the situation in 1994.[19] Two hundred and seven sites on some twenty-seven thousand acres were under Volunteer Stewardship Network management, and almost four thousand volunteers had collectively logged fifty thousand hours of work. Chicagoland was becoming the prototype for citizen-led ecological restoration. This form of volunteering, Ross argues strongly, brings great benefits to human as well as natural communities. The sense of engaging "tangibly" and "positively" with the environment, and with other volunteers, creates a sense of deep significance in people's lives—a point echoed in other restoration experiences across the world. Her article is also studded with unabashed references related to spiritual experience: people find elements of the "sacred," "redemption," "transcendence," "kinship, awe, respect, and even love" in their new relationship with nature.

She also stresses the strong development of citizen science among the volunteers. She concedes that the quality of their scientific knowledge varies, and that it is passed on more as a craft skill than as a systematic education. However, she cites with pride the high degree of local expertise in areas as diverse as lichen, sedges, plant propagation, and mapping that individual volunteers have acquired. This rich intellectual and practical capital is widely recognized today, even among those who disagree with the environmental politics of the movement. One must, however, wonder if the movement was not beginning to suffer from hubris. In her conclusion, Ross approvingly quotes one volunteer at a dinner to honor members' work: he "drank in the information and ideas he was hearing . . . his eyes shone as he said, 'I'm sitting in a room full of greatness.'"[20]

A Vacuum at the Heart of the Enterprise

But where, you may wonder, were the municipal authorities while a number of their preserves, all of them public land, were undergoing this radical transformation by what was, however you cut it, a self-selecting group of unelected citizens?

"The FPD were not doing anything, they did not know anything, they were not interested,"[21] Packard recalls bluntly. He remembers that they sent out representatives when the volunteers were conducting controlled burns. That was as far as their participation generally went in the early years, though some FPD staff became deeply involved in restoration at a later stage. The managers were, it seemed, quite happy to hand over control of sites to the NBRP, once Packard and his colleagues had identified them as suitable for restoration, and to rubber-stamp the requisite permits for their activities. In retrospect, this was a quite unsustainable situation in social and political terms, though most of those involved seemed to have genuinely regarded it as perfectly normal at the time. Volunteer activists were not only changing the nature of public parkland without any broad public debate, they were burning and poisoning brush and burning prairie in small parks right in the middle of large and heavily populated suburban developments. The fact that no significant accidents happened is a tribute to their care and discipline—and to the nature of fire in a generally damp environment. But the concerns such practices later raised are hardly surprising.

The vacuum of clearly constituted authority at the heart of the enterprise sucked in the storm that followed. The NBRP regarded themselves as contributors to their local communities, but they had not asked those communities what contributions they wanted. Other citizens began to question how this private group could create and implement policy for public land. Perhaps the most surprising thing about the emergence of opposition to the NBRP is that it took so long to become a coherent and visible force. Indeed, Packard himself recognized that some aspects of his project were under criticism from an early stage and mentioned this in the opening paragraphs in his 1988 article. His response, then as now, is breathtakingly confident, given the complexities of the issues involved: "Our objective was clear, however. It was to restore these tracts to their original, natural condition. . . . People responded quickly to the purity and grandeur of the vision."[22]

His bald use of problematic phrases like "original, natural condition," coupled with the bold attachment of emotive words like "purity" and "grandeur" to his own vision, suggests that, with his apparent rediscovery of tallgrass savanna, he already had drawn a line in his once wide-open mind. In rereading his articles, the interviews he has given to others, and transcripts of my own interviews, I am constantly struck by how he and the volunteers who have remained close to him have expressed the same views, often in almost identical sentences, over many years. Such changes as there are seem to reflect more a strategic shift to deflect criticism than any real engagement with an ongoing process of reassessment based on experience. The tablets have been handed down from on high, and they are set in stone. Trial and error, the rough testing of hypotheses, was still being applied to the detail of North Branch restorations. But the big picture had been painted, and its basic composition was no longer open to question. The story of the rediscovery of prairie savanna was too good to be spoiled by the emergence of new facts or lines of enquiry. This was true not only in terms of scientific knowledge, but also in terms of response to public criticism. From a very early stage, the NBRP took the stance that if people challenged what they were doing, that was either because they did not understand it, or, later, when things got really hot and heavy, because they were pursuing some hidden and malicious agenda.

The battle lines of the great argument to come already stood out clearly in a snapshot captured in William K Stevens's excellent account of the early years of the NBRP, *Miracle under the Oaks*. Packard told him how he was conducting a controlled burn on Sand Ridge Nature Preserve when a local resident rushed in to put out the fire. Packard told him that fire was necessary to restore the ecosystem, and the man actually started to cry, exclaiming, "This is my favorite place; you can't burn this up." But it is Packard's response, as recounted to Stevens, that is really telling: "He didn't understand the ecology."[23]

It seems to have never struck Packard—it still doesn't—that some people, acting in good faith and with equally strong convictions about environmentalism, might fully understand the ecology (or rather, his interpretation of it) and still object to what he was doing.

What's Not to Like about Restoration?

The objections that would ultimately emerge are manifold, ranging from the (relatively) trivial to the profound, but each one of them is a challenge to the assumptions, often unspoken or even unconscious, of well-intentioned restorationists everywhere. People who have walked in the preserves since childhood are often attached to familiar features, regardless of their ecological function. One person's patch of invasive aliens is another's long-cherished grove. In small parks, in an urban context and on public land, who has the right to decide?

The preference for woods—whatever their native/alien makeup—over savanna, and especially over prairie, is widespread in Euro-American culture. Most of us can appreciate the stark beauty of winter woodland landscapes. But a prairie in autumn and winter can appear dreary and somehow untidy. I have learned to find great and subtle beauty in such landscapes over the last five years. But for most us this is an acquired taste. My first visit to the International Crane Foundation was in November, and I was quite taken aback by the unkempt appearance of the restored prairie communities around the crane pens. "Why don't they clean up that mess?" I heard other visitors ask. So the restoration of prairie on public land may also require the restoration of a taste that has probably never been widespread in our recent culture in the first place. And on a purely practical level, buckthorn patches are welcomed by many residents for the shade they provide on hot days and for screening out traffic noise and ugly urban vistas.

Even if the general public appreciates—or can be led to appreciate—the end result of restorations in the preserves, no one questions that the process of restoration creates long periods where the landscape will offend almost everyone's aesthetic preferences. Again, Packard was clearly aware of this problem from a very early stage. "Oddball Species" has a how-to-restore panel that includes the following advice: "Don't expect quick perfection. Be ready to be embarrassed for many years by charred bush, bare ground and weeds. Prepare the public . . . for this transition period."[24] Ironically, it was the NBRP's best efforts to cope with this problem that proved to be the fuse that ignited the controversy. The counterintuitive aspects of restoration raise legitimate questions in a public mind sensitized to environmental issues by ubiquitous reports on deforestation and, more recently, climate change. Isn't it a big mistake to cut down *any* trees when we are being told to plant entire forests to absorb urban and industrial pollutants, act as carbon sinks, and so on? Do we not need to prioritize vital ecosystem functions in the present crisis rather than indulge in the luxury of attempting to restore lost communities of plants?

Even those who grasp and share Packard's passion for restoration of the North Branch's prairies and savannas might question his ruthless approach to "native invasives," his unshakeable conviction that he and the other NBRP stewards have worked out which species, in which proportions, are appropriate to each site. When I first step onto Somme Prairie Grove with him in 2007, I almost immediately trip over the stump of a tree that was still resprouting resiliently, despite the evident application of chainsaw, fire, and herbicide. "Buckthorn," I say with the eagerness of the new pupil. "No," he corrects me, "native green ash. Native invasives are like cancer cells that kill off most everything else. These species would not have been here two hundred years ago, or two thousand years ago. In the neighborhood, yes, but not here. This place burned. Fire raced across the prairie, and swept up onto the moraine. You can see in the old pre-settlement notes from Thomas Jefferson's surveyors that native ash and elm were in the valley down by the river, but not here."[25]

It's not just native invasive trees that get the chop. On the same visit, I am taken aback by Packard's declaration that there is "much too much" tall goldenrod at one point on the trail. Tall goldenrod is a characteristic plant of both prairie edge and savanna, and you might think that, if it flourishes on a restored site, even at the expense of other native species, then that is what "the ecosystem is expressing," to use a favorite phrase of Packard. Otherwise, surely we are not talking about restoration, but about gardening in the limited sense of the word, a system where humans, and not the interactions of the system itself, determine what plant should be in which place? The distinction, however, between gardening and restoration is not as clear cut as you might initially think. Restoration is a process needing long-term (and probably, in most cases today, perpetual) management. The traditional aim—which Packard shares—is to reach a moment when the natural interactions are working as the restorationists believes they did in the historical reference ecosystem. Then the restorationist can "let go" of the project. "When planting seed or planning fire, every restorationist knows the ecosystem will respond in unpredictable ways that rise out of itself. That is precisely what we want to liberate . . . comparing restoration to gardening she [a volunteer] said: 'Gardening implies control. Restoration implies surrender.' The goal of restorationists is precisely to set in motion forces we neither fully control nor fully understand."[26]

But Somme Prairie Grove has clearly not reached that stage. Given its size and isolation, it may never do so. So a considerable degree of ongoing intervention is necessary. Packard, in a 2007 article on the grove's website, describes the continuing phase of restoration there as analogous to the medical interventions against infections that follow major surgery: "One of the worst 'infectious' plants is tall goldenrod, a native weed, which uses its own herbicides and its thug-like competitiveness to eliminate other grasses and wildflowers. But it has no symbiotic relationships with the many species of a healthy ecosystem. So soon this thug itself succumbs to shrubs (in part because it doesn't carry the fire that would control the brush). Like the worst of disease organisms, it kills the patient. From tall goldenrod, to shrubs, and the grassland is gone."[27]

So Packard, with his scythe, is the surgeon slicing away new infections while trying to avoid cutting down too many of what he calls "quality plants" as collateral damage. This is a lighthearted piece in style, but it deals with a serious issue, and Packard's language gives easy ammunition to his critics. To categorize one plant as a "thug" and another as "quality," when they both belong in the same ecosystem, is at best problematic, though it may be useful as shorthand in educating volunteers. But the underlying question remains: who decides which native plants should be subject to human control in a restoration project, and for how long?

Paul Gobster, a Forest Service social scientist who has made a major contribution to the debate about the North Branch, is insightful on this issue: "These are small sites, so fragmented that they [volunteers] must control natives or it will get out of whack, and they won't have the kind of representation they think is ecologically best. But there is also a human value component, the same as you would find in any other gardener, that they have a feel for what harmony and balance and beauty should be in the landscape. If you point that out they would deny it though."[28]

Informal Democracy or Self-Selecting Elite?

This brings us to the final point in the charge sheet against the NBRP. From the outset its modus operandi has combined apparent openness to newcomers with an organizational structure that appears, in the eyes of many outsiders, to foster the self-perpetuation of a like-minded group of cognoscenti in all the key positions. "The North Branch is not incorporated, it does not have by-laws," says Laurel Ross. "We do have a fair number of rules we made up to guide us; when a problem comes up we make up a rule to solve it, and then someone tries to remember that rule, so yes, you could say we run it by custom and practice."[29]

Such a setup sounds very liberal, but informality can come to cloak, however unintentionally, a power structure all the more insidious for being hard to define. The process by which stewards for particular sites are appointed—and whether they can be removed—remains opaque to outsiders, and even to many insiders. Candidates can "step forward" of their own volition, but they must be mentored by an existing steward until they are deemed fit to take on a site of their own. But the criteria for "fitnesss" are never clearly stated. Bearing in mind that each site is on public land, this is hardly a system likely to engender confidence outside an inner circle. It appears less an informal democracy and more a self-selecting elite. Many lovers of the preserves wonder why their local site has a "steward" over whose appointment they have had no say. Even long-standing associates of Packard concede that they have seen new and younger members drift away after years of hard work because they feel their input on restoration and public policy and organization is neither sought nor welcomed.

None of this is to say that the leaders of the North Branch have not acted in good faith and out of the best motives. Any firsthand encounter with the depth and breath of their commitment to restoring some form of ecological integrity to their local sites would impress even very skeptical observers. But it is precisely because this is the case that their story is both so instructive and, at times, so heartbreaking. They have become emblematic of what happens when good restoration goes wrong.

The Mendelson Challenge

The most direct challenge to Packard's position from the scientific community came in a paper published in *Restoration & Management Notes* in 1992.[30] Three Illinois scientists led by Jon Mendelson questioned almost every aspect of NBRP theory and practice. But their main thrust was their most radical one. Far from assisting nature by attempting a restoration to the pre-white-settlement ecosystem on the forest preserves, they argued, Packard and his colleagues were actually preventing nature from undertaking a "wonderful recovery"[31] from postsettlement disturbance through natural succession. Buckthorn in the region should thus not be stigmatized and removed as a human-introduced intruder, but should be welcomed as the pioneer and vanguard of a new ecosystem forming before our eyes, which we should not interfere with. They dismissed notions that restoration assisted nature to express itself more truly, insisting instead in a scathing sentence that the NBRP practiced "a form of landscape architecture limited by the imagination of the manager."[32]

This article, along with a more philosophically based attack on restoration by Eric Katz,[33] prompted Packard's response with his 1993 article "Restoring Oak Ecosystems."[34] This is an impressive defense not just of the NBRP, but of a new and broader concept of nature that he identifies as emerging in the conservation movement. He sees off the Mendelson argument that the buckthorn-dominated preserves are a valuable expression of nature's resilience by making a cogent distinction between natural succession and "artificial succession." This leads him, through a neat piece of dialectic, to argue very fluently against the traditional conservationist view of nature as "animals and plants unaffected by people."[35]

Human influence, he points out, as long as it does not "eliminate species from any long-evolved community," is actually often beneficial to biodiversity. And, citing Native American burning, he continues that "the loss of people from a natural system in which we have played an essential part can be as destructive to the functioning and survival of that community as the loss of a key predator, pollinator, herbivore, or any other key species."[36]

Packard's argument is a strong one. To ignore the human influences in shaping "natural" ecosystems is either naïvely idealistic—the nostalgia for a nonexistent pristine paradise—or disingenuous. Indeed, if we do not include human impacts in our concept of "nature," a clear-sighted view of ecological history forces us to the opposite semantic extreme, which is to describe most landscapes—and all of them in the anthropocene epoch of climate change—as "cultural." And if we ignore the impact of abandoning a human cultural practice as ecologically influential as Indian burning, especially where IAPs are present, the outcome is indeed likely to be the impoverishment of biodiversity, not the flourishing of "untamed nature." Indeed, Mendelson's approach prefigures the argument that it is misguided to attempt restoration in "novel ecosystems," which we will discuss in chapters 8 and 14. Packard's appreciation of the Native American role in forging the presettlement landscape is, however, oddly limited. He writes that the Native Americans were "as much a part of nature here as . . . the beaver, bear, and bumblebee,"[37] a very strange phrase I have heard him repeat as a given in conversation many years later. He appears uncomfortable with the view that Native Americans exercised a *cultural* influence on the landscape, and that cultural influences, like ecological ones, can (and did) change many times over relatively short periods. This blind spot ties into his reluctance to consider that specific pieces of land, like Somme Prairie Grove, may have shifted ecosystems several times since the last Ice Age, depending on both human cultural practices and broader ecological influences like climate change. In one breath he recognizes that ecosystems are dynamic, but in the next he makes sweeping statements about what was happening on a particular site five thousand or even ten thousand years ago. The lure of a stable, eternal past, which might be recovered through restoration in the present, exerts a powerful magnetism on his vision.

And so he falls back on the idea that the pre-white-settlement ecosystem is a "natural" one, as opposed to the cultural ecosystems he recognizes in Europe, and indeed in the eastern United States:

> We have very focused, original, ancient ecosystems here. This is an ancient heritage. If I was doing restoration in Massachusetts or Pennsylvania or North Carolina, I wouldn't have that option, they don't have natural ecosystems surving there, it's closer to what you

> find in Europe, cultural landscapes, though a lot more recent. By Thoreau's time, nature was gone from Massachusetts, it had all been utterly changed there, it was not what we could call a natural area. What the pre-settlement survey of 1838 told us was equally true in 1738, and 1538, and in 2038 BC. The most important thing for most of us is what is there [on a given site now] now. If you have got some of the original woodlands, if you have got some of the original prairie, you assume that some of the original microbes in the soil, nematodes, funghi are also there.[38]

The prairie is five million years old, he continues, created by lightening strikes long before there were Indian fires. The biological communities, he says, were changed by the Native Americans, but not created by them. "If you have got something which goes back 5 million years and you have the last of it, maybe you should keep it. . . . It's almost like finding parts of DNA of a mammoth and trying to put it back together."

His sense that the remnants on the preserves really are the last intact traces of a system that has vanished almost every where else is what drives his passion to save it. But it is also what makes his critics fear that he wants to impose a universal template on the preserves, and that that template is the landscape exactly as it was before white settlement. He denies this.

> I never said it is important to put this back as it was in 1830. . . . Somme Prairie is amazing, you don't want to walk on it, it is so full of stuff now so rare, once so commonplace. [And in fact it does seem that he has shifted from metaphors of restoring the past to metaphors of restoring health in recent years, perhaps under precisely the pressure of criticism that his vision is based on nostalgia.] The prairie and savanna, it is an injured thing, we are just trying to heal it and allow it to spread and increase. . . . *[I]f someone came to me and said this [SPG] was a forest in 1830 I would say I don't care.* . . . [B]iodiversity is a precious and wonderful thing, the future will appreciate that we saved or resurrected some of it, and if under modern conditions the [biological] community behaves in a different way, I don't much care, that's all right.[39]

In a 1990 review of Bill McKibben's *End of Nature*,[40] Packard effectively countered the author's argument that environmental degradation was disconnecting people from nature because "we usually don't choose new friends from among the terminally ill. . . . Our experience is the opposite," wrote Packard. "It's an honor to be among the first to have a nurturing experience with wild nature." Concluding his reply to Mendelson and his colleagues, he pitched the rhetoric higher: "Restoration is helping redesign our species' relationship with the rest of life on this planet."[41]

Where Packard's redesign would prove to be weakest, however, was in the NBRP's relationship with other members of his own species in Cook County.

Chicago Wilderness Is Born; the Crisis Erupts

The crisis erupted in 1996, the very year in which the NBRP's influence was at its zenith. This may not, of course, be a coincidence. Chicago Wilderness, an innovative and potentially very influential conservation consortium for the broad Chicago region, was set up in April. Chicago Wildnerness brought together a remarkable array of thirty-four partners, ranging from landowners like the forest preserves (in five counties) to major NGOs like The Nature Conservancy, as well as local, regional, state, and federal agencies from the Chicago Park District to the US Forest Service and the US Fish and Wildlife Service.[42]

The early Chicago Wilderness could be seen as a fairly direct offspring of the NBRP, though it was to grow very significantly and quite rapidly beyond these roots, and develop into a very different kind of organism. Veteran North Branchers like Packard and especially Laurel Ross had been key figures in the skilful politicking that brought the partner groups together. And the first round of projects that the consortium endorsed included restoration in the forest preserves and related endeavours. The organization was launched on a flood of increased funding from local and federal agencies. The statement issued at the launch had passages that might have been—and probably were—drafted after a North Branch restoration workday: "We have hopes that the Chicago Wilderness initiative will become a model both for citizen participation and for inter-agency cooperation in conservation. This effort has . . . the elements of a new environmental ethic, one that recognizes human beings in a metropolitan area as important and necessary components of a thriving natural system. We envision the work in Chicago moving like a prairie fire, igniting the spirits of people in other places and inviting others to take, like sacred fire, this idea home to their own communities."[43]

To some, it looked as though Packard's volunteer movement had succeeded in transferring its template from the Cook County forest preserves to the whole Chicago region, and was aspiring to transfer it to the rest of the nation. The problem was, some of the human beings in the metropolitan area were deeply suspicious of the NBRP agenda, and they were about to find a voice.

That voice belonged to the late Raymond Coffey, then a respected and influential columnist with the *Chicago Sun-Times*. Ironically, his first target was not the NBRP itself, but a restoration project in neighboring DuPage County. The irony is double, because in DuPage the Forest Preserve District was firmly in charge of its own patch and its own policies. The volunteers simply implemented them. Only in Cook County did the volunteer tail, as it were, wag the FPD dog. But the forest preserves in the collar counties had indeed largely bought into the restoration model the NBRP had developed. DuPage had raised an $11.6 million bond to restore 7,000 acres to oak savanna and tallgrass prairie. Since such bonds are raised on the basis of local referenda, the democratic endorsement of restoration in this part of Chicagoland seemed secure. But Coffey's sharp antennae had picked up a strong dissident current, which he helped to mobilize with the first of his many dramatic headlines on restoration: "Half Million Trees May Face the Ax: DuPage Clears Forest Land to Create Prairies."

The trigger that detonated the wave of dissent was the clearing of buckthorn where it had previously screened residents from traffic noise or unsightly urban vistas. To add to the irony, this kind of insensitive clearing was something the NBRP itself had generally been careful to avoid in Cook County, allowing screens of alien brush to continue to surround the preserves. But a multipronged antirestoration agenda was now emerging, epitomized by a new group called ATLANTIC—the Alliance to Let Nature Take Its Course—and by another called Trees for Life. Whether their members had read the Mendelson paper or not, they tended to agree with its general thrust: "natural" (that is, currently existing) landscapes in the preserves should be left alone and not be actively managed. If buckthorn was squeezing out oaks and associated prairie/savanna vegetation, well, that was just evolution in action. This thinking was shared by animal rights activists in the county, who had long opposed the culling of deer by the DuPage FPD.

The most radical position was taken by Bathsheba Birman, a leading member of yet another group, the Urban Wildlife Coalition. She rejects the whole theory of invasive plant biology, and is an animal rights activist. "If you want to drop a prairie on a vacant lot," she told me, "well, rock on. But to remove existing living things and maintain a system indefinitely with artificial methods, that is ethically suspect. Species that have adapted to new circumstances in the way that nature intended are being penalized. Every layman knows these are bad environmental practices. These restorations look like the deforestation of the Amazon. And there are health and safety issues: a regular citizen can't burn leaves in their gardens, yet these people [NBRP volunteers] can burn brush literally in peoples backyards."[44]

What happened next is, inevitably, deeply contested. In an interview in 2007, Packard told me that the whole controversy came about because North Branchers had previously testified very effectively, on behalf of the forest preserve authorities, in hearings on deer culling in other counties. Angry animal rights activists had regrouped, he said, and spearheaded a retaliatory campaign against restoration in three counties, including Cook. While he had some empathy with people who could not stomach the killing of deer—"they are like vegetarians"—he gave the opposition to his form of restoration no credibility as environmentalists. "These are not people who love nature, these are not people you see at natural history meetings." Three years later, he made long and weary pauses before replying to questions about the controversy, which obviously irritated him as a topic. He sidestepped a question about whether he felt the North Branch should have done anything differently in the years leading up the crisis. He explicitly rejected Alexander Hamilton's reminder that in every dispute there are good people on both sides. "There are not people of goodwill on both sides, except the deer people. *There are not two sides.*" He insisted that the whole row boiled down to "one little neighborhood in Edgebrook and Sauganash where there is still antagonism," where politically motivated and self-publicizing people sought to make trouble. He was still convinced that "education" can—indeed already has—solved the problem. "You can easily educate people who care about trees and birds. There really is not an argument any more." The issues that moved the political authorities of his county to ban restoration activities on the sites he had championed had simply passed him by.

It is clear, however, that Coffey was articulating, albeit hyperbolically, the views of a broader constituency than a tiny group of "agitators." And he was goaded on by something that would drive any good journalist—he found that he could not get straight answers to simple questions from the FPDs, especially in Cook County. His almost-weekly columns on restoration-related issues very rapidly built up a tremendous head of steam and precipitated a series of heated public hearings on restoration in the preserves, which raised many of the questions we have considered above.

A mere four months after Coffey's first article, the Cook County Board president issued an executive order putting a stay on all restoration activities in the county's forest preserves. This moratorium on the work of the volunteers was an extraordinary reversal, overturning the policy of the county for the previous two decades. How on earth did it come to this? For Laurel Ross, the answer was simple: "Ray Coffey's columns *were* the problem. From the forest preserve people's point of view, they were afraid that every week there would be another of these devastating pieces that were designed to humiliate, destroy, challenge, and ridicule [their policy]."[45]

So, fourteen years later, the messenger still gets blamed by a sophisticated senior protagonist. This is surely a sad illustration of how much of the controversy over restoration in Chicago has been, and to an extent remains, a dialogue of the deaf. Ross concedes that the FPD's own internal "disarray" at the time—she asks me to keep her precise comments on the agency at that period off the record—fed the fire. No doubt the FPD was also hypersensitive to bad press. But she still holds Coffey responsible for the crisis, rather than looking at the underlying issues that had drawn his attention to the project. His columns were undoubtedly strident, and appeared aimed to inflame. Many of the "trees facing the ax" were in fact buckthorn stems, barely bushes, for example. But his positions were also nourished by awkward facts about the volunteer project. The restorationists *were* acting without a public mandate, and some of their practices did appear devious to outside observers. As we continue to discuss the issues, Ross very frankly admits that her own assessment of Packard, a close personal friend and mentor, is mixed. She describes him repeatedly as "brilliant," and says there is no one from whom she can learn more during a day's fieldwork. But she added that he is "difficult," that he "alienates people": "if you know Steve, you know he manipulates. That word sounds like a really bad thing, but we all do it. Especially if you are trying to do something hard, like playing a game of chess. . . . Isn't it the right way to do something, not to call attention to it, not to upset people?"

"But don't people become much more upset when they find out?"

"Exactly right. But we would not be talking about it today, had a major national newspaper not run thirty-plus of those columns, some of them on the front page."[46] Once again, we are back to blaming the media. The real question is surely whether it is valid, in a democracy, to do what you believe to be good in the public arena, without consulting the public who will be most directly affected by your actions.

The moratorium was lifted, quite swiftly for some sites and some activities. But most or all restoration work at six key North Branch sites was halted for ten years. In some cases even the pulling by hand of the notoriously invasive garlic mustard, was forbidden for a period. The full ecological cost has not been counted yet, says Laurel Ross. At Bunker Hill, one of the most affected sites, an FPD ecologist estimated in 2003 that the moratorium had resulted in invasive woody plants increasing on restored savanna by a factor of ten.[47] Packard says the damage to Somme Prairie Grove, where the moratorium was only briefly in effect, is limited and has been reversed. But Ross believes that the biggest costs were in time lost going to endless hearings, and above all in the damage to the morale of volunteers and full-time restoration workers. Some, including many of the handful of employees of the FPD who had taken on specialized roles related to restoration, were brokenhearted and walked away or resigned. "They could not believe that they had put thirty years into this work and were being treated by the county in this way," she says. Hard-earned skills were lost.

Even today, with restoration work once again being permitted again on all sites, the atmosphere at some of them is sour and tense, as antirestoration campaigners continuously photograph the volun-

10. **A sour and tense atmosphere.** Opponents of the NBRP protesting against prescribed burns and photographing restoration volunteers at Bunker Hill Forest Preserve after the moratorium on restoration was lifted in 2009. (Photograph courtesy of John and Jane Balaban, Master Stewards, Forest Preserve District of Cook County.)

teers while they clear, cut, burn, and spread herbicide. "Working at Sauganash is like being in the Twilight Zone, it is completely surreal," says Ross. This is a far cry from the idylls of restoration workdays in the past, and she feels for volunteers who are treated like trespassers on land they have come to love. Meanwhile, contentious activities like burning and spreading herbicide are now much more strictly regulated (theoretically, at any rate) throughout Cook County FPD than they were previously. The practice of girdling trees—slowly killing a standing tree by cutting a strip of bark around the trunk remains outlawed on all preserves in Cook County despite the ecological benefits: a girdled tree does not have to be burned or chipped; it provides perches for birds, whose defecations seed the ground beneath, and is also a prime habitat for insects and reptiles as it decays.

Nevertheless, the lifting of the moratorium has released a remarkable revival of volunteer restoration activity. Both Ross and Packard are sanguine about the current situation. The Cook County FPD has reorganized its departments and now has many more staff trained and dedicated to restoration than before the controversy. A substantial budget is available to bring in contractors to do the heavy lifting—and a lot of the burning. And these contractors work under the direction of stewards—including a number who were appointed by the NBRP before the crisis. "The stewards have more authority than ever," says Packard. On the surface, things have returned to business as usual, but the opposition has not gone away, as we shall see. The crisis has left deep scars in the restoration movement, and there are many lessons to be learned from it.

Concealing the Truth, or Being Considerate to the Neighbors?

The issue of girdling—and how the NBRP attempted to conceal it from the public—had been another spark that ignited the antirestoration movement. In a further ironic twist, the source of evidence was the book that is an eloquent, mostly positive, and always well-informed hymn to the North Branchers, William K. Stevens's *Miracle under the Oaks*. Its publication in 1995 was the first account for general readers of the NBRP's restoration work—after nearly twenty years. For many Chicagoans, it was a revelation that set out clearly how and why their local preserves had been undergoing such a remarkable transformation. And not all of them liked, or like, what they learned.

Stevens is an excellent writer and reporter, and his generally celebratory tone does not play down discordant notes, especially those sounded around Packard's complex personality, charismatic to many, deeply divisive to many others. However, he does buy into the idea that only people "uninformed about nature" would object to their restoration practices. But in making this case, he lifted the lid on an NBRP tactic that even he describes as "sneaky": "they removed unwanted invading trees (always called brush by the North Branchers, to deflect criticism) by girdling them below the vegetation line, where the killing cut could not be seen. Then the tree would slowly fade away and no one would notice. The whole idea was to keep people from becoming upset about destroying 'nature' so that nature could actually be restored."[48]

Many people who read this passage, of course, became very upset indeed, not least because their concept of nature was being dismissed within scare quotes, while the North Branch concept was respected as inherently authentic. The passage was repeatedly cited by opponents of the NBRP at public hearings and has underpinned the perception by its critics that the movement is devious in its dealings with the public. This allegation stings North Branch veterans, says Laurel Ross. "We were the good guys, all we cared about was that the land was not being managed, we found a way to interact with it in a really positive way, we wanted to make a contribution. It was very shocking to the North Branch, who felt ourselves as people of goodwill, to be perceived as of bad will somehow. We were pretty weirded out by that."

She argues that much of the responsibility for the fact that the volunteers were flying under the radar lies with the county authorities themselves. "The FPD general superintendent, Joe Nevius, really did not believe that the county should talk to the public," she says. "When the Forest Service made a grant of $879,000 to the preserves for restoration, we actually begged him to hold public hearings. He refused, and he had all the authority." But what could the North Branch Restoration Project have done differently? "That's a tough one. I feel as if we were so dumb, so innocent." Their first instinct, she says, was indeed to conceal some of their practices. But when challenged, they acknowledged this tactic openly, because their motives were good and they did not grasp how such concealment would be perceived. "People would ask us why we called what we were cutting 'brush,' and we would just tell them the truth: 'We call it "brush" because if we call it "trees" people get upset.' Well, that sounds as if we were lying, and we sort of were, but we were saying what we were doing. We thought we were being considerate of people's feelings."[49]

Ross, a person who combines gentleness and strength with remarkable grace, is patently sincere in saying this. But what she is perhaps missing is that the North Branchers' underlying assumption was that people *needed* to be protected from their own feelings, because those feelings were ill-informed and because the restorationists knew best. This position was crystallized in a very telling bullet point in a North

Branch educational pamphlet at the height of the controversy: "Informed people will not resist."[50]

Karen Rodriguez, a North Brancher of similar experience to Ross, and a very senior figure in the Society for Ecological Restoration, remembers that efforts to conceal girdling and leaving screens of brush to hide clearances were aimed at "hiding the visual impact, not hiding what we were doing." There is, in fact, a long tradition in similar practices on US public land. One National Parks textbook from the 1930s recommends pouring chemicals on rock faces, where they have been blasted for road works, for example, so that they appear aged and weathered—and therefore appropriately "natural" and pristine—to the first group of visitors who drive past them.[51] This can be traced to the traditional view of the arts, where an exhibition only takes place when the necessary mess of creation in the studio has been tidied away.

Philosopher Bill Jordan takes a very different position, arguing that restorationists should deliberately expose the public to the "shame" inherent in their project, because a clear and unromantic vision reveals that "shame" is inherent in nature itself, in the messy chaos of evolution, in the "shame of the hunter, who is aware that his life depends on his deliberate destruction of his fellow creatures." He advocates, therefore, that girdling should become a kind of public sacrificial ritual: "If restorationists systematically—and publicly—ritualize the killing that is part of all restoration efforts, might resistance to such killing fade away, revealing the genuine caring and the sense of shame behind it?"[52]

Rodriguez also recalls that there were regular discussions at early North Branch meetings on the need to communicate with communities where restoration was taking place. Indeed, Packard's own writings repeatedly advocate "extensive public education" for neighbors of restoration sites.[53] Undoubtedly, many efforts at outreach were made. Unfortunately, however, NBRP members had become so convinced of their own rectitude that they conceived of public education not as a dialogue, but as the one-way transmission of a paradigm they no longer believed was open to question by people of goodwill. And yet the fact that they felt a need to dissemble about the nature of some of their activities inevitably fuelled suspicion among genuinely enquiring minds. They were caught in a vicious circle of their own making.

Of course, good two-way communication with the local community is no guarantee of a positive outcome, for restoration, or for any other cause. A community may resist proposals for a whole range of reasons, good and bad, even when all the available data has been clearly presented to them. But rejection in a democratic forum, however painful, is surely far preferable, both on ethical and pragmatic grounds, to attempting to by-pass such debates, even with the best of intentions.

The "Paul Gobster Arguments"

Forest Service sociologist Paul Gobster is one of the shrewdest observers of the crisis that followed the moratorium. He had been an enthusiastic fan of the North Branch restorations, to the extent of naming his daughter Savanna in honor of the rich carpets of native flowers he loved to walk through on the restored site near his home at Bunker Hill. But his professional training taught him to maintain a critical distance, a characteristic that did not endear him to the North Branchers as he persisted in teasing out the implications of the controversy without fear or favor.

His first response was to be shocked at the harsh treatment people he respected and liked, such as Packard and Ross, were getting in the media and at public meetings. He was distressed at the prospect of a ban on restoration. But he was equally taken aback by the North Branchers' reaction to criticism in general, once the controversy erupted. Instead of presenting a listening, respectful ear, and seeking dialogue, they tended to use *ad hominem* arguments, often lumping all their opponents together as "a bunch of loonies."

"I found," he said, "that where anyone questioned them they took a very arrogant defensive posture and went into attack mode, as in 'you don't know what you're talking about.'" Their position was—and sometimes still is—"like a mother bear's in defense of her cubs."[54] As soon as the moratorium came into effect, he set out to do a survey of attitudes to restoration in the forest preserves. Insofar as the debate had been (and remains) extremely polarized, the results were surprising.

"I found," he wrote, "that opposition was far from absolute, and there was a good deal in common between ... 'opponents' ... and 'proponents' of restoration."[55] His research suggested that very few people were absolutely opposed to restoration, per se, but that many people were concerned about one or more specific restoration practices, whether the use of fire or herbicides, or the removal of trees in the name of environmentalism, and so on.

Stressing throughout that the debate was about restoration on public land, to which members of the same neighboring communities ascribe multiple and sometimes conflicting values and uses, he argued that a more sensitive response to public concerns by the restorationists might smooth the way to a resolution. Ten years later, he saw little sign that that had happened, and told me sadly in 2007, "I've yet to see an open-minded dialogue about restoration take place here. If only the restorationists could make some concessions, like chipping cleared brush instead of burning it. Or even showing on a spreadsheet what carbon gains, and carbon losses, there are from restoration practices like clearing, burning, and planting." Gobster went on to coedit a collection of essays on the philosophy and sociology of restoration, which deals elegantly with the issues underlying the Chicago controversy in much more depth and detail than I can accommodate here.[56] I suspect that most readers would find the book essentially sympathetic to the restoration cause—Gobster wears his restorationist's heart on his sleeve in the introduction—but the response from the NBRP was that he was giving credibility to the opposition by taking them seriously.

I first arrived in Chicago to research this chapter in 2007, shortly after the moratorium on restoration had been lifted from all North Branch sites. I found that the two sides in the controversy were still separated by apparently irreconcilable differences. What I found most disturbing was the utterly dismissive way in which veteran (and some newcomer) North Branchers still treated their critics. Repeatedly I was told that the entire opposition to their work was made up of ten or twelve people, a maximum of twenty, a troublemaking minority. "The squeaky wheel gets the grease" was a phrase I heard, repeated like a mantra, when I asked how so few people could have so much political impact. There was a sad irony in hearing this kind of argument from a group including many who had participated in the 1960s antiwar movement, when similar arguments were used against them by the US establishment. Similarly, I was struck by how Packard used the term "community" for the restorationists, but "agitators" for the opposition. There is something particularly poignant about poachers turning gamekeepers, as it were, but retaining the moral conviction that they are on the side of the angels.

On that visit, I attended a town hall–style meeting in North Park Village, in which both sides appeared to use procedural obstructionism to avoid debating the real issue at all. Afterward, I put it to a group of North Branchers that the aldermen would not have imposed the moratorium unless they believed that many or most of their constituents had real concerns about the way restoration was being carried out. One of them told me that I sounded like I was "using a Paul Gobster argument," and the others nodded sagely, as if that reference alone was enough to disqualify an entire line of thinking. No further response was forthcoming. Just as any strict application of ecological science to favored restoration projects was often suspect on the North Branch, so it seemed was the application of social science to the situation in which the North Branchers were embroiled. I drove up to see Bill Jordan the day after this encounter, still exercised by the question of how people of good will and intelligence could have got themselves locked into such a cultish corner. "Ah," he said sadly when I told him the story, "now you have found that there is fundamentalism among restorationists, haven't you?"[57]

A Walk in the Woods—and Prairie and Savanna—with the North Branch's Critics

In 2010, emotions in some areas remained as high as ever. Two people regularly quoted in Coffey's series of columns as key critics of the NBRP, Petra Blix and Carol Nelson, invite me to take a walk with them at Bunker Hill, along with Jackie Boland, a prominent supporter of the moratorium. Bunker Hill is one of the North Branch sites where restoration was closed down totally by the moratorium, but which has been the object of quite intense NBRP activity since the ban was lifted in 2006.

It happens to be one of those spring days in the city when nature makes its presence felt everywhere, as cottonwood seeds drift like patches of fluffy snow even past the skyscrapers downtown. At Bunker Hill, in the leafy northern suburbs, they carpet the paths along the river in an impressive demonstration of natural regeneration. It is also very hot, but that does not stop Nelson, a feisty seventy-five-year-old with a tousled mop of hair and sparkling eyes, from leading us energetically through territory with which she is intimately familiar. She has monitored insects, plants and birds here for sixteen years. As she identifies an eastern tailed blue butterfly that I can hardly see at twenty meters, comments on the abundance or absence of prairie bethany or rattlesnake master at particular points, or notes the confusing hybridization between oak species from their leaf shapes and patterns, I cannot help thinking that I might have been walking with a soul sister of Steve Packard.

But then we reach an edge of the restored savanna where eighty quaking aspen have been removed by the NBRP. This is a native tree, and the stand was beloved of many who walked or cycled by the grove. According to my guides, the steward, Jane Balaban, simply decided that these trees "did not belong here." Balaban, with her husband John, is one of the core restorationist group that has worked with Packard since the 1970s. There was no debate with local people. Blix and Nelson claim that a single NBRP volunteer who questioned the decision got short shrift from the steward and subsequently left the organization. Out in the open areas, the midday heat is becoming leaden, oppressive. We step in under some buckthorn, and the refreshing coolness is palpable. The point hardly needed making—the shade from an invasive alien is a lot better, as many people see it, than no shade at all.

The trio guiding me do not all sing from the same hymnsheet. Boland and Blix belong to Trees for Life; Nelson is unaffiliated. Boland takes the hardest line against the NBRP and restoration generally. She accuses the volunteers of making a "land grab" once the moratorium was lifted and believes they remain unaccountable to local communities or the FPD in practice. She accuses them of ignoring safety guidelines in using fire and herbicides, and of a stubborn refusal to accept any response, even from the environment, which does not fit their plans. "Their attitude is, 'if we don't see what we want happening, then we should burn more, cut more, herbicide more, until it happens.' Surely the question should be, 'Why burn? Why herbicide in the first place?'"[58] She is happy to pull garlic mustard wherever she finds on her local preserves, but has deep misgivings about any tree clearance whatsoever. All three women also claim that senior NBRP figures, whom they name, are burning and poisoning recklessly, and in violation of their own guidelines, now that they once again appear to have a relatively free hand. North Branchers deny this, but a single spot-check of my own after a volunteer workday in May 2010 revealed three unattended burnt brush piles on a single site. There were substantial clusters of red-hot embers still glowing at their hearts. Perhaps this is a safe practice, but it is easy to see why a neighbor might fear that a change in the wind levels would ignite a forest fire next to their backyard.

Blix and Nelson repeatedly stress that they are not opposed to restoration as such, but they question both the science and the politics behind NBRP's modus operandi, and what they believe is its uncritical adoption by Chicago Wilderness. Nelson claims that her figures show a serious decline in some species, especially birds, at Bunker Hill, since restoration started there. However, the data she gave me in fact mostly covered the period when the moratorium was in operation, and senior Chicago Wilderness figures have cast doubts on her methodology. Be that as it may, it is often hard to get satisfactory monitoring data from the restorationists, and Nelson's general point is well made: "I am not in favor of stopping restoration," she says. "It still brings me moments of magic. I have a life list of 612 native species for Bunker Hill, 5 miles from my house, and know that I will see more insects, more fungi, more flowers, every year—that makes me give a thumbs-up to what they are doing. On the other hand, what they are doing to the birds and butterflies, to maintain the flowers, is not so good, and for that they get a thumbs-down from me. Do you see where I'm at? They are not handling all the elements of the ecosystem equitably."[59]

Restoration inevitably involves this kind of triage—always painful for all concerned—and these issues can really only be resolved in a very broad context, taking into account the always vital question of scale. Ecosystem management across a region can balance the gains and losses of native species, prioritizing threatened ones and maximizing overall biodiversity. This is surely exactly the kind of referee role a consortium like Chicago Wilderness could and should be playing. But Blix and Nelson remain deeply suspicious of Chicago Wilderness at the time of writing. "Chicago Wilderness is fully behind restoration as it has been practiced here, and will not rock the boat," says Nelson. Blix claims that Chicago Wilderness will not publish data that challenges a positive view of NBRP and other restoration work, and that scientists are afraid to speak out because they fear losing funding for their projects. Students working on restoration projects have told her, she says, that questions that don't fit with the NBRP vision are not welcome.

Blix, who has a PhD in molecular biology, feels deeply frustrated by what she claims is the lack of open scientific debate in Chicago Wilderness. "The site selection is arbitrary, there are no baseline surveys. I have nothing against taking out buckthorn, or culling deer, where appropriate. The challenge is to choose appropriate sites for restoration, to manage sites appropriately, and to use science as the basis for that management."[60] She says that the volunteers are "so enamored of certain rare prairie plants" that they repeatedly attempt to restore prairie and savanna at inappropriate sites. Or, as Nelson puts it, "They should find out what's using a site before they change it. If there are birds nesting there, it may be best to leave it alone." She says her own decision to support the moratorium came when she found that nesting habitat for four species of warbler was being destroyed by a restoration project at Palos Fen. But she has developed doubts about the outcome of the course she then chose. "The moratorium was a mistake," she now believes in retrospect. "It went on for too long. We should have put a shorter time limit on it, and then invited everyone back to the sites to do surveys, using them as the basis for future management."

She obviously regrets the personal rifts this has caused. "Jane and John Balaban will never forgive me for supporting it," she says. "I know it sent twenty years of their work down the tubes. And for all for our disagreements, I want to say on the record that Jane is the most knowledgeable person about the native plants of this area I have ever met."

"Yes, but nobody ever anointed her queen, you know," Boland cuts in tartly.

For Blix, the fundamental issue, after science, is trust. She says that in the early 1990s, FPD officials actually denied that any trees were being cut down, until she took them to sites and showed them the stumps. "What has bothered me always about restoration in Chicago is the lying. If something is that good, [why do] you have to lie about it? Trust is completely contaminated in this situation." And, for her at least, this contamination extends to Chicago Wilderness.

I heard from a number of individuals not associated in any way with support for the moratorium that Chicago Wilderness was indeed a cold house for people outside the volunteer movement, or simply with independent spirits, in its early days. Questions from scientists and laypersons alike were brushed off by veteran members when criteria for site selection, the use of particular indicator species to measure progress, or the collateral damage that might be caused by frequently utilized herbicides were raised by newcomers. They talk about sensing an unspoken code of values and practices that did not have to be made explicit or defended. No doubt this closed consensus was heavily reinforced by the defensive armour donned by many Chicago Wilderness founders in response to the shock waves created by the moratorium, in a spiral of escalating hurt familiar from so many adversarial political conflicts.

A Vision Can Make Things Happen

That context, however, appears to have changed quite dramatically in recent years, and it seems likely that critics like Blix may be fighting against positions that have already been largely abandoned by Chicago Wilderness. In fourteen years, the consortium has grown from thirty-four to more than two hundred and fifty partners, many of whom may never have heard of the NBRP. Many lessons appear to have been learned from the controversy by those who do remember it, though the protagonists with a deep history in that era do not necessarily express it that way. Packard himself no longer plays a high profile role outside his beloved Somme Prairie Grove. Laurel Ross's thirty years of hard labor and commitment as a North Brancher make her reluctant to concede any points to the opposition. However, speaking of Chicago Wilderness, of which she is vice-chair at the time of our 2010 interview, she says, "We don't really think you can conserve biodiversity in an area of nine million people if they don't want it, if they don't value it, and if you don't do it in such a way that it *is* of value to them, it benefits them. It is our job as conservationists to put people in a positive relationship with nature, and it doesn't matter if that is because that is where they want to walk their dogs or ride their bikes or watch birds. This is a metropolitan area with some really important natural resources and some innovative and successful approaches to managing it as a consortium, and I think it is good example for other fragmented landscapes."

She cannot talk about Chicago Wilderness without producing maps, without a visual demonstration of the scale of the project, sketching how it sweeps down around Lake Michigan from Chikwaukee Prairie in southeast Wisconsin and the northern collar counties, through the city itself—a founding member of Chicago Wilderness—on through the vast industrial rustbelt of around the Calumet River and east to the Indiana Dunes along the lake's southern shoreline.

She highlights the paradox that more native vegetation is secreted in the preserves of this metropolitan region than exists in the countryside of the entire state of Illinois, where indigenous plants have been almost totally erased in the name of agriculture.[61] She sees the brownfield sites of the Calumet greening up, restored by trainees initially financed by President Obama's stimulus funds, perhaps the embryo of a Civilian Conservation Corps for our times. She speaks of Chicago Wilderness's four great goals—combating climate change, developing green infrastructure, the No Child Left Inside Campaign, and the "restoration of the health of local nature." This is indeed a strong sapling to have grown from the acorn of the NBRP, though of course it will be tested many more times as it strives towards maturity.

"It's more a vision than a reality at present," she concedes, "but a vision can make things happen." Regarding the selection of sites for restoration, Ross says she has "no problem with pragmatism, they are so altered already, we have to compromise, we must say: what is the best conservation solution here? The approach, I would say, is to look at a piece of land, and say: 'What is it now? What was it? What can it be?'"[62] As a succinct summary of the tasks of restoration, that triad is pretty hard to beat, and it appears to meet the concerns of critics like Petra Blix and Carol Nelson. Meanwhile, the Chicago Wilderness Science Team is forging ahead with three remarkable programs that promise to address, from different angles, almost all the issues raised in this chapter.[63] If they are successful, they could be enormously valuable, not just within the region, but as flexible templates for restoration schemes almost anywhere.

11. **Participation, education, inspiration, and vision.** Laurel Ross (center foreground) leads a college group on a field trip to the restoration site she stewards at Somme Prairie. (Photograph courtesy of the Field Museum.)

The team, led by Lynne Westphal of the Forest Service, Liam Heneghan, and David Wise of the University of Illinois at Chicago, has initiated research ranging from the very specific (the ecology of buckthorn invasion and reinvasion), to the very general (the biodiversity outcomes of diverse restoration practices on a wide range of sites), the sociological relationships involved in restoration planning, and the response of communities to restoration outcomes. It's a systematic investigation of restoration in Chicago, from soil to society, you might say. These extremely ambitious programs—one of them immodestly proposes to span a century!—represent a massive advance on the situation Liam Heneghan found as a young ecology professor arriving at his teaching post in DePaul in 1998, though the groundwork for them was already being laid by others, including Lynne Westphal and Paul Gobster.

Ironically, when he first came to Chicago he took it for granted that the key research on restoration there must already have been done. In a city with a long history of producing distinguished ecologists, he says, "I assumed that this was a very well-plucked bird."[64] He was surprised, therefore, when his departmental head, Tom Murphy, suggested that the relationship between buckthorn and soil was "a problem I might like to think about." His surprise turned to amazement when he searched the scientific literature and found that not a line had been published on why buckthorn had outcompeted all natives to become the most common woody plant in Illinois. Like many others, Heneghan guessed that the tree somehow transforms the soil, so that it is very difficult for native plants to reestablish spontaneously in areas where it has been cleared, but very easy for buckthorn to reinvade. There was, however, no experimental research, no data, to confirm this hypothesis.

He found that the volunteers, though keenly aware of, and deeply frustrated by, the persistent problem of reinvasion after twenty-five years of restoration clearances, were not asking scientific questions about it. Instead, they resorted to attitudes like "we are not doing it with enough purity of heart, if we just roll up our sleeves and work hard it will go away."

A Respectful Companionship between Scientists, Planners, Restoration Activists, and Communities

Heneghan became determined to "develop a suite of citizen science projects, where research questions would emerge from volunteer activities, and we would use the best type of ecological experimental design" to answer them. Ten years later, with the support of many colleagues in science, land management and citizen conservationism, he believes that the three programs Chicago Wilderness has established offer a unique opportunity to create a "very respectful companionship" between scientists and restorationists. "We are taking the phenomena that they [restorationists] are creating in the world and we are asking: if you aggregate enough of them, in a fairly systematic way, can you get enough replication to start answering questions about what works where and when?"

"The companionship part of it is that most of the scientists I am working with don't have the botanical skills that these amateur naturalists have. They simply don't 'make things' the way these people [restorationists] do. But professional ecologists know how to think about data, design experiments, and analyse data. They are different skill sets that overlap." His vision is that scientists should provide a "sympathetic critique" of restoration on the ground, constantly asking the questions: "'Are these things working, and under what conditions are they working?' The sympathy part is that we are also conservationists, we share the vision that restoration is the best option for conservation in the region, and we want this to work out. The critique part is 'show me the money, show us the results.'"

Developing this companionship will be a delicate task, but the other great axis of the science team's work treads even more sensitive territory, the relationship between planners (whether volunteer stewards or professional land managers), the biodiversity outcomes their plans produce, and the response of the broader community to those outcomes. "We are explicit that we are not studying the moratorium, or the 'Chicago controversy,'" Heneghan says carefully, insisting the focus will be on the future, not the past.[65] "We want to study this loop—how do people plan, what are the implications for biodiversity, and how do they translate into response by the community? Our thinking is that if planning has gone well, and if the biodiversity outcomes are positive, whatever that might mean, there will be a receptivity in the community for more conservation management. If the planning is poor, if people are excluded from the process, if the plan when put into action has a highly negative impact on the community, then there will not be that receptivity. Our research is rooted in the idea that democratic planning has positive outcomes for both community and environment."

He is heartened by developments over the last two years, as the science team has developed these proposals: "People are learning each other's disciplines, each other's language. I hope we can bring together people who have been mutually suspicious."[66]

Including people like Petra Blix and Carol Nelson, historically associated with the "opposition"?

"As long as I can have a pint with someone, and a conversation, that is fine," he says evenly. Speaking of a recent discussion Blix and Nelson had attended, followed indeed by an amicable beer and a burger, he says that he was "intrigued that Petra's lack of ease with restoration practice," her "fascinating questions," related to "the same things that interested me as a researcher."

Other figures in Chicago Wilderness tend to assume that the conflict that sparked the moratorium is now ancient history, and many new to the movement are simply unaware of it. But while the controversy was both low-key and localized when this chapter was going to press, Heneghan is far from blandly sanguine about the future: "There is no guarantee that restoration is here to stay in Chicago. We must learn the lessons of the conflict, or our projects will not be resilient."[67]

You might imagine that Heneghan has enough on his hands with the prospect of collaborating in the research and development of a restoration project which reaches four states and includes nine million people. But his restless mind insists that there is another branch that must develop if the complex restoration tree is to mature and flourish: "Think of some of the questions we are asking: What is nature? What is it to manipulate nature? And what is it to step back from nature? These are philosophical questions. They bring in so many aspects of the human heart and mind that we cannot think adequately about conservation without getting the best game from our poets and our artists and our philosophers." He is already fund-raising so that the Institute for Nature and Culture can launch a journal that will help to fill this gap.

Can restoration in Chicago really carry all this weight? Can it what? Isn't it a city with big shoulders?

6 *Keeping Nature Out? Restoring the Cultural Landscape of the Cinque Terre*

We get e-mails from the park about international conventions on dolphins and bees, but no real communication about what matters to us. * MASSIMO EVANGELISTI, *private wine producer, Vernazza*

There is a deep lack of democracy here. We know nothing about the park's budget. This is a sore point. * BARTOLOMEO LERCARI, *ecologist and private wine producer, Vernazza*

We are a strange national park, a new concept of national park, we are responsible for many things outside the remit of a normal national park. * MATTEO PERRONE, *Cinque Terre National Park agricultural and environmental office*

The conservation value of ecosystems shaped by human use . . . is particularly notable in the Mediterranean Basin. . . . [This is] one of the world's biodiversity "hotspots" . . . partially as a result of the long-term human presence and related activities. * MIGUEL N. BUGALHO, MARIA C CALDEIRA, JOÃO S. PEREIRA, JAMES ARONSON, AND JULI G. PAUSAS, "*Mediterranean Cork Oak Savannas Need Human Use to Sustain Biodiversity and Ecosystem Services*"

I believe that the terraces make a significant contribution to biodiversity. . . . Abandonment and degradation of rural systems, and homogenization of landscapes, could cause species associated with these habitats to decrease. * RICCARDO NARDELLI, *ornithologist for the Italian Institute for Environmental Protection and Research*

* *

Italy's Cinque Terre is very small. Its five villages cling to less than twenty kilometers of steep coastline, a little to the south of Genoa. Its territory extends only to the top of a six-hundred-meter ridge, which never wanders more than a few kilometers inland. Yet this miniature region has been so tightly crosshatched with stone-walled terraces that they would be longer than the Great Wall of China if they could all be laid end to end.

Today, most of these terraces are abandoned. They were painstakingly created over many centuries for the cultivation of vines, fruit, and vegetables, but most are now overgrown with a variety of classic Mediterranean plant communities, from arid scrub (*macchia*) to full-blown holm oak forest (*lecceta*). They are often visible only as faint ridges on the sometimes tortuously twisted skirts of the slopes. These vestiges are more or less imperceptible to the inexperienced eyes of many of the 2.5 million tourists who pass through every year, mostly walking a single, spectacular, but heavily eroded trail. Where the terraces do still show their rocky faces, they are often in the process of collapse. First they are eased apart by the roots of invading vegetation, fingering their way through the tiny gaps between the stones in their persistent search for nourishment.

After roots, pigs. The terraces are gouged by the exploding population of wild boars. Emboldened by a 1990s ban on hunting when the region was declared a national park, the boars trespass from their advancing forest refuges right up to the outlying houses of the five villages which give the Cinque Terre its name, the "five lands." Young boars often seem to make havoc in old terraces just for the hell of it, though their juvenile exuberance probably teaches them a lot about food sources. Older boars have a finely tuned nose for grapes. According to local lore, possibly apocryphal, they seek out the very best varietals on terraces that are still in use, to the despair of cultivators.

After boars, water. A complex hydrological system of small channels, designed long ago by illiterate but savvy peasants, once irrigated each individual terrace. When storms hit the coast, as they often do, the same channels allowed floodwater to skitter harmlessly down the ridges, and into the streambeds that lead to each village, and through them to the sea. Today, these channels are increasingly plugged with debris, if they have not been entirely uprooted or tusked and snouted away. When the sky falls now on the five lands, water flows wildly and unpredictably, accumulating such critical mass that when it slaps up against a weakened terrace it may sweep away hundreds of years of devoted labor in less than a minute.

Let this happen often enough, and entire terraces will mass into a more deadly flood of stone, hammering down hillsides into their neighbors, and tumbling them on down like cascading dominoes. The unstable sandstone beneath them then begins to shift, and finally the whole slope may slide, carrying rocks from the upper slopes all the way to the Mediterranean waves, and threatening every human habitation in its path. I had finished what I thought was a final draft of this chapter when just such a disaster, long foretold, occurred.

On October 25, 2011, unprecedented rainfall over two hours triggered massive mud and rock slides that devastated the heart of Vernazza, the loveliest of the villages, and smaller parts of another, Monterosso. Vernazza's entire main street and piazza were buried under more than three meters of debris. Three lives were lost, many homes and businesses were ruined, and the village had to be entirely evacuated for several months, since water, gas, and electricity supply systems had been ripped apart. What I found most disturbing when I returned in June 2012 was not how much, but how little, visible damage there is to the hillsides above these villages. The landslide scars look almost insignificant in the context of the whole landscape. The point being that there is an awful lot more trouble stored up there, where the last disaster came from. We have a fatal tendency to underestimate scale in nature. A very little bit of mountain can hammer a village.

The villages themselves were restored remarkably quickly and, on the surface, remarkably well, in just eight months—a tribute to the renowned hard work and resilience of the Cinque Terre's citizens. Yet one had to wonder why the children's playground in Vernazza was being rebuilt exactly where it was before—right in the path of the little river that carried much of the mud and rocks into the village. This is surely a case where restoring to the immediate past, instead of learning from a much longer historical trajectory, is an act of blind faith, bordering on folly.

Some earlier landslides have been much more devastating to the hillsides, but have fortunately not found villages in their path. You can find a dramatic example by taking a vertiginous mule trail from

Vernazza initially toward its neighbor, Corniglia. First you pass a tower built to protect the coast from Saracen raiders in the Middle Ages. Suddenly, scintillating views open up of the Golfo dei Poeti, where Percy Bysshe Shelley drowned in a sudden storm in 1822.[1] But you must now turn your back on this dreamy panorama to scramble along some banded sandstone shelves. Then you must plunge up an even narrower and steeper track between a few still-cultivated terraces and the cypresses that mark the boundary of a small villa. Within two hundred meters, the path almost vanishes into the dense *macchia* engulfing the terraces that once stretched from here almost unbroken to Riomaggiore, the southernmost village, and beyond. They are masked now by showy yellow globes of spurge blossom, by thickets of bushy heathers and broom, by myrtles and mastic bushes and a rich variety of herbs and grasses. The ubiquitous helichrysum gives the air a spicy tang.

Trees—Italian buckthorn, maritime pine, holm oak, cork oak, and sweet chestnut—have found few footholds so far on this particular slope, but they are common elsewhere, regaining the forest territory they lost to agriculture as much as a millennium ago. On more arid slopes, there is the sparser vegetation characteristic of *gariga*, among patches of bare ground.[2] Almost every step here sends big grasshoppers, vividly liveried in lurid greens and yellows, rocketing away and disturbs a small flurry of other insects. Lizards rustle constantly in the undergrowth, while occasionally a disappearing tail betrays the presence of a snake. Sardinian and Dartford warblers set the dry air rattling with their harsh calls and songs, while less easily identifiable small birds chant tantalizingly similar notes from the dense scrub. High above, two kestrels dive-bomb a buzzard in a vain attempt to dislodge the rodent in its talons. If you are lucky, you may see a short-toed eagle slide down a thermal over the ridge. Abundant biodiversity here, then, and since we are on one of the more obscure and demanding trails, tourist footfall makes very little impact on the condition of the land.

So this is spontaneous ecological restoration, right? Indeed it is. The classic Mediterranean plant and animal communities are reassembling themselves where homogenous terraces of vines, lemons, and olives—underplanted with lettuce, basil, and zucchini—once dominated the landscape. Surely a national park could only be delighted to see such healthy natural grow-back, and do everything in its power to encourage it?

Wrong. The Cinque Terre National Park has—or had—a very counterintuitive mission, which might be summarized in the edict "Keep Nature Out!" In this chapter we will be looking at the restoration of a quintessentially "cultural" landscape.[3] When I was drafting the first outline for this book, Dave Egan, a former editor of *Ecological Restoration*, told me bluntly that he saw no place in it for a chapter on the Cinque Terre, that it bore no relationship to the other projects I was researching. Other restoration ecologists take a different view. In a 2006 article for the *Journal for Nature Conservation*, Francisco Moreira and his colleagues make a strong argument for restoring cultural landscapes, which often include biological niches, and for restoring traditional agricultural practices, vital for the survival of many animals, and which therefore have multiple values for human societies: "Cultural landscapes are increasingly threatened, and some, like species, face extinction. Their eventual disappearance will be a double loss, in both natural and cultural terms."[4] And while they were not considering the Cinque Terre specifically, their words have a resonance for the region: "In some situations, it may even be necessary to heavily modify or destroy valuable ecosystems to 'restore' cultural landscapes, such as clearing forest to reveal former abandoned agricultural terraces or stonewalls."[5]

In a very real sense, every landscape in the era of anthropogenic climate change is a cultural one. The human footprint falls implacably today on every Arctic glacier and on each remote rain forest ecosystem, even though no actual human being may have ever set foot there. And well before the advent of climate change, many or even most landscapes were "human shaped," even if the human impact is often not easy to spot at a casual glance.[6] Taking this argument further, if *Homo sapiens* is part of nature, then the New York subway and the slums of Mumbai are as much ecosystems as the Doñana wetlands or Yosemite's high meadows. However, there is clearly a broad spectrum of landscapes, running from relatively "pristine" glaciers, deserts and (some) isolated rain forests to zones where the impact of human culture is evident everywhere.

The Cinque Terre case is certainly very much toward the cultural end of this spectrum. And as we will see, the national park has indeed paid precious little attention to strictly ecological questions to date. Its founder even christened it "Man's Park" to emphasize that this is a national park with a very big difference. When I asked a park technician about biodiversity in the region, he corrected me: "This park is about *cultural* diversity. And not just about the diversity of the Cinque Terre from other regions, but also about the diversity between its very different villages."

"Protection of the environment here is protection of agriculture," another park official told me. Despite this, the work here has a lot in common with many ecological restoration projects. Restoration often requires a return to earlier, traditional agricultural practices, and this is being done systematically in the Cinque Terre. True, in most ecological restoration ventures the priority reason for this strategy is the restoration of biodiversity, but it is also often linked to the well-being of the local human population.

Collapsing Biodiversity

The distinctive feature of the Cinque Terre project is that it is restoring these practices not to replace intensive monocultural agriculture or reverse urban degradation, but to *prevent*, quite explicitly, the spontaneous restoration of natural processes. However, we will see that many of the same issues arise here as in more typical ecological restoration projects. And we will also see that ecological questions, though very inadequately addressed at present, are deeply relevant to the park's present actions and future development. Regardless of its organizers' intentions, the partial restoration of the terraces may well be having a beneficial effect on biodiversity in the region. In short, the atypical example of the Cinque Terre may tell us as much as many mainstream examples of ecological restoration about the complex relationship of humanity and nature.

Climb a little further on our path from Vernazza and you will find another reminder of this relationship, as one of the region's possible futures stretches beneath you. Below the tiny cluster of houses at San Bernardino, site of the sanctuary church for Corniglia, the slope has fallen away completely where a series of terraces had collapsed. The result is a still-unstable scar of nearly vertical landslide, which occurred after natural revegetation had taken place. The scree supports very little significant plant life. As we saw earlier, when the terraces are abandoned, the process of natural regeneration that follows is often ultimately a destructive one, leading to a rather literal kind of ecosystem collapse. Look down further, and you will see plenty of evidence of earlier landslides in the rocky debris piled up along the narrow shoreline at Guvano. While the Cinque Terre's steep topography and unstable geology undoubtedly make it historically prone to landslides in the best of circumstances, the existence of the terraces presents environmentalists with a paradox. Their carefully calibrated structures made most of the slopes significantly more stable than they would have been without human intervention. But in their present abandoned condition, the terraces both increase the risk of landslides and may greatly exacerbate the damage done when they occur. Given that these slopes dominate the villages, and the trails between them so beloved by tourists, the social motivation for reversing abandonment is clear enough. Sooner or later, abandoned terraces are likely to lead to another catastrophe that could threaten both the Cinque Terre's many visitors and its citizens. The periodic closure of trails in unstable condition bears witness to this real and present danger.

So the "pure" ecological restorationist is faced with a dilemma here, familiar in dealing with many landscapes impacted by agriculture, and different only in the extreme form it takes in this context. You could try to return the landscape to what it was like before the terraces were built, removing them and using the chestnut and holm oak forests elsewhere on the Ligurian coast as a valid historical reference system. It is most unlikely that this herculean task would be feasible, however, since the stone walls are deeply embedded in the shallow soil on the volatile slopes. The likely outcome would be a massive increase in landslides, leading to the transformation of today's rich plant communities into the sparse vegetation that can find a brief foothold on capricious piles of detritus.

Or you could let unmanaged nature take its course, leaving the area to spontaneous restoration of *macchia* in the hope that this would ultimately lead by natural succession to the complete return of the chestnut and, lower down, holm oak forests that once cloaked the slopes almost all the way to the sea. But instability caused by collapsing terraces would result, in many cases, in what we saw under San

Bernardino: an increase in landslides denuding the slopes of their biodiversity. The cost-benefit analysis of these options is never simple, however: the first postmortems on the October 2011 landslides suggest that the vegetation on spontaneously restored terraces *slowed* their momentum in many cases. This runs counter to the assumptions on which park restoration policy was previously decided. It is likely that every slope would need to be assessed on a case-specific basis to achieve the best outcomes for stability.[7]

But there is another important argument against totally restoring this landscape to its preagricultural ecosystem: even if we allow that the restoration of chestnut and oak forest to their maximum historical extent on these slopes were indeed possible, this achievement would not meet one of the key criteria of ecological restoration as I understand it. This is the imperative to increase or at least maintain biodiversity in restored areas, on a planet where biodiversity is taking a drubbing almost everywhere. While chestnut and oak forest forms an important ecosystem, it is in fact relatively poor in species, compared to the rich biodiversity to be found in the patchwork landscapes so typical of ecosystems in the Mediterranean. This patchwork pattern is mainly due to millennia of cycles of clearance, cultivation, and abandonment by humans; one of those instances where our species impacted positively rather than negatively on biodiversity. In any case, chestnut and holm oak forest is also still relatively common in the whole Mediterranean region, so there is no compelling biodiversity argument for restoring them here—and depriving the local people of both agricultural and tourism income in the process.

On the contrary, there is an excellent ecological argument for *partially* restoring the terraces to agriculture, though curiously, the park has hardly deployed this argument so far, and never very forcefully. The mosaic of partial restoration would almost certainly support more biodiversity than a complete restoration of the precultural landscape would do, assuming the latter were even possible. We will return to this point at the end of the chapter. But first it is important to understand the history of the terraces, and of the extraordinary battle which Franco Bonanini, the park's first president, unleashed to reverse this process.

Bonanini is a consummate politician, an ambiguous quality that inevitably has brought both positives and negatives to his contribution to the restoration of the Cinque Terre. However, he is one of those rare politicians driven by genuine passion for his people, for their culture (in the broadest sense of that word), and for the unique and special place they occupy on the planet. As mayor of Riomaggiore, he saw people, culture, and place confronting an unprecedented crisis as more and more terraces were abandoned in the 1980s. And he had had the vision to turn this crisis into a remarkable opportunity for the region. As president of an agricultural cooperative in Riomaggiore, he had seen young—and not-so-young—people desert the hard labor of the terraces for softer and better paying jobs. He hated to see an ancient and sophisticated system of cultivation fall into disuse. He dreaded even more the prospect of villages in the region becoming ghost towns full of mostly empty second homes for city people, or tourist traps that offered foreigners bad versions of their own cuisine—sauerkraut for Germans, bacon and eggs for the British, hot dogs for the Americans. Tourists were welcome, he thought, but they should be encouraged to enjoy the region's highly sophisticated cuisine, based on first-class local products, and they should learn something of its other cultural traditions. Somehow, he thought, it might be possible to find a common solution to all these problems.

The flight from the land that Bonanini lamented in the 1980s was not new. However, the Cinque Terre's geographical isolation had slowed the impact of the economic and social changes that have swept northwestern Italy over the last hundred and fifty years. The region's local tradition of fierce independence has grown out of its long inaccessibility to the outside world. Even the writ of the Roman Empire is said to have run only with difficulty on these slopes. The Cinque Terre is less than ninety kilometers from Genoa as the crow flies, and the region participated in the commercial life of that city-state's great trading republic, which dates from the eleventh century. Monterosso's anchovies were a staple item on the capital's tables, and Vernazza's skilled seamen navigated for Genoa's great armadas and merchant fleets. Politically and socially, however, the citizens of the five villages lived in a different universe from that bustling, cosmopolitan port. Some of them went to Genoa, but Genoa, generally speaking, did not come to them. Its feudal hierarchies were alien on these slopes. The people were poor—often desperately—but they were imbued with the pride of small landowners in their self-sufficiency and individual liberty. In Vernazza, this was expressed until quite recently by contemptuous references to inland folk, even as close as the village's own sanctuary hamlet of San Bernardino, as "people who bend the neck," that is, as serfs and slaves.

Today it is hard to imagine, when the tiny village of Manarola swarms with tourists from as far away as California and Queensland, just how isolated the Cinque Terre used to be. That's in summer, of course. Visit during the winter, and you can clearly hear a lonely echo of centuries of solitude. Indeed, you can still hear that echo even in summer, if you venture off the most-trodden trails into the deeply layered folds of the slopes. It was not until 1874, only four years after the final unification of Italy, that a railway connected Genoa to the Cinque Terre, en route to the nearby military-industrial boomtown of La Spezia.[8] Roads came much later, and never carried much traffic. Before the railway, the villages of the Cinque Terre were almost as remote from each other as they were from the outside world. Each village maintained a pride in its own identity, and retained—and to some extent still does—its own dialect. Almost the only opportunity for social mixing between the little towns took place during religious pilgrimages to the sanctuary churches. These rituals, whose routes can often be traced to pre-Christian practices, have a rich repertoire of traditions related to cycles of the year. On Good Fridays, you will find urns of eerily white wheat, apparently albino, surrounding the altar of every church. It has been grown in windowless cellars, and this is its first exposure to the light. By Easter Sunday its deathly pallor has been replaced by vivid greens, an appropriate symbol of resurrection for the faithful and of natural regeneration for the rest of us.

The railway itself could only connect the villages by long tunnels beneath the slopes, whose excavation significantly exacerbated the region's inclination to subsidence. To this day, a daylight train ride through the area consists of brief flashes of sea blanched to silver by blinding sunlight, framed repeatedly by minutes of immersion in blank darkness. And until the first steam engine traversed the route, the only links between the villages—and between most of the region and the outside world—were the sea and the mule trails that snake through the terraces.

One set of trails links the villages directly, usually close to the sea, though sometimes forced by cliffs to climb many meters above it. This is the now-famous Sentiero Azzurro (a.k.a. Route Two), along which tourists walk by the thousands every day of the season. Other trails climb rapidly and exhaustingly to the ridge route that defines the region's inland limits, passing by the sanctuary churches en route. A further set links the sanctuaries, roughly parallel to the ridge route and the coastal route and roughly midway between them. Those are the trails you will see on tourist maps, neatly set out with distances, times, and graded levels of difficulty (beware, the latter sometimes bear scarily little relationship to reality). The really fascinating paths are the subtrails that run—or rather ran—at all angles within this basic grid. These were made by peasants to access their highly fragmented land holdings. It would not be unusual for a family to have four terraces together by the sea, and five at the level of the sanctuary churches; several hundred daunting meters higher up, they might have yet another one alone among myriad neighbors' properties. Situation and soil dictated which terraces were best to maintain a variety of vines, olives, lemons or vegetables.

"Agriculture in a Pot"

Walking even the easiest of these trails on a hot morning, unburdened by anything more than a light backpack and binoculars, one begins—but only just—to appreciate the brute labor that constructing and maintaining the terraces required. First, loads of stone had to be shifted, using baskets balanced on human shoulders or slung across a mule, up, down, and across the slopes. Then each stone had to be carefully placed, the biggest at the base. These must be roughly flat and roughly rectangular, narrow end toward the hill, and overlapping. The angle of the front of the terrace must not be perpendicular, but lean gently inwards toward the slope, so that gravity secures the whole structure and does not tip it over toward the sea. Usually no mortar is used, which demands that the stones fit very snugly indeed. A hammer is employed to sculpt them, crudely but effectively, to fit the next gap, and then to tap them firmly into place.

"There are no wrong stones," says one of the builders, who has been building dry stone-wall structures for twenty-five years and now works for the park. "It is just a matter of finding the right place to put them." Smaller stones must be packed in, almost poured, to plug the space behind the front wall. These pebbles ensure that the soil the terrace contains will neither filter out when dry nor leak out while wet, and they also draw off its excess moisture. Even the soil itself is a human creation here, carried down from the ridge or the valleys inland, rich humus from the chestnut woodlands to supplement the thin layer of poor clay on the slopes.

"This is agriculture in a pot," says Matteo Perrone, a forester with the national park, in a phrase that neatly captures the way the soil is held within the terrace.[9] But it is also the agriculture of communicating vessels, since every terrace must be constructed in equilibrium with all the others around it. Each forms part of the spider's web of small drains that is the real genius of this system. So while the work of terrace building requires rugged individualism, it also requires a high degree of collaboration between landowners. Robert Frost's memorably ambiguous line, "good fences make good neighbors," would sound less ironic in the Cinque Terre than in New England.

It is no accident that this region, where cooperation is survival, gave more than its share of communist and socialist partisans to fight Mussolini's Fascists. Every second street seems to be named for a freedom fighter, half the piazzas dedicated to those "fallen for liberty." The Cinque Terre still votes largely for Italy's fissiparous parties of the hard and center left today. But this socialist tradition seems to sit easily enough with the entrepreneurship of families who proudly run their own businesses. After the terrace has been built, the ground must then of course be planted, and for the majority of the Cinque Terre terraces this meant planting vines. The tradition here has been to train the vines through a *pergola*, an overhead trellis structure. This means that every terrace's mini-vineyard was a dense mass of small tunnels. "Overhead" gives the impression that you can stand upright within the *pergola*, but in fact you have to crouch beneath a low-level lattice to get to the grapes. The work of harvesting was, and still can be, almost literally backbreaking. And then the crop has to be carried, again on shoulder or on muleback, along steep and narrow paths down to the village.

"The *pergolas* maximized quantity in production, but minimized quality, which was rational in the past but would be irrational in today's market," says Bonanini, who favors the simple row cultivation that is becoming more and more common.[10] "*Pergolas* also sheltered the grapes from the wind," adds forester Samueli Lercari, "but the cost in additional labor would be much too high today."[11] As president of the cooperative, Bonanini introduced other labor-saving innovations, the most important being the *trenino*, the "little train." The sinuous steel tracks for these miniature cable cars can now be encountered on most of the slopes. Today the loads of stones and baskets of grapes are transported up and down by the power of a small petrol engine, not by human sweat. In the least accessible corners, great slings of stone are lowered in by helicopter. Silvio Benedetti's glowing mural in Riomaggiore railway station, *History of Men and Stones*, celebrates the region's heroic rural labor in a style that might be described as socialist realism plus sunshine. Now it refers more to the past than to the present.[12] Get up early enough, though, and even today you can see callused hands, bent backs, and furrowed, leathery faces working on the relatively few productive family terraces remaining.

These laboring people are as often women as men, as Benedetti's own epic work illustrates. Historically, in Vernazza, women were much more likely to work the terraces, because a legendary tradition of seamanship in the village kept the men away for long periods. A kind of matriarchy ensued, and if men were needed, the women hired them from among the even more impoverished "bent necks" inland. As you walk the mule paths of the Cinque Terre today, you can still see impressive stretches of venerable terraces in continuing cultivation between the villages of Monterosso and Vernazza at the northwestern end of the region. There are also surviving patches between Manarola and Riomaggiore, at the southeastern end, and higher up around Manarola's sanctuary church village of Volastra there are extensive and impressively maintained terraced vineyards. But these are only fragments of a landscape that sculpted the entire coast a century ago, and still dominated much of the region as recently in the early 1980s.

Four distinct events have reduced the hectares under private cultivation from 1,400 in the 1870s to well under 100 today. The arrival of the railway reduced the distance to rapidly industrializing La Spezia from several hours' demanding walk to a few minutes' comfortable ride. Young people began to choose commuting factory jobs with set hours over endless days in the terraces in all weathers. However, the railway also made access to wine merchants in the cities much cheaper and more reliable, and production continued to expand in the early twentieth century. The seepage of labor from the terraces became a flood in the 1920s, when even the isolated Cinque Terre was reached by a globalizing biological force. *Phylloxera vitifoliae,* the insect that had been devastating American and then European vineyards since the mid-nineteenth century, rotted the vines from the terraces. As elsewhere, the crisis was finally overcome through grafting European vines to native American rootstock resistant to the pest. Many of the varietal grapes that made the region's wine distinctive were lost, apparently forever.

Expanding industries following the Second World War further weakened traditional agriculture, and abandoned patches began to outnumber cultivated ones on many slopes. But even in the 1980s the area between Volastra and Manarola was almost totally cultivated, as impressive photographs from the period show. Then the most recent, and most extensive, flight from the land began. This was the point at which Bonanini noticed that a market for second homes was opening up in Riomaggiore. Families found that they could buy two apartments in La Spezia for the price of their home in a Cinque Terre village. And as elite regional tourism threatened to evacuate the villages of their people, mass international tourism began to flow into the heart of the region by train and boat and spill out along the Sentiero Azzurro.

The Cinque Terre lies to the southeast of the classic high-end tourist meccas of the Ligurian Riviera, like Portofino and Rapallo. By the 1950s, tourists fleeing the extortionate prices of these resorts discovered Monterosso. But they tended to stop there, because they wanted to spend their holidays on beaches, and the remaining villages offer only very limited access to the sea, and hardly any to sand.

Gradually, more adventurous tourists discovered the spectacular views encountered when walking to Vernazza via the Sentiero Azzurro, and the charming seclusion and culinary delights of all the villages, though very few restaurants were operating. By the 1990s, the word was out internationally that this region was a very special place to visit. The vacation networks attracted ranged from serious hill-walking groups (often senior citizens) from northern Europe to young backpackers in search of a romantic location and cheap wine. Even today, when the Piazza Marconi in Vernazza fills up with hikers from the trails and disoriented tourists off the latest train or boat, there is a curious sense that the Sierra Club has just collided with Party Central. The demand for bars, restaurants, and accommodations shot off the scale, and every other house in the villages seemed to be offering some service or other to tourists. This development hit a weakened traditional agriculture with a one-two punch. First, many of the ground-floor rooms converted into restaurants had previously been wine cellars. In an area where storage had always been at a premium, many producers found they had nowhere to put their products—and stopped growing grapes. "Every street door lost as a cellar meant a plot of land lost to production," says Matteo Perrone.

Second, many of them no longer wanted to be producers in any case: it was much more profitable—and much less demanding physically—to sell meals to tourists on the spot than to endure hard labor to sell wine or lemons to Genoa. And if tourists wanted sauerkraut and burgers, that is what they would get.

Terraces: The Core Brand of the Cinque Terre

Enter Bonanini. First, he began to advise his constituents not to sell to second home buyers. "If outsiders are willing to pay a lot for our houses," he used to say, "that is because they are worth a lot more than they are paying us. Hold on."

This was a pragmatic tactic to buy time for a much more ambitious strategy. Bonanini knew that the intensely terraced agriculture of the Cinque Terre was unique. Terraces are common on Mediterranean slopes, but nowhere else were they packed so densely, in such a dramatically beautiful setting. Nowhere else did they produce so many high quality specialist products in such a small space. (The basil from these slopes is said to make the best pesto, a sauce that Liguria claims to have invented.) The wines had been highly rated for centuries, possibly as far back as ancient Rome. He was convinced that it was the terraces that had attracted the tourists in the first place, the one thing that gave the area both a distinctive landscape trademark and a reputation for first-class gastronomy. But now, the more the tourists came, the more the terraces disappeared. Soon there would be no terraces, no local produce, and then there might be no tourists.

As it stood, the tourism money tree was poisoned at the root, and things were going to get worse. For a start, the secluded spaces early tourists had sought were now as thronged as the New York subway at rush hour. The local population had dropped from 9,000 to 4,500 over the previous century. Now this tiny citizenry was hosting hundreds of thousands of tourists annually, rising toward millions. The Sentiero Azzurro was visibly crumbling under this incessant footfall, and the atmosphere in the villages was asphyxiating. No wonder the restaurants were turning into fast food joints and the bars losing their local character. Worse again, the abandoned terraces were beginning to shift in earnest. Riomaggiore and Manarola in particular were directly in the path of potential landslides. Bonanini foresaw a future with not only no tourists, but no villagers and no villages.

The solution had to be radical, and it was. Bonanini has been accused of many things, but never of thinking small. And so he appealed to the United Nations. He had studied the criteria for the UN's World Heritage Site program and decided to ask UNESCO to recognize the territory as one of "Outstanding Universal Value." At the same time, he applied to the Italian government for national park status. Both applications were successful. Having pulled off a remarkable national and international coup, Bonanini took office as president of the new park—and faced straight into daunting challenges at home. At the most basic level, the region's new double status—and his own interpretation of it—required him to impose numerous new regulations on his fellow citizens, and they were far from universally popular. All hunting was banned within the park, in a region where a shotgun was almost as familiar a tool as a spade. Outside the modern part of Monterosso, the villages were entirely pedestrianized, with only deliveries, national park buses, and emergency vehicles excepted. In an attempt to restore some tranquility to the villages, at least at night, the construction of hotels was prohibited. Overnight stays would be limited to bed-and-breakfasts and small hostels. These would have to meet high standards, including architectural and ecological criteria, to receive the park's quality brand. Likewise, only restaurants that served local produce in local styles would receive official endorsement. Planning permission for any new dwellings or extensions in the villages must now conform to rigorous new environmental and aesthetic norms.

His core vision, however, was focused on the restoration of the terraces and how that restoration related to tourism, and this is also where there is most contention today. Bonanini liked to repeat that "for every euro from the terraces, there will be a hundred from the visitors." But because he believed that that 1 percent contribution, insignificant in financial terms, is in fact the ultimate source and *sine qua non* of all the rest of the income from tourism, maintaining that 1 percent from the terraces in terms of quantity and quality is critical for the success of the whole Cinque Terre enterprise. He also argued that the revival of a vibrant and productive culture on the terraces will in time change the nature of the tourism in the Cinque Terre, attracting a much greater component of "cultural" and "ecological" visitors, eager to both respect and explore local traditions, and reducing the numbers who come in search of cheap wine in exotic surroundings. But the only way to maintain, improve, and increase production on the terraces is to persuade people to work there again. Reversing the flow of people from the land is difficult anywhere in Europe, flying in the face of powerful economic, social, and demographic trends. In the Cinque Terre it has been, and remains, a steep uphill battle in more senses than one.

By the time the park was established, the number of hectares under terrace cultivation had shrunk to less than a hundred. Bonanini's target was the restoration of a relatively modest four hundred hectares, but he must sometimes have felt a little like Sisyphus, as he watched more terraces succumb to the *macchia* and begin to crumble even as restored plots inched across the landscape. Figures on progress toward this target are extraordinarily hard to come by—the park appears to be keeping no reliable records. The best estimates suggest that the park has restored approximately fifteen hectares, private owners probably much less than that.

"I am still pessimistic," Bonanini admitted after a decade, "but not so pessimistic as I was ten years ago."

Some of the reasons for this painfully slow advance are obvious and come, as it were, with the territory: "The minimum viable plot [made up of multiple terraces] in the Cinque Terre is three thousand square meters," says Matteo Perrone. "And to farm three thousand square meters here requires three hundred days of one person's labor, because machines can hardly be used at all. To farm three thousand square meters of vines on the plains with mechanization requires one day's labor," he adds.[13] So it is necessary to produce very high-quality products, at premium prices, to compensate for the cost of such intensive labor. Cinque Terre wine acquired the coveted DOC quality assurance label status in 1973, but now a wide range of products, from honey to pesto, can also command higher prices because they carry the park's own quality endorsement symbol. In many cases, this also indicates organic production. Eventually, Bonanini hoped to extend organic criteria to all products, including wine. There is an increasing focus on the production of *Sciaccetrà*, a highly regarded sweet dessert wine unique to the region, which is by far the most profitable product per square meter planted.

However, profits from these products alone fall far short of the costs of further restoration and maintenance of the terraces. So the park has been tapping into the potentially vast reservoir of tourist income to make up the difference. Visitors who hike the lower trails are now charged a fee of €5 a day. Daniele Moggia, who works for one of the park's tourism cooperatives, told me that up to 1 million tourists, 40 percent of the total, now pay these fees, which would represent a very substantial income.[14] Bonanini claimed the park's restoration program is already self-financing.

But it is impossible to get details or confirm these figures, because the park does not publish its accounts. When I ask why, I hear the same cryptic phrase—"This is Italy"—with only minor variations from park employees and local citizens alike. Transparency and accountability have never been obvious virtues of Italian public finances. Since the majority of voters endorsed Silvio Berlusconi's bizarre contempt for ethics in public life for most of the last decade, things are worse than ever. Bonanini's ideology is closer to old-fashioned socialism than to the rightist politics of Berlusconi, but the Italian Left is not exempt from financial scandals. I was saddened—but not altogether surprised—to learn, after I had researched and written most of this chapter, that he had been removed from his position after being charged with misuse of public funds. These charges were still pending in June 2013. Some citizens of the Cinque Terre, especially in Vernazza and Monterosso, feel excluded from the patronage of the park. Long before Bonanini's arrest, I heard people openly question where the income from the Cinque Terre card and other schemes was going.

"The park is not sensitive to the needs of the community, it does not communicate with us," said accountant and private terrace restorer Massimo Evangelisti of Vernazza. "It would be good to understand what the park is doing. If we don't understand, then it is likely that we will think poorly of it."[15]

Whatever his flaws, the first park president was certainly a man with an attractive vision. But he did not communicate it effectively; it was often not at all clear if he was talking about what had actually happened already or about what he would like to have happen.

The Roots of a Vision

To see where Bonanini's vision has put down some real roots, you have to climb up from either Manarola or Riomaggiore to Corniolo Hill, where most of the area restored by the park is concentrated. Rank after rank of cleared and rebuilt terraces have been planted in vines, so that the brow of the hill has regained that striking similarity to a neat cornrow hairstyle that once characterized the whole region. In some places the terraces march in almost unbroken lines most of the way down to the edges of the village. In others, patches of *macchia*, sometimes already crowned with mastic trees and maritime pines, still interrupt the procession. Overall, the restored terraces at Corniolo form an impressive sight, and one that had been almost entirely lost to this locality ten years ago. But occasional pock marks on the hill's face show that even brand new terraces are subject to collapse when boars, or simply heavy rain, put too much pressure on them. Restoration can have a target in space and time, but maintenance and repair may be needed in perpetuity.

As you ascend the steep stone steps between the terraces on Corniolo (or take the easier option of riding the *trenino*), the first impression of monocultural rows of vines gives way to a wealth of detail. Herbs are cultivated on some terraces for culinary and cosmetic uses—the park is marketing an increasing range of soaps and perfumes. Massed ranks of basil soak up the sun. A long row of rosemary shimmers in the heat, the rippling air dancing with half a dozen different butterflies—though no one can tell me how many species occur in the park. On the ridge itself, the principles of synergetic agriculture—using wild plants to maintain fertile soil among cultivated crops—are being experimentally applied to a variety of vegetables.

Over on the opposite slopes to the southeast, between Riomag-

12. **Like a cornrow hairstyle on the landscape.** Agricultural terraces, some still in use (or restored), others overwhelmed by *macchia* and woodland vegetation, on the ridges above the village of Riomaggiore. (Photograph courtesy of Daniele Virgilio.)

giore's sanctuary church at Montenero, the cemetery, and the village itself, other significant stretches of restored terraces can be distinguished by their fresh greens and pale browns and light grays, amid both *macchia* and quite large areas of darker terraces which have been in continuous cultivation. Here, most of the restoration is private, apart from some olive groves, but the Corniolo terraces are almost entirely the work of the park. The park itself, however, actually owns very little land. Most of the terraces it is currently restoring have been leased from private landowners for twenty years. At the end of this period, they will be offered back their land, on the condition they maintain it in traditional agriculture. If they fail to do this, it may be confiscated in perpetuity by the park.

Where a terrace is not maintained today, the park can offer this deal to the landowner, and if the landowner refuses to take it up, and does not restore the terrace him/herself, then again the park has the right to take it into public ownership in perpetuity. All this, at least, was Bonanini's intention, but the temporary administration in place following his removal has raised many question marks over these regulations. I write "the park" here for convenience, but I should mention in passing that, under Bonanini, the park had a very odd structure, with no direct employees except for the president, the director and, according to one account, the president's wife.

Basically, the park subcontracted work to a variety of cooperatives, one of them devoted to agriculture and two to tourism. While this might seem like a good measure to keep local people involved in the park's structures through organizational modes with which they are familiar—co-ops have a long history in the area—it also has negative impacts. Everyone in the co-ops is employed on short-term contracts. The park has no permanent staff accumulating experience and feeding that experience into policy. Policy has therefore been entirely in the hands of the president. This lack of feedback and Bonanini's personal monopoly on policy has had numerous negative impacts. The park undertakes to assist local landowners to recuperate their family terraces. Since the vast majority of cultivated terraces are in private hands, one might think that subsidies to this sector was the royal route to restoration in the Cinque Terre, rewarding those who have maintained their land in production over these difficult years, and offering incentives to others to revive their family traditions. But it does not seem to be working out like that.

The Evangelistis, son Massimo and his retired father Angelo, are one such family in Vernazza. Together, they have painstakingly restored ancestral terraces over the last few years. "We don't do it as a business," says Angelo, "we do it as a passion, and out of respect for what previous generations created here. If we are a UNESCO heritage site, we should all do what we can to preserve our heritage."[16]

"I do it to find a place to escape the tourists," says Massimo, smiling quietly, "and my wife says I spend too much time up on the terraces. There is no money in it for us. But the best thing is the moment you pour yourself a glass of wine you have produced yourself."

They think that the number of families recuperating old plots is still very small but growing, and that—for those with appropriate south-facing terraces—rising sales of *Sciaccetrà* will change what has become a hobby back into a business. Neither of them feels the park has either inspired them to restore, or assisted them much in their efforts. "Let's be diplomatic but realistic in what we say here," Massimo continues evenly. "The one time we asked the park for stones when a wall collapsed they delivered unsuitable ones. I don't want money from the park, and I certainly don't want them to do everything, but I do want some help, and I want the right stones when I need them. We get emails from the park about international conventions on dolphins and bees, but no real communication about what matters to us." Nevertheless, they both agree emphatically that Bonanini deserves full credit for having had the region denominated as a national park and UNESCO site. Speaking before his arrest, they told me: "For sure, it would not have happened without him. And he has to run the place, he can't listen to everyone, and some people are impossible to satisfy. But he should listen to a wider variety of people."

Two people who believe they should have been listened to a lot more are Bartolomeo Lercari, also a native of Vernazza, and his Danish wife Lise Bertram. Both are agronomists. He is also a plant biologist who took a degree in ecology early by Italian standards, in 1969, and has many publications in international science journals behind him. Like many local people, they referred to Bonanani as "Pharaoh," because he seemed so inaccessible to them, so high-handed in his actions.[17] "To be independent-minded is a problem with the Pharaoh," says Lercari. "I don't want to fight him, but I believe he is getting bad advice, and that competent people should be critical. I think he is a little afraid of people with a university background."

The Lercaris certainly know what they are talking about in practical terms. With their own hands, they have restored much of the hill overlooking Vernazza railway station over the last ten years. The light green vines on his family's land stand out vividly in the June sun with the duller colors of the *macchia* that occupies some of the neighboring terraces. Even the stones in their terraces seem to gleam, pristine in the brilliant light. Such appealing appearances often count for little in agriculture, but if you order a bottle of their crisp Cheo wine in El Capitano's or Giani Franzi's restaurants on the Piazza Marconi, you will find that the quality of the Lercaris' work carries all the way to the table. Their revived business is very successful, even profitable—as long as you don't count the countless hours of their unpaid labor which support it.

"The park has been positive for the Cinque Terre, but I don't like the way it operates," says Lercari. "You are not told what money is available, how to apply. It is almost impossible to get a meeting with the Pharaoh. Then one day we met him in the street here in the village, and in two minutes he agreed to give us cellar space rent free and a corking machine." They greatly appreciate this support, but they don't think it should have depended on a chance encounter. "If he likes you get help. He is very positive, very active, but there is a *deep* lack of democracy here. We know nothing about the park's budget. This is a sore point."

They agree that good restoration work has been done at Corniolo. The team there has worked from the top of the slope downward as systematically as possible, given the fragmentation of landholdings and the vagaries of reaching agreement with landowners. Here the park has followed a key principle for restoration, whether (agri)cultural or more purely ecological: always work at the appropriate scale. As we have seen so often, because all natural processes are interlinked, if you cannot control every significant element around your restoration site, there is a real danger that your restoration may fail. And so it is better to focus on recuperating all the interlinked qualities of one such area—often a watershed, or at least one slope within a watershed—rather than carry out isolated patches of restoration in unconnected areas. Private landowners in the Cinque Terre can rarely work to an appropriate scale, because their holdings are spread out among a mosaic of neighbors' terraces. Thus the Lercaris have painstakingly restored the channels which draw rainwater in to irrigate their terraces, and which will also carry excess water off before it gathers the momentum of a flood. But this labor will be in vain if the terraces above theirs have not been restored—or have been restored badly—and more, or less, water than their terraces were designed for flows down onto them.

"Hard Work *Plus* Book Work"

Likewise, the Lercaris' efforts to eradicate insects harmful to vines through organic methods are often frustrated because surrounding patches of unrestored terraces provide nearby incubating grounds for the pests, and they are forced, against their environmental instincts, to use insecticides. They feel that the park's emphasis on cultural restoration tends to produce an uncritical respect for tradition, for what is often described elsewhere as local or traditional ecological knowledge.[18] "People say 'I am going to do it like my grandmother did it,' and then bad traditions go unchallenged. Hard work alone is not necessarily good work. We need hard work *plus* book work." The Lercaris are keenly aware, from Bartolomeo's own family background, of the grinding poverty that went along with the old traditions. They show me a photograph of his father as boy, his legs like spindles, his large head out of proportion to his body. "That was due to the kind of hunger we see in Africa today," says Lise. "Obviously, no one wants to restore those conditions." They both recognize that the park has made a significant contribution to creating new employment on the land, though as private producers they are also critical of the park's low productivity in its restored areas and, through subsidies, its "unfair" competition with their products.

That kind of tension is probably inevitable in any enterprise where public and private ownership coexist. And in fairness to the park, the restoration at Corniolo shows no sign of blind adherence to tradition, with innovation very evident. But then the Lercaris have never been invited to see what is happening there close up, and discuss techniques with the co-op staff, something they would gladly do on a voluntary basis. The park's failure to foster communication with local people is reprised by the Evangelistis, and a number of other people I spoke to. "What they are doing at Corniolo may be great work, but we can't be sure, because the park never explains itself to us," says Massimo. And then he adds a point which is commonly made in Vernazza, and in Corniglia and in Monterosso: "The park has a good image in Riomaggiore and Manarola, where it gives people work. We are waiting for something similar here."

The park runs into a dilemma familiar to all restoration projects with public funding, reflected here as much as it is on the Working for Water program, or in Chicago's North Branch Restoration Project. Restoration principles advocate tackling problems strategically, concentrating work on where it is most needed or will be most effective. The park can justify focusing its limited resources on Corniolo, as we have seen, on the grounds that a single large scale restoration is more effective than a lot of small ones. But people in other villages inevitably wonder why they have had to wait for a slice of the park's financial pie. They question whether this ridge was really selected because the University of Genoa recommended that it was most at risk from life-threatening landslides, or whether it was chosen because the villages beneath it are Bonanini's long-term political power base.

Matteo Perrone gives several reasons why Corniolo was selected as the first big restoration project by the park. He insists the priority criteria are public safety and scenic impact, but does not deny that local politics may have played a part in the decision. "Landowners leased land to us more willingly there, because the park's offices are in Riomaggiore and Manarola, and we employ most people there, so there is a lot of goodwill." But it was also chosen on grounds of relatively cheap and easy access, he says, since its terraces are right beside one of the few major roads near the Cinque Terre. Other sites will require much more labor to restore, and sometimes the costly use of helicopters to ferry in material. Perrone says that there is a park strategy to restore terraces above the other villages, but cannot give details about when these projects will start, and where exactly they will be. Perrone was in an invidious position during our first interview in 2008. He is clearly both highly educated and highly skilled in his profession of forestry, and has worked on park projects for eight years. But he had had no real input into park policy, and no job security. "I am the functionary in charge of the park's environmental and agricultural office, but for the moment that position is not defined. We are all in the same position, it is unbelievable, it is an Italian problem. The park has a plan for fourteen official positions, but they are all vacant, covered by temporary contracts. But I am working where I am born and I am satisfied with my job and"—he gestures elegantly toward the sea from where we are having an espresso at the spectacularly situated Bar dell'Amore—"there is always a beautiful view."

He accepts some of the criticisms I had heard. But he points out that these issues had arisen in a very challenging context: "We are a strange national park, a new concept of national park, we are responsible for many things outside the remit of a normal national park. Our priority has been to create a new economy in the region, based on traditional agriculture and on tourism. We are now reaching this goal of a self-financing park." Progress toward other goals, he admits, has been slow, especially on nonagricultural ecological issues. "The protection of the environment here means the protection of agriculture—and its protection against tourism, or better, how to manage tourist fluxes." The contradiction at the heart of the park, he agreed, is that it must both attract tourists and minimize their impact. Despite the park's organizational connection to a marine protected area of prime biodiversity importance,[19] it is really only in the area of managing tourism that it has even paid lip service to ecological issues, as most people would understand them, on land.

From the outset, the park has recognized that the footfall of hundreds of thousands of visitors annually on the highly scenic Sentiero Azzurro has major environmental costs. "The risk is collapse," says Perrone. "And the paths *are* collapsing, right now. In medieval times, everyone who used the paths was obliged to maintain them. And that was for only a few thousand people at most."

"Tourism in this sense is a sort of pollution," concedes Daniele Moggia, whose job is to attract visitors. "We have gone beyond the limit, there are too many tourists already," Bonanini told me in our first interview back in 2005. "We want cultural tourism, not mass tourism."

Despite recognition of this threat, the park has been reluctant to close the Sentiero Azzurro, except for short periods during the hyper-busy holidays around Easter and May Day, probably because it is a prime source of income.[20] Instead, there is rather vague talk of attracting a different kind of tourist. The theory is that such tourists will be more interested in culture and nature, and thus in exploring the dolmens, churches, terraces and wild places in the hinterland of the villages, so taking pressure off the coastal route.

The park has made significant practical efforts to facilitate such tourists by improving access to the higher paths through the provision of "eco-buses." These can drop you off at trailheads, saving hours of sweaty ascent. The upper routes offer more adventurous hiking, and some spectacular views, as well as access to the sanctuary churches and to active agricultural villages like Volastra. They also yield greatly increased opportunities to enjoy the region's rich flora and its fauna (rather poorer), of which you can only get a sketchy impression from the most popular route. Despite this investment, however, the buses mostly run empty, except for a handful of locals who live or work in the countryside. Some of the higher trails are now wildly overgrown with brambles, many are poorly marked, and you can walk the whole of the ridge trail almost entirely in your own sole company in high season. This is an added attraction for those who prefer their nature in solitude, but it is an indication of the abject failure of the policy of attracting "cultural" tourists to a sufficient extent to relieve pressure on the lower paths. The park's information offices and website do not highlight these opportunities and can provide precious little information to those who want to avail themselves of it. The Cinque Terre is a national park without rangers or eco-guides. Requests for information about birds or plants, beyond the basic identification sheets on sale, and some rather good poster-panels on plant habitats, are met with puzzled stares by information office staff.

A Willful Blindness to Wild Wealth?

In fairness, a new path has been created from Riomaggiore to a botanical garden at the Torre Guadiola, a good spot to observe migrating birds and cetaceans. Visiting in midsummer, I suggested to a park employee that there must be quite a lot of wildlife in such rich scrublands throughout the year. He was adamant that there was "nothing" to see outside migration seasons. Yet a Sardinian warbler—a delightful bird with red eye-rings standing out like fancy spectacles against its smart black balaclava—was singing from a nearby bush at that very moment. A ten-minute walk in the grounds rewarded me with a view of large snake slipping into the bushes. This almost willful blindness to the park's wild wealth might seem to justify the concerns of those who argue that the restoration of (agri)cultural landscapes has no relationship—or indeed a negative relationship—to ecological restoration. Those arguments, however, are based partly on the assumption that entirely "natural" landscapes exist anywhere. This postulates places where humanity stands, or can stand, outside the ecosystem, and where the role of restoration would be limited to simply assisting it back to self-sufficient functioning where we have degraded it. In our "anthropocene" era, it is doubtful whether any landscapes, anywhere, meet this criterion.

But it is no more helpful to proclaim that "everything is cultural" than it is to declare that only the "purely natural" is worthy of the attention of ecological restoration. The distinction between natural and cultural landscapes is useful, and reflects real and often radical differences, but they are on a spectrum, not in separate silos. In the

Americas and Australia, understanding indigenous peoples' cultural practices and their impact on a landscape often require forensic skills in archaeology, history, and anthropology, as well as ecology. Whether or not such cultural practices should be revived depends, as always in ecological restoration, on what kind of a future we want to restore a landscape to, and why.

In a European setting, cultural influences are generally much more obvious to most of us. In the Mediterranean region, where we have been making and remaking landscapes for millennia, restoring to a landscape at the "natural" end of the spectrum raises a number of issues, as we saw at the beginning of this chapter. The varied mosaic of habitats created by agriculture in the region, and the biologically rich ecotones[21] between them, often support a much greater variety of species than occur when the fields are abandoned. As Miguel N. Bugalho and his colleagues put it, "The conservation value of ecosystems shaped by human use (hereafter termed 'human-shaped ecosystems') is particularly notable in the Mediterranean Basin, where numerous societies have shaped natural ecosystems for more than 10,000 years . . . generating many cultures and land uses that have contributed to the landscape diversity now present in the region. . . . Indeed, *the Mediterranean Basin is one of the world's biodiversity 'hotspots' . . . partially as a result of the long-term human presence and related activities*."[22]

In the Cinque Terre, therefore, we have to ask several hard questions about the comparative value of restoration through spontaneous regeneration of the dominant native vegetation, and restoration through the recreation of a cultural landscape. Which will produce greater biodiversity? Are there rare or even endangered species which may be gained—or lost—as a result of either method? If so, which species among these should we prioritize? Finally, are there important ecosystem services which may be gained or lost, and what is the cost-benefit analysis between them?

We have seen that the abandonment of the terraces to natural processes may lead to landslides on many slopes, with huge local losses to biodiversity, though assessment of this risk needs to be fine-tuned after the 2011 disaster has been fully analyzed. Such landslides would also threaten critical ecosystem services, in terms of soil stability, water supply, and productivity in terms of wood and natural fruits, to say nothing of the threat to human dwellings and human lives. We

13. **After the deluge.** The disastrous floods, mud slides, and rock slides of October 25, 2011, cut the ground from under a house near Vernazza. It is still an open question whether terrace restoration minimizes the impact of such extreme events, or, at least in the short term, makes them worse. (Photograph by Paddy Woodworth.)

also saw that where something approaching a typical natural plant community for the area—closed canopy holm oak and chestnut forest—could indeed be restored, biodiversity would probably also decrease relative to the abundant variety of species attracted to the current patchwork of terraces, *macchia*, *gariga*, and several types of forest.

But a further question also arises: why has the national park to date shown little or no interest in evaluating the full ecological impact of either form of restoration? I sent a series of questions about this impact to two senior park employees, both of whom felt that they were not equipped to answer them. They recommended I contact Riccardo Nardelli, an ornithologist who works for the Italian Institute for Environmental Protection and Research. He comes from the Cinque Terre and has done some research in the area in a personal capacity. His main area of interest is the conservation and ecology of Mediterranean bird communities, especially agro-ecosystem avifauna. His responses, given before Bonanini's arrest, were illuminating.

He prefaced them by pointing out that Italy lags far behind countries like the United States, Britain, and Germany in gathering this kind of data, even in national parks: "In particular, the Cinque Terre National Park is driven by the best politician of Eastern Liguria (Franco Bonanini), a very "practical man," who cares for the heritage of our terraces and for human (economic) activities in this coastal context. Unfortunately, slight attention is paid to monitoring the changes occurring in fauna related to environmental dynamics, like the abandonment or the restoring of terraces. Few studies have been carried out, and the best contributions have come from volunteers."[23]

"Obviously, the abandonment of terraces and soil degradation must be priorities in a management plan for the park, but more activities concerning the natural environment are needed. The risk of the park becoming 'another Florence' (a lot of tourists, less and less indigenous people) is real."

Nardelli organized a number of bird counts on a volunteer basis in 1995: "This was the first study to relate the presence and abundance of bird species to habitat. Fifteen years later, it would be interesting to see what is changed! But birds (and other animals) are not part of Bonanini's mindset." He went on to address my questions as best he could from his personal observations. He reckons that *Sylvia* warblers, like the Sardinian—which is ubiquitous in the Cinque Terre—have indeed benefited from the abandonment of the terraces and the extension of *macchia* and *gariga*. He expects the same is true of forest birds. No bird or mammal found in the park is endemic to Italy, he points out, but there are several endemic amphibians of conservation concern, including two salamanders, a toad, and a tree frog.[24] He feels that not nearly enough research has been done to determine whether any of these species benefit significantly from the restoration of the terraces, but points out that one of the salamanders, *Speleomantes ambrosii,* "lives in the holes of terracing-walls, between the stones, in damp micro-climates. This species could be [affected by] processes connected to abandonment."

A number of bird species clearly benefit from the open spaces created by cultivation, and especially from the ecotones between vineyards, *macchia,* and woodlands. They have done so in some cases since Roman times, he adds. Stonechats, spotted flycatchers, cirl buntings, and red-backed shrike all exploit the stakes in vineyards as perch sites to catch insects and/or for courtship or territorial display.

"Sometimes, the dry stone walls are chosen by hole-nesting bird species, like tits, rock thrushes, and hoopoes. Lizards and snakes—especially the dark green snake—store food in the terrace crevices."

He has also observed an increase in some top avian predators, like the peregrine and the short-toed eagle, which probably indicates the increasing ecological health of the region as a whole. As to whether climate change has affected the composition of the park's flora and fauna, he simply cannot say due to lack of data. In general, some species will benefit from the restoration of the terraces, and others will lose territory. But Nardelli sees no significant threat to any species, given the relatively modest target of restoring only 400 hectares of the 1,400 originally cultivated. Indeed, he believes that both fauna and flora will benefit from the mix of cultivated and uncultivated land, perhaps becoming more biodiverse than they have been for many centuries: "I believe that during the eighteenth century, when vineyards were at their maximum spread and the slopes were intensively exploited, the total number of species in Cinque Terre was probably lower than it is today. Biodiversity here results from the interpenetration of natural and human habitats, which promotes environmental variations and the 'ecotone effect.' I believe," he concludes, "that the terraces make a significant contribution to biodiversity."

It seems reasonable, then, to regard the Cinque Terre National Park as a case of cultural restoration that could produce many of the benefits of ecological restoration—reasonable, but far from proven. The huge number of variables in any ecosystem means that there are many unanswered questions about all restoration projects. What is disturbing about the Cinque Terre is that most of the key questions have not even been asked by the organization in charge. It is very understandable that the park authorities should have focused on social, economic and agriculture issues during the first fifteen years of its existence. But it is surely now time to recognize that the park's mission offers a priceless opportunity to study the relationship between ecological and cultural restoration in a Mediterranean setting, with lessons that could probably applied to many other regions.

The expertise needed does not have to come from within the

park's structure, though some of it is already there and just needs to be properly deployed. The park has been very successful in tapping into the resources of universities—some local, some from as far away as Australia—for studies on everything from the provenance of grape varietals to the economics of tourism to landslide risk assessment. It could do the same in the fields of ecology and biodiversity studies. Its relatively small scale makes it particularly attractive as an open-air laboratory for experiments ranging from climate change to the relationship of its varied "natural" and "cultural" habitats to species populations. Such studies could also inform the future strategy of the park. Is four hundred hectares the ideal maximum expansion of restored terracing in terms of biodiversity? What are the costs and benefits—ecological and aesthetic—of controlling populations of alien plants like agaves, which are closely identified with the popular image of the Cinque Terre landscape today, but which contribute to destabilizing the soil?

Our exploration of the Cinque Terre National Park touched on many features common to other restoration projects: issues of organization, democracy, and accountability; of the role of the charismatic and visionary leader; of biodiversity and monitoring; and of the relative merits of local ecological knowledge and scientific methods have all been raised as we walked its paths. Many similar issues, problems and pleasures arise in restoring a cultural landscape and other forms of ecosystem restoration. But something fundamental is clearly missing here, so far.

"A study of the ecology connected to the Cinque Terre terraces would be fantastic," Nardelli writes wistfully. Indeed it would.

* * *

The post-Bonanini reorganization of the park was still in flux on my most recent visits, in June and September 2012. Matteo Perrone has been appointed head of scientific research, and is committed to undertaking projects along the lines suggested here. But he also expressed fears for the future of the park in the context of the cutbacks imposed by the crisis-ridden Italian economy. The most attractive aspects of Bonanini's vision for a cultural restoration of the terraces may, he fears, simply not be sustainable in the new situation, and the park may simply revert to much more traditional environmental management. In September, he was more optimistic about park policy under a new full-time director, but expressed deep concerns, which I heard echoed everywhere, about the flaws in the regional authorities' response to the aftermath of the October 2011 disaster. There is a budget for rebuilding damaged roads, but not for work on the "hundred points of instability that remain above Vernazza."

"I would hope," Perrone told me, "that the park and the authorities will prove capable of closing the circle between nature, agriculture, biodiversity, economy and landscape."[25]

7 *The Last of the Woods Laid Low? Fragile Green Shoots in Irish Forests*

How long does it take make the woods?
As long as it takes to make the world.
The woods is present as the world is, the presence
Of all its past, and of all its time to come.
WENDELL BERRY, "1985 V"

In 100 years' time, will there be oak woodland still present at Brackloon, if it is not managed? Good question! * PAUDIE BLIGHE, *former Coillte forester*

When you propose change, you get irate feelings rising up. People say, "That is a lovely walk in the woods, and now you are going to fell it on us." * SEAN QUEALY, *Coillte restoration project manager*

You can nudge nature, but she may not go the way you want her to. * SEAN QUEALY

You should not try to improve nature too much. * DONAL CORKERY, *Kerry farmer and restorationist*

* *

It is notoriously difficult to read a wood, and very easy indeed to misread this seductive landscape. It is no accident that getting lost in a tangle of trees is a recurring motif in global folklore. Even scientific specialists get disorientated here from time to time, and laypersons need to learn to look twice—and then several more times—at what appears to be obvious.

For most of my life, I believed that the mossy oak woods that cloak most of the southern slopes above the twin lakes of Glendalough, in

14. **Ancient forest or secondary growth?** The woodlands surrounding the well-loved medieval monastic site of Glendalough have gone through many more changes than most visitors—or residents—realize. (Photograph by Paddy Woodworth.)

my native County Wicklow in Ireland, were an "ancient forest" without ever thinking much about what that meant. I certainly thought of them as much older than the early medieval monastic churches and round tower that attract visitors to the valley from all over the world. It was a rude shock, in the early stages of researching this book, to see nineteenth-century photographs that showed the same slopes stripped bare of trees. (One such photograph, admittedly a rather blurred one, had been hanging on the wall of my kitchen for years, but I had misread it completely; I had seen woods where I had expected to see them). Those green and shadowy places, so evocative of magical fantasies in my youth and of equally fanciful and vaguely conceived notions of pristine nature as an adult, turned out to be secondary growth, little older than the century in which I was born.

Does that matter?

Not much on one level: the imaginative and environmental pleasures they gave me were real enough. Whether I was stalking migrant warblers or courting girlfriends, these woods fostered an abiding sense of connection with the natural world. Eventually—and I admit to being a very slow learner in this—that sense of connection has led me to become curious enough to see a little beyond the apparently obvious, and to seek to read an oak forest at a less superficial level. That can be a painful process, of course, because it involves a loss of innocence, an awareness of Aldo Leopold's "world of wounds," when you see through to the ecological injuries which lie behind the skin of many attractive landscapes.[1] There is no quick balm for that existential discomfort, but restoration, with its promise to liberate the extraordinary recuperative powers of altered and damaged ecosystems—the very powers that created Glendalough's new woodlands—offers a route to new and richer ways of engaging with nature as it is. And that is in many cases, and perhaps today in most cases, nature as generations of humans have reworked it.

But still, seven years after I began to shift my focus in this direction, reading a wood with any accuracy remains damn difficult. This is hardly surprising, when experts like Oliver Rackham and George Peterken can spend a lifetime on this enterprise and still argue with each other, and indeed with their own earlier versions of a woodland's history.[2] In any event, especially when looking at woods in the process of restoration, I like to have a guide with me who can separate the layers of ecological—and human—history that lie beneath the surface view. Sometimes, however, I find it is a useful exercise to poke around on my own for a while, at least after doing some initial research, and attempt to piece together what is happening under the canopy. As it happens, I was unable to link up with guides on my first visits to two of Ireland's best-known woodland restoration projects, and had to make of them what I could with the help of a book, some old field guide notes, and a newspaper article.

Brackloon Woods, nesting among rocky hills above the coast of County Mayo near Westport, has been a testing ground for two important projects. It was chosen as the prototype for an Irish Ecological Monitoring Network in 1996. The network, sadly, has failed to spread across the land, but the intense and systematic research done by soil scientist Declan Little and others at Brackloon made it an ideal pilot site for the Native Woodlands Scheme (NWS).[3] The NWS encourages and supports landowners to restore existing native forests and to create new native forests on greenfield sites. Brackloon is also the subject of a beautiful and informative book by Deirdre Cunningham.[4] It is an ancient, seminatural woodland, that is, it has been wooded since at least 1660, but it has also been repeatedly disturbed, exploited, and modified by human use. Among its characteristic trees are oak, birch, and holly.

We are very lucky to be able to study and enjoy it today, because in the 1960s it nearly disappeared altogether. The state forestry service felled fifty of its seventy-four hectares and replanted them with alien conifers.[5] The remaining mature woodland was fragmented, the largest single unit left being only eleven hectares. Brackloon's survival and restoration is a case history in the shift of the culture of Irish forestry in recent decades. The entire wood was recommended for conservation and restoration as early as the 1970s, but it was only in 1996 that the current owners, the state-owned commercial forestry company Coillte,[6] designated it for biodiversity management.[7] In 1999 the whole site was declared an EU Special Area of Conservation (SAC).[8] Coillte then opted for full-scale restoration, ultimately using NWS criteria. The research under the monitoring project provided the project with a quantity and quality of scientific information that is probably unique in Irish restoration and would be envied by many projects worldwide. Using seed collected exclusively from Brackloon and nearby woods that were germinated locally, Brackloon was replanted with native trees after the conifers were cleared.

An Aura of Robust Ecological Health, Some Sour Notes

Walking into the wood on a late afternoon in September, eleven years after restoration began, it is very hard to believe I am on the right site. Remember, two-thirds of this wood had been converted to conifer plantations some forty-five years earlier, and they had been clear-cut less than a decade before, with some areas left to regenerate naturally, and others augmented with planting. I am walking the loop trail, which offers an apparently comprehensive tour of the wood. But I never see an obvious clear-cut gap, nor even any of the telltale plastic tubes that protect recently planted seedlings from grazing. Late-blossoming wildflowers—goldenrod, devil's bit scabious, rosebay willowherb—raise their heads confidently in open areas along the main trail. Stepping randomly off that trail into the woods, the forest floor is thick with hard ferns and studded with at least two of the more elegant *Dryopteris* fern species, all set in a bed of mosses. Thriving willow and birch trees thicket the many damp hollows—the sound of water from streams feeding the nearby Owenee River is omnipresent after two days' heavy rain. There are plenty of fine, upstanding ash trees, heavy with their "bunch of keys" hanging fruit, and holly and hazel flourish in the understory.

Mature birches are richly decorated with fantasy maps of mosses and lichens and liverworts. Occasionally, rarities leap out at me: lungwort among the lichens, and a heather graced with sparse but striking bell-shaped flowers, purple-pink and outsize. This turns out to be St. Dabeoc's heath. Another rarity, the narrow-leaved helleborine, an orchid of damp woodlands and one of the trophy plants of Brackloon, evades me completely—not surprisingly, since it would have stopped flowering at least two months ago. Every so often, fine individual specimen oaks rise like keynotes to confirm this aura of robust ecological health. Toward the end of the circuit there are groves of them; clearly this is one of the sectors that had never been cleared, at least in recent centuries. A good mix of seedlings and saplings beneath them heralds another generation of local giants.

There are sour notes, too: an occasional rhododendron finding a new foothold despite intensive clearance during restoration, an alien spruce popping up as a reminder of what had gone before. There is evidence of past human use of the woods—a large pit beside the trail, already well camouflaged by mature vegetation, must have been used for charcoal production. Overall, then, the impression is of a flourishing, long-established native woodland. If I had not seen a Coillte sign on the entry, I would have been convinced I was in the wrong wood. How could so much destruction and disturbance be so little evident, so soon after the restoration had begun?

I decide to make a more thorough exploration the next morning, and ring Deirdre Cunningham en route, almost begging her to tell where I might find some obvious concrete evidence of the dramatic changes that her book describes. She suggests I look for new plantations to my left just after I entered the wood. Sure enough, looking a little more closely, I can see one or two pale tubes just a few meters from the path. Plunging into the thick undergrowth, I find dozens of ash, birch, and willow sheathed in wire-reinforced plastic. Some are still struggling to emerge, but others are doing well, already twice or even three times the height of their tubes. The main function of the tubes is to protect the young saplings not only from grazing but also from the competing vigorous vegetation on the forest floor. Gradually, as I criss-cross the off-trail woods as thoroughly as I can, I come across more traces of human influence on the structure and composition of the wood.

Beech: A Charismatic but Problematic Alien

Several times I encounter small stands of mature beech, often mixed in with the oak. Cunningham writes that "veteran" beech trees will be retained for "recreational value," and that their rot holes are good for biodiversity.[9] Younger beeches will be removed, though this is "very contentious" with local people.[10] Extensive consultation, however, has prevented this controversy from reaching the kind of ugly impasse I saw over similar issues in Chicago.

"A lovely stand of beech with bluebells surrounds a souterrain cave," says Paudie Blighe, a retired Coillte forester who led this restoration project. "It is going to be retained because it has scenic/amenity and wildlife value. No significant beech regeneration is observed at the site, so there is little sign of the beech encroaching on the native woodland."[11]

Beech is certainly a very familiar tree in Ireland, and a much-loved one. It was first introduced, probably by Norman colonists in the thirteenth century, has been widely planted since the eighteenth century, and has naturalized seamlessly into our woods. Indeed, many Irish people are not aware that it is an alien. Given the scarcity of native woodland communities, however, it is a problematic candidate for inclusion here. The biodiversity argument about such charismatic aliens cuts both ways, and the beech loses out on balance: its dense canopy shades out almost all midstorey and understory plants. Cunningham argues that, if not controlled, it will outcompete and replace oak on the wood's most fertile soils. Little thinks that it should be watched very closely indeed wherever the aim is to restore native woodland.[12] The souterrain Blighe mentions, associated with a nearby ring fort, dates from as early as the third century; both are part of the evidence for fairly intense human habitation locally and within the wood in that period. Less obvious evidence shows that humans have been present here for three millennia.[13] The pollen record confirms that many—but not all—areas were cleared up to the thirteenth century at least. This opened up already-poor soils to increasing rainfall that leached away many of the remaining nutrients and minerals. All these factors contributed to the disappearance of tree species like Scots pine, aspen, juniper, and yew, and the decline within the community of elm, hazel, oak, and ash. The soils in the wood are now quite varied, which obviously influences what trees and plants can be restored in what areas. But the research has shown that the parent mineral materials for Brackloon's soil are more or less uniform, so that the variation is largely due to human impacts—a good example of how subtle but substantial anthropogenic changes can influence the development of what we see as "natural" woodland.[14]

Intrusion of grazing animals from neighboring fields was commonplace right up until the conservation order was put in place. To maintain the natural regeneration of the woods, it is vital that grazing should be limited, and a stout fence now rings the perimeter. But I find that it has been breached and trampled flat at one spot. The trespassers were probably human, given the adjacent remains of a campfire, a scattering of beer cans, and a shirt hanging rakishly from a tree nearby. Happily, the presence of many regenerating seedlings suggests that not many straying grazers have found the gap yet, but it is a reminder that the restoration's price, like liberty's, is constant vigilance.

What I find most surprising, and most pleasing, is the way in which evidence of the conifer plantations has vanished so quickly, at least to the untrained eye. Even in the very occasional remaining openings, you have to kick around a lot in the mossy forest floor to find the decaying stump of a Sitka spruce or a Douglas fir. For those who very understandably want to see quick results from restoration, Brackloon is an inspiring place to start. As a pilot site for the whole NWS, this project displays many more features than I could spot unaided, or indeed, than I have space to describe here. But they include the retention of standing and fallen deadwood as wildlife habitat, the former especially for long-eared owls and several species of bat. Bat boxes are also used. Another basic strategy has been to reconnect, through active planting and natural regeneration, all the fragmented oak-dominated communities.

On both visits I see a large female kestrel hunting just outside and then over the woods. As I exit for the last time, my gaze is pulled skyward by the angry yickering of a juvenile peregrine, flashing across a gap in the trees. The presence of these two significant predators could be taken as another sign of the health of this project.

Small Questions, and a Very Big One

Nevertheless, several relatively small questions—and one very big one—loom over its future. Some of these can only be asked because of the research and monitoring that preceded the restoration and now runs alongside it. Declan Little has a double involvement in Brackloon, first as one of the prime movers in developing the prototype monitoring site there, and then as project manager with Woodlands of Ireland, the initiator of the NWS. These roles give him a keen insight into questions about its history and restoration potential. He has set them out in a paper on the challenges facing the future management of the woods.[15]

To put a smaller question first, the research established that the dominance of oak in the woods might be more due to human management than original soil types. Little stresses that the privileged position we tend to accord to oak in our seminatural woodlands is often not "natural" at all, but the outcome of centuries of our *cultivation* of this species, with rigorously planned coppicing cycles and the concomitant exclusion of those native trees our ancestors often considered "weeds"—alder and birch, for example.[16] So while soil types are used as a guide to determining where to plant oaks, down the years it may prove that oaks will not flourish everywhere they are planted—unless there is active management that could be detrimental to other native species. This raises a familiar conundrum: are we restoring toward the community of species we imagine *should* be there? This is often an impression based on what happens to be present in remaining native woodlands today, which is itself the outcome of generations of human intervention. Or are we restoring toward the community of species that the ecosystem, left to itself, can support? Are we restoring so that nature can express itself, or to express our own desires? Or does the answer always lie at some point on the spectrum between these two motivations?

But these are only preludes to a bigger version of the question at Brackloon: according to Little, the soil in some areas has been so impoverished over the centuries by leaching caused by clearances, that fertilizer may now be needed to maintain it as a woodland at all.[17] The idea of a restoration permanently maintained by artificial fertilizers somehow does not ring true, and this leads to a bigger question still: what is the ultimate trajectory of the woodland likely to be? Devoted though he is to restoring woodlands, Little gives some credence to Oliver Rackham's view that the natural ecosystem trajectory of Ireland's species-poor Atlantic woodlands tends ultimately toward their transformation into blanket bogs, even without human intervention. If this is the case, Little writes, the best option might be to refrain from attempting to hold back the tide of natural succession with lime-based additives to the soil, and to instead allow "the woodland . . . to develop without intervention, which may result in the woodland being succeeded by blanket bog in the long term."[18]

Little stresses that this is a very complicated issue, and the outcomes will vary from region to region, from site to site, and even within sites. He argues that the best strategy is to expand the forest area, but to let those parts of it that cannot support trees in the long term shift to bog.[19] Blighe readily concedes that we cannot fix a future trajectory for these woods with any certainty: "In 100 years' time, will there be oak woodland still present at Brackloon, if it is not managed? Good question!"[20]

And so Brackloon, for all its remarkable and rapid success, or perhaps because of it, poses in a clear and acute form one of the great restoration issues: are we restoring to preserve particular communities of fauna and flora which we cherish—perhaps for the best of reasons—or are we restoring ecological processes so that, independently of our wishes (and therefore, perhaps, more "naturally") they will continue toward new destinations we do not attempt to determine, and perhaps cannot even clearly imagine?

"I go for the latter," says Declan Little.[21]

The Raw Mechanics of Restoration Laid Bare

My next stop was Clonbur Woods, on the border of counties Mayo and Galway along the southern shores of Lough Mask—very close to where John Ford created a highly influential—and deeply sentimentalized—image of Ireland when he filmed *The Quiet Man* in Cong with John Wayne and Maureen O'Hara. This 300-hectare estate is the biggest of 9 sites, totaling almost 550 hectares, restored jointly by Coillte and the EU LIFE-Nature program[22] between 2006 and 2009 as part of Coillte's commitment to manage 15 percent of its estate "primarily" for biodiversity. The program aims to restore four types of native woodland, all now rare in Ireland and across Europe: alluvial woods, bog woodland, yew woodland, and woods associated with limestone pavement. These are all priority habitats for conservation and restoration under the terms of the EU Habitats Directive.

The Clonbur Woods often grow straight out of limestone pavement and form the largest area of this spectacular karstic habitat in Ireland outside the Burren,[23] which makes this a very special place. But it also includes a remarkable range of other habitats: alluvial woodland, heath, grassland, alkaline fen, and low-nutrient lakes and ponds. There are some old areas of yew woodland with a few trees surviving, and their seed will be used to restore it over larger areas. There is also a lot of mature woodland on calcareous soil, which had been heavily planted with conifers and beech. Given its size and diversity, it is not surprising that Coillte has nominated Clonbur as one of three "demonstration sites" in the program.

If a walk through Brackloon is to experience an apparent miracle of rapid regeneration, a stroll through Clonbur is to see the raw mechanics of restoration laid bare, a work in progress. This is not always easy on the eye. However, signage at a small parking lot gives the lay of the land and sets out the general outline of the restoration program clearly and attractively. Coillte specified public education as an essential part of its restoration program. I enter the site through unrestored areas that remain part of the company's mainstream operations. There is a dense plantation of mature Douglas fir on the right, without a glimmer of green on the dark needle blanket that stretches beneath them. Rosebay willowherb has massively colonized a swathe of clear-cut on the left, beyond which lie more stands of monocultural conifers. There is perhaps no harm in being reminded of what commercial forestry looks like before reaching the restored zone.

The first evidence of change is another zone of clear-cut, but here there are ample signs of replanting with native trees, mainly birch and ash. Spindly birch trees that were spared during the clearance are the only tall trees, and the site has an untidy, unfinished look, as you would expect at this stage. This is a younger restoration than Brackloon, of course, but only by a few years, and the comeback seems to be slower here so far. One problem is evident at once. Japanese knotweed has taken a firm grip alongside the trail and looks set to perform expand like a tsunami if it is not rapidly controlled.

This is a site with a lot of history. Ballykine Castle, a heavily fortified family residence rather than a major military installation, has long been unroofed, but all its stout main walls are still standing, thickly coated with mosses. Most of what remains was built during sixteenth-century wars as a reward to Scottish mercenaries who assisted the British Crown against native rebels. But some of it has much older features. I am rarely sensitive to intimations of the paranormal, but the place has a decidedly spooky feel. This is probably because it is almost hidden in a dark beech wood where a pair of dry branches are creaking ominously. Most such castles originally stood on cleared land, the better to see approaching enemies. More recent landowners here, the Guinness family, responsible for Ireland's best-known black stout, planted the beech trees around it. No doubt they felt the wood would make a more romantic setting for a ruin. As in Brackloon, the beech pose a dilemma for the project, as extensive local consultations showed that people are very attached to them. To clear or not to clear? Coillte has decided to leave them be for the moment, save for a little thinning, and to reduce them gradually but not necessarily completely. This decision was bolstered ecologically by other attachments to the beeches, some of them literal: they provide a local stronghold for three rare species—the lungwort lichen, the parasitic bird's nest orchid, and the lesser horseshoe bat. And another fine alien has escaped the ax for the time being: towering high over even the high beech canopy, veteran silver firs, which do not not pose a regeneration threat, will be left to die of natural causes.

"The people of Clonbur have long historical memories," Sean

Quealy, Coillte project manager for the EU LIFE-Nature woodland restorations, told me later.[24] "I call trees like these firs and beeches 'cultural' trees. People remember that their great grandparents worked on the estate when these trees were saplings. They value them for the same reason people like to go to a graveyard and rub the family grave." He found the consultations at each restoration site different, but always challenging. "When you propose change, you get irate feelings rising up," he says philosophically. "People say, 'That is a lovely walk in the woods, and now you are going to fell it on us.' So at least some of the trees local people love, like American oak and beech, will not be felled within twenty meters of a forest path. Quealy also points to the function of the beeches as "nursery trees" for lungwort, but stresses that these woods will regularly be thinned, and that oak and ash are already springing up in the gaps. This lichen is not fussy about its host species, and will spread quickly to the new growth.

A Zen Space Sculpted by the Ecosystem

The trail back out to the lakeside runs back through the first big cleared area. Mulch from the thinned beech is protecting the roots of many young saplings that sit among rotting conifer stumps already vanishing in the rising sea of pioneer plants on the site. I cannot find the quadrats which have been set up for monitoring, or any of the yew planted in this area. "The yew did not strike universally the first time around," Quealy concedes when we meet in his office in the midlands. "There is an educational aspect to this program; we have learned a lot about propagating yew and will go back in and plant more."[25] As for the quadrats, he pointed out gently that I had misread the map and missed them. They remain, it is good to learn, part of an annual monitoring system.

After a good stretch of mature native woodland, including some venerable Scots pine, the limestone pavement becomes visible underfoot. And pavement it seems to be, its characteristic clints (flat slabs) and grikes (crevices) often appearing curiously man-made. As it opens up, small ponds and fens appear, and the bordering grikes are rich in lime-loving plants like wall-rue, maidenhair spleenwort, rusty-backed fern and biting stonecrop, while the sparse soil also gives sustenance to small trees and shrubs, spontaneous bonsais if you will, blackthorn, pines, willows, and alder. This intimate landscape, midway between the road and the lake, has an exquisite tranquility about it, a Zen space sculpted by the ecosystem.

"You can stand at one point on this pavement," Quealy tells me, "and see every native Irish tree, bar aspen, as you turn through 360 degrees." Conserving this extraordinary level of biodiversity on the limestone pavement may present challenges to futures managers of Clonbur, he suspects. Some species, probably fir and hazel, will tend to monopolize it, especially if there is no grazing. So do we take out a native species, he asks, or do we lose the pavement's current gift of variety? He answers his own question with a line that might be a motto for all restoration projects: "You can nudge nature, but she may not go the way you want her to."

I have to admit that I never see anything close to the whole range of Irish trees from one spot. But every time I move the composition shifts again: I stumble on two ancient yews, perhaps five hundred years old, and among the parents of the new plantation, plus some guelder rose and buckthorn, heavy with red and black berries, respectively. And then the path deepens and darkens into a hazel wood, their multiple stems sending roots deep down into the grikes. There are more areas of clear-cut conifers with some evidence of natives beginning the long process of natural regeneration, and then, through a gap, the lakeshore suddenly opens up, its waters softly slapping in the wind against the bare limestone. Here we are on the edge of the restored site, and conifers stand in industrial ranks on a nearby island, linked to the mainland by a narrow isthmus. The plan for the future is to remove commercial trees on short rotation in this border zone, rather than clear-cutting, and slowly replace them with native hardwoods, though still exploiting them commercially.

Unrecognized Sites of Humungous Biological Diversity

The whole site comes alive again for me when I talk to Quealy three months after my visit to Clonbur. And so too do other Coillte-LIFE sites I have never seen, like Durrow, Hazelwood, and Camcor. Quealy is wired with energy, his piercing eyes scanning photos from these woods on his desktop and on my laptop for an image that will illustrate points he wants to make, ticking off each point in his sentences with a rhythmic "yes" in a torrent of affirmation James Joyce might have envied.

"A forester's view is a long view," he declares. "My aim is to kick-start restoration, yes, remove aliens, yes, allow natural regeneration; because we do not have native grazers, we will have to manage in perpetuity, yes." Quealy is a passionate educator, taking children out on field trips to show them fungi and damsel flies, in his own free time as much as on Coillte business. He exudes positivity, but he is frustrated by the "philistinism, the lack of environmental awareness, in Irish culture."

"These are sites of *humungous* biological diversity, but they are not recognized, they are not appreciated by the public. But a forester's view is a long view, yes, I know what they will look like, yes, I know they will be gems."[26]

Clonbur is, as I have said, very much a restoration in progress and it will be decades before any firm conclusions can be reached about its success, especially regarding the yew. But its exploration of a variety of management techniques for restoration over the site should make it a valuable laboratory for the future. In order for restoration to navigate its way to a future that is both viable and in some way authentic, it needs a compass from the past. I asked Declan Little where I might find a forest less managed over the centuries than Brackloon or Clonbur, closer to what might be a historical reference system for native woodland where *Homo sapiens* had not played the lead role. He brought me to a valley where what remains of the past, and what may be the future, can be found side by side.

A Commercial Forester's Despair, a Bryophyter's Delight

Glaciers carved out the long, broad-bottomed valley of Gleninchaquin thousands of years ago, leaving it rich in lakes and rugged topography. At the head of the glen, tucked in under the summit of Dish Mountain, the corrie lake of Cumeenaloughan feeds a waterfall; it spills anarchically over a broad shield of blackened sandstone, varying from a modest trickle to a raging spate, depending on the weather. In turn it feeds a stream, which winds down the barely perceptible gradient of the valley floor, spreading out into three more lakes where it levels, before a final short and steep descent to the Atlantic. We are on the northwestern edge of the Beara peninsula in County Kerry, one of Ireland's more biodiverse areas. The ancient oaks of Uragh Woods border the entire southern shore of Lough Inchaquin, the uppermost of the lower lakes. Most of the valley was wooded before the arrival of the first human settlers in the Bronze Age, but today Uragh's eighty-seven hectares are by far the largest remnant of forest on its southern side. The northern slopes are entirely bare of trees, apart from a few clusters, huddled in small ravines where farmers could find no grazing for their sheep.

The sinuously twisted oaks of Uragh, their bark often invisible beneath dense mats of moss and thickets of epiphytic ferns, look as though they have been there forever (or at least since the last Ice Age). But the pollen record shows that Scots pine dominated the woods until it declined sharply approximately two thousand years ago, and had virtually disappeared three hundred years later. A combination of selective felling and an increasingly wet climate are thought to have led to the extinction of Ireland's only native pine in Uragh, which was one of its last western strongholds, though it lingered in the midlands for a few more centuries.[27] Uragh's name yields another clue to changes on this land, because *iúr* is the Irish word for "yew." No yews can be found there today, though impressive stumps periodically surface in the surrounding bogs, providing dramatically suggestive shapes and vibrant tones for artists to carve and polish. The strawberry tree, one of the treasures of Ireland's "Lusitanian" species,[28] and relatively common on the lower edges of some of the nearby forests in Killarney, has been exiled here to an island in the lake.

There have been many other changes. The steep gullies formed by the capricious folds in the underlying old red sandstone, and the presence of many erratic boulders dumped by the glaciers, make the woods inhospitable terrain for casual visitors today. But this did not deter earlier and hungrier generations from exploiting them over the centuries. You can find many telltale multiple trunks on the oaks, indicating extensive coppicing, and you will find frequent traces of charcoal if you sink an auger into the peaty, acidic soil.[29] "Ancient" in this context, therefore, as in so much of Europe, does not mean "virgin" or "primeval." Technically, like Brackloon, these woods are "seminatural." The oaks thrive on the drier, free-draining sandstone ridges, while birch, willow, and alder dominate in the damper troughs. Ancient oak forests in the west of Ireland, influenced by the Atlantic climate like their counterparts in Scotland, Cornwall, and Denmark, are dominated by sessile oak. They are often found, as in Uragh, on highly acidic soil, whereas most European oak woods feature the pedunculate species, and stand on much more fertile, base-rich soil. The oceanic climate also fosters the almost subtropical luxuriance of ferns and mosses here: a commercial forester's despair, perhaps, but a bryophyter's delight.

This place combines magical ambience and significant scientific value. It is designated as an Special Area of Conservation under the EU Habitats Directive, and managed by the National Parks and Wildlife Service (NPWS) as a Nature Reserve. But rulings in Brussels and Dublin do not guarantee ecological health. The understory is as rich as the trees in rare and scarce ferns, like the endemic Killarney fern, the hay-scented buckler fern and several species of filmy fern. But it is dominated by their overbearing relative, bracken. Goat and deer droppings at our feet show why. Bilberry, grasses, and woodrush are grazed almost to ground level. Seedling trees can hardly sprout from the forest floor before they are nibbled back to their roots. So bracken, immune to grazing, takes over, and its shade then further inhibits diverse understory growth. The forest is fenced, and trespassing grazers are culled, by the NPWS, but clearly more needs to be done.

Despite these signs of degradation, and the consequent and disturbing lack of natural regeneration, Uragh remains a place of great beauty. This makes the visual shock that awaits us all the greater, when we leave the woods and drive on up toward the head of the valley. Almost all tree cover has been recently and abruptly removed from the eastern shore of the lake by clear-cutting. Just a few skinny and straggly birch still stand uncertainly among the raw stumps of Sitka and Norway spruce and larch, while great stacks of logs await the journey to the sawmills. The birches had reached beyond their own grasp to find the light in intimate competition with the fast-growing conifers; now they are in danger of keeling over without their competitor's support. "The foresters might have been better to coppice them even at this stage," Little comments. "They would still sprout from a stump, but they don't look like they will last very long as they are now. The wind will have them sooner or later."[30]

This site had been planted with the alien conifers by the Forest and

Wildlife Service in the late 1960s and became Coillte property when the company was set up in 1988. The site is adjacent to Uragh but according to Coillte it was not wooded at the time of the plantation, but was "probably . . . an open wet heath or blanket bog."[31] Part of it was cleared in 2007, and has already been planted with oak, birch, alder, and Scots pine, as will the bigger sector recently clear-cut. This is a restoration of native woodland at Coillte's own expense, but it fits well with the preservation of Uragh—it should form a biodiverse buffer zone on this side of the wood. It is also in tune with the state-assisted private restorations on this side of the valley under the auspices of the NWS, one of which we are about to visit. If successful, these restorations will spread a rich mantle the valley has not seen for centuries. The scheme, along with the preservation of Uragh and the Coillte restoration, is stimulating discussion about whether the whole watershed could eventually be managed for biodiversity under the EU Habitats Directive guidelines. But there are a lot of hurdles to cross before that vision can be realized.

And even limited restoration, even to those who understand it well, is often painful and problematic in the short-term—which can be rather long-term in the context of a human lifetime. "My forest is gone," laments Peggy Corkery, only half joking, as she greets us. We are now looking down at the raw scar that the Coillte clearance temporarily forms on the glorious valley view from the farm and visitor center she runs with her husband on the slopes under the waterfall. "The German tourists are going mad about it."[32] Her husband, Donal, coming in from the next field to greet us, is more sanguine, more conscious that restored native forest will hopefully heal that scar with a much more varied green skin than the spruce plantation had provided. He knows he will not see it come to maturity, but he has a long perspective. The Corkerys are deeply committed to the NWS project on their own land.

"They Tell Me It's Like Heaven"

Corkery is a trim septuagenarian, with thick and close-cropped dark hair just beginning to show hints of gray. He is remarkably light on his feet as he strides later through tussocks of purple moor grass on the unforgiving slopes above his few green fields. The upper valley bottom offers decent prospects for cultivation and some cattle grazing, but until recently the extensive uplands that form much of his property were good only for a scattering of hardy black-faced sheep. Donal Corkery's family moved into the valley in the late nineteenth century. The Great Famine of the late 1840s, along with ruthless landlordism, had cleared much of the land of the people whose ancestors had originally cleared most of the trees. His grandfather brought back Scots pine to the area as a single, small, and tightly packed windbreak plantation, but the broad and boggy slopes between the waterfall and Lough Inchaquin remained bare apart from a few stunted rowans, holly and birch among the heather, gorse, bracken, and rushes. That is where his part in the restoration project comes in. It is part of a process whereby the Corkerys have expanded their concept of land ownership from farming to a modestly profitable form of stewardship.

Twenty years ago, he and his wife realized that these slopes were also good for tourism and began to develop unobtrusive hiking trails and charge a modest fee for access to the spectacular views and rich flora his farm enjoys. Twenty thousand people a year visit his holding now. "They tell me it's like heaven," he says wryly, his voice rippling fast in Kerry's characteristic sing-song accent. "I tell them it would be heaven indeed, if I could just finish work as early as six in the evening. Scenery never fed anyone." In a sense, though, scenery feeds him now, and in any case he is far from insensitive to its pleasures. Speaking later of a patch of larch he has planted under the cliffs, he waxes lyrical about this unusually deciduous conifer, how its powerful palette of colour changes through the seasons. Then he adds that he has been careful not to obscure certain views of the rocks with the trees. "As long as it keeps my eyes happy, that is the main thing," he concludes.

A logician might say that his take on the landscape is contradictory, but in fact it expresses the complexity of a life dependent on soil and weather. Like many Irish farmers, he lives well now but is close enough to poverty, and the folk memory of hunger, to retain a no-nonsense pragmatism about priorities on his land. Food on the table comes before poetry and painting. But his vision is binocular. The dimensions of beauty, natural resources—and biodiversity—are all clearly visible to him. His eye sweeps effortlessly from the tiny blue milkwort flower he spots at his feet to the dramatic glacial erratic, perched like a Neolithic satellite antenna on the horizon, which gives

15. **Burgeoning biodiversity.** Male fern, widespread in native Irish woodland, flourishing in the restored forest at Brackloon. (Photograph courtesy of Declan Little.)

Dish Mountain its name. In the spaces between, he is not embarrassed to remind us that the woodland he is restoring will bring him additional income.

Kerry people have a reputation for "cuteness," which in Hiberno-English means playing your cards very close to your chest. Corkery is no exception, and he must be doubly careful. An explicit espousal of environmentalist values can raise hackles in Irish rural communities. "Greens" are often regarded, sometimes with some justice, as ignorant, interfering townies or gentry insulated by wealth who know nothing and care even less about the struggle to make a living from the land. He is also a naturally reserved man, so he opens up slowly to his visitors, though he has met Little at an earlier stage in the project, and loses no opportunity to ask him shrewd questions about soil, and about which tree species might best be planted where.

Gradually, however, he shares confidences, his smile all the warmer for being slow to spread over his weathered face, his initially guarded eyes brightening to a glitter. "I've loved trees all my life," he says, as he confesses the pleasure he took as a boy, relaxing among Uragh's oaks and alders, when he slipped off from wearisome tasks like stacking turf. "I remember a wet day when I took a kettle in there under a rock, and made a fire and made the tea. Waiting for the rain to clear off it was magic, like, you would hear the sound of the birds. . . ." His voice trails off at the recollection.

He admits that he would have made no big distinction between native oak forest and introduced larch or lodgepole pine until fairly recently. "I suppose I wouldn't have, but somehow the whole thing seemed completely natural to me when I got wind of the plan to restore native woodlands . . . and it would give us an extra bit of an income alongside the farm." Like a good Kerryman, he heard of the scheme before it even started, and was one of the first to put his name down for a grant.

So much so that he found that local foresters were still being trained for the scheme, as they had never worked with ecologists before. The scheme required an ecological survey of his land, which yielded a rich and detailed report on his land's diverse flora and fauna—and provides much attractive information for his website.[33] So there was a happy synergy between the restoration scheme and his tourism business.

"The Natural Way Will Put a Hump in Your Back Very Quick"

He discusses the background to the scheme over coffee in the visitors' center, often gesturing up the hillside to underline his points. There was a shortage of oak seedlings when they started planting, he says, but this was a blessing in disguise. The oaks he has planted are struggling to survive. It may be that the pioneer trees like alder, willow, and birch need to get established before the king of the Irish forest, as the old Gaelic lawmakers called the oak, can find a foothold on soil that has not nourished its roots for such a long time.

I find myself rather confused, however, because apart from a few patches of holly and a lozenge of conifers, all aliens, I can see no new growth at all on the slopes he is pointing to. Even when we go out, he takes us first up a back road, with panoramic views over his land and the whole valley, but with precious little sign of the new forest he has been talking about, though I later realized we were overlooking it all the time. Meanwhile, he points out the sites of charcoal pits, dating back to the seventeenth century. On the brow of the northern horizon, he guides our eyes to parallel ridge marks, which we take at first to be "lazy beds," signs of abandoned potato cultivation. "Much too widely separated," he says. He thinks they may be late Neolithic cereal fields, on land the early settlers had cleared or was above the tree line, and could have sustained such crops when summers were warmer than they have been long since. Perhaps this was the same culture that built a mysterious circle of standing stones on the shore of Lough Inchaquin below us.

On the southern slopes of his own land, he indicates a one-room cottage he is restoring to remind visitors that human life was less than cosy here in the colonial period, dogged by evicting landlords, savaged by Atlantic wind and rain, and always haunted by the threat of famine. Political independence did not bring great improvements for many of the few who remained. Only the subsidies that came with entry into the European Union have lifted Irish small farmers in these hard valleys above subsistence levels. Ironically, it is that same Union's environmental regulations that often infuriate some members of this newly prosperous class today. Corkery walks a thoughtful line between the rapacious over-exploitation that has degraded many farms like his, and any starry-eyed version of environmentalism. He recalls wryly that a man from the government agricultural advisory service had once told him he should clear the bracken and heather between the cottage and his bottom fields and make one big meadow. "What would have been the consequences?" he asks himself. "The grass would have died, I think, and there might have been erosion. You should not try to improve nature too much." Apropos, he remembers the time he overeagerly altered the course of a mountain stream to speed up drainage and then spent wakeful nights fearing his farmhouse was going to flood.

But then, without a pause, he reflects pragmatically on the human cost of organic farming: "It's a good idea, of course, but doing things the natural way will put in a hump in your back very quick."

We are now further than ever from Corkery's woodland restoration scheme, which in fact lies in the vicinity of the restored cottage, but from this high vantage point he wants to show us something even higher. Tucked into a near-vertical gully on the cliffs above a second corrie lake, we can just discern a tiny patch of mature trees. "Would it not be a good thing," he asks, "to connect that patch to the Uragh

Woods below? I have suggested this to the NPWS, I think it is their land, and I can't understand why they haven't taken it up." Declan Little stares at the vertiginous slope sweeping down to the upper edge of the lakeside forest, and ponders. Trees probably extended all the way up from the lakeshore to these pioneers a long time ago, he ventures. But could they ever find purchase and nurture there again, after centuries of erosion, and an absence of the leaf litter that held back the soil from becoming too acidic even for Irish oaks? And Little comments that the removal of the ancient forest probably left the shallow soils very vulnerable to erosion. That is, a threshold may have been crossed that makes woodland restoration impossible here in practice.

It is the kind of question that haunts many restoration projects, but no one could accuse Corkery of not thinking big. He says he is willing to put in fencing on his side of the putative forest to keep out deer, goats, and his own wandering sheep as a voluntary contribution. It is a generous offer, because fencing is not cheap, and the labor involved in bringing it up such gradients would add to the hump in anyone's back. The issue is noted for further discussion with all concerned. On the high slopes right around us, Corkery has long noticed—and been puzzled by—occasional odd hillocks with little depressions apparently scooped out beside them. He wondered if they were ancient human structures. Then he read Cunningham's book on Brackloon and realized that he was looking at the aftermath of centuries-old windthrow.[34] The lay of the land here, once you recognize it, is dramatically visible evidence that the forest did indeed climb to these heights before clearance.

The Irish poet Patrick Kavanagh fiercely satirized romantic visions of rural life in his poem, *The Great Hunger*. The title transfers the vernacular phrase for the potato famines of the nineteenth century to the spiritual and cultural famine of the mid-twentieth century. "Clay is the word and clay is the flesh," Kavanagh intones, and there is no love lost in this instance between the poet and his "stony grey soil." Corkery belongs to a more fortunate moment, and his engagement with the soil does not drag him toward insensibility, but fascinates and energizes him. Looking down across the majestic sweep of the land under the lough, he expresses wonder at the depth of its blue boulder clay left by retreating rivers of ice, layered as thick as ten inches in places, a sterile barrier to roots. "But when you puncture it," he says with delight, "you find the most beautiful red earth that was ever known of."

And at last he turns the car around, and brings us to what we have come to see. But the preparation was wise. We will see it in its context now.

A Long and Uncertain Experiment

We have seen on several occasions that restoration can be ugly in its early stages, but it can sometimes be very easy to miss. You almost have to trip over this project to realize it is there, and from any distance it blends into the variegated shades of the hillside. The first sign is little cluster of alders, perhaps a meter high, comfortably dominating the thick carpet of soft rush and purple moor grass along the side of the track. As you get your eye in, the richer, deeper green of dozens of similar alders emerges from paler rushes and grasses for hundreds of meters up to the fenceline. Occcasionally willows, birch, and Scots pine pop into view among them, and something else also, so small and yellow that at first I take it to be goldenrod. Sadly, the yellow flecks turn out to be the leaves of little oaks, still not much more than seedlings, struggling to rise above the level of the rushes, their vivid yellows indicating their severe unhappiness with the waterlogged soil in which the alder and birch are thriving. Whether they will come into their own once the pioneer trees have fully established, drawing much of the moisture from the land and enriching the soil with their leaf litter, will not become clear for decades. Like almost all worthwhile restorations, this will be a long and uncertain experiment.

As we climb toward the shores of Cummenadillure,[35] the lower and larger of corrie lakes above the project, a profusion of white plastic tubing encases many of the small trees, and this indicates another challenge to their survival. Quite frequently here, seedlings shoot up only to be chewed back until only a snapped-off stick remains. The fencing keeps out goats, sheep, and deer, but this problem is caused by native fauna.[36] "This is the work of our friend the hare," Corkery remarks philosophically. Unlike the deer, the hare population has not exploded and is still subject to natural predation. So it will not wipe out the new plantation, and can mostly be tolerated. Occasional culling, coupled with tubing the seedlings in the worst-hit areas, could cope with the problem until the new woodland is established—but this raises issues of triage, as hares are protected in Ireland.

The flora is rich here, and Corkery pauses to check the bees pollinating the goldenrod, greeting them with satisfaction as hard workers from his own hives. The rusty-colored fruiting spikes of bog asphodel are ubiquitous; the Virgin Mary–blue of milkworts still in flower are either more occasional, or at least less visible. From a rocky outcrop, we can see most of the project, and it is clear that its growth rates are very mixed five years after the first planting. Some quite big patches are almost entirely bare of visible trees, others are already taking the shape of bushy little thickets, and may even need thinning. Sometimes the alder are already fecund with the "smelted emerald," as Seamus Heaney puts it, of the green cones that will become, in winter, "so rattle-skinned, so fossil-brittle."[37] Elsewhere they are still scrawny, their few branches barely leaved. The uneven development does not faze Corkery. His intimate knowledge of his land offers convincing explanations. Dense growth under a large rock is not just due to wind protection for the young trees today, but to the fact that his family's sheep have sheltered there for many generations, ensuring a rich fertilisation of the soil with their excrement. Another flank of the mountain shows remarkable variations, with no tell-tale shelter spaces. This is the result, says Corkery, of a landslip a century ago, which redistributed topsoil and sometimes tore away the boulder clay, exposing the red earth horizon he loves so much. Elsewhere, differences in minerals flushing through the bog offer clues. In any case, the NWS allows for open spaces of up to 20 percent of new native woodlands for biodiversity enhancement.

As we walk, Little and Corkery often kick at a bank or a lose boulder and discuss the constantly changing composition of the soil. Farmer and ecologist use different but complementary approaches. Little tells me afterward that he is awed by Corkery's awareness of the impacts of small variations across his land. Ecological restorationists who rush into projects without taking full account of what lies beneath their feet might do better if they consulted the local ecological knowledge of those whose families have cultivated the land for centuries.[38]

We turn back by the lough, which still holds Arctic char, a fish with distinctive wine-red markings that is among Ireland's oldest fauna.

The char possibly survived the last Ice Age by spawning in the meltwaters of the southern glaciers in Kerry, but is now threatened by acidification and climate change. Eagles, both golden and white-tailed, have had even less luck locally, persecuted to extinction by zealous gamekeepers in the nineteenth century. The name "Eagle Rock," on the rugged cliffs that tower above the darkly glittering water, recalls their presence here. The recent reintroduction of white-tailed (sea) eagles to Kerry has been controversial, with farmers fearing for their lambs. There have been a number of poisonings, possibly deliberate. No eagle has yet turned up on Corkery's loughs, but he would welcome them. "The hoodie [hooded crow] does much more damage to sheep," he says, "and it adds far less to the drama of the countryside." On our way out, he shows us a sheep blinded by the crows.[39]

Just before we leave, he takes us up one last path, where the few old trees huddled in a gully and along a hedgerow are raining seed among the new arrivals, raising hopes for some early natural regeneration. He pauses to admire a bank cloaked with a patchwork quilt of Sphagnum mosses, studded with the star-shaped leaf-whorls of one of the region's endemic plants, the great butterwort. This is one of that small group of insect-eating plants that give a whiff of tropical exoticism to Irish bogs. "It's a beautiful bank in May," he says, "when the butterwort's blue flowers are in bloom." Corkery's eye for flowers is not restricted to natives. We had been surprised to see a patch of rhododendron on his driveway, and a single plant, still small, perched above a gully stream that bisects one of his sheep pastures. He confesses that he likes its flowers a lot, but assures us that he is "keeping an eye on it." That the same person should be restoring native woodland, and flirting with one of the most vexing invasive plants in the locality, is perhaps not as contradictory as it seems—or simply reflects a contradiction at the heart of our human nature. Corkery may be inconsistent, but he is aware of his inconsistencies—and he gets a lot more done than many right-on environmentalists endowed with ideological purity but often hopelessly ignorant about their native places.

In any case, Little points out that he can avoid the risk of infestation and still enjoy a rhododendron's flowers—all he has to do is change the species. Only *Rhododendron ponticum* is invasive in Ireland, but unfortunately that was the plant that ninetheenth-century landed gentry decided they wanted for their gardens, and also offered to their laborers as a quick-growing hedge to shelter their cottages from the wind. Gleninchaquin is lucky in that the shrub has never gained more than a toehold here, and one must hope that Corkery and his neighbors can keep it like that. Just a few miles inland, among the fabled vales and lakes of Killarney, the rhododendron has gone on a rampage that came very close to reducing some of Ireland's finest and most beautifully situated woodlands to a virtual monoculture dominated by a shrub from the other end of Europe. Only a long-running voluntary clearance campaign, coupled with major investment of scarce resources by the NPWS, has turned this tide, and the battle is far from over, as the most cursory visit to this iconic national park will demonstrate.[40] Alien invasive plants are, as Little stresses again and again, the biggest threat facing conservation and restoration of woodlands in Ireland.

People's Forests for the New Millennium

> *Cad a dhéanfaimid feasta gan adhmad?*
> *Tá deireadh na gcoillte ar lár*
>
> (Now what will we do for timber,
> With the last of the woods laid low?)[41]

The great eighteenth-century Gaelic lament, *Caoine Cill Cháis,* expresses the essence of the potent Irish nationalist charge that English colonists had destroyed our native forests along with our indigenous aristocracy. It is ironic that it is only after researchers like Oliver Rackham exposed that accusation as more symbolic than real that we are finally taking practical steps to restore some of those woodlands to something approaching their former glory. Most of Ireland was forested after the glaciers of the last Ice Age withdrew, but a combination of human activity and climate change relentlessly drove the trees back long before the English arrived. Of course, the settlers did continue the clearances, so that only 3 percent of the country was wooded by the middle of the seventeenth century, and much less than 1 percent remained under timber two hundred years later, giving Ireland the sad distinction of being the European country with the least forest cover. Most of the remnants were in fact themselves secondary growth rather than primeval forest, and their indigenous integrity was compromised by the spread of successful exotic species like rhododendron, cherry laurel, sycamore, and beech.

In the twentieth century the newly independent Irish state belatedly embarked on an ambitious reforestation program, but almost all state investment went into monocultures of nonnative conifers. The work provided by these plantations was sometimes crucial in staving off total collapse in desperately poor rural communities in the short term, but impoverished the environment both biologically and aesthetically. The turn of the current century provided a focus for a shift in policy, and the People's Millennium Forest project took shape. This was an initiative developed by Woodlands of Ireland, with a lot of input from NGOs, such as Crann, and managed by Coillte in partnership with Woodlands of Ireland. The result is impressive, both in terms of environmental success and public impact. Every household received a certificate for a tree planted in its name in one of sixteen native woods restored or created throughout the island. These include two in Northern Ireland, and cover a total of fifteen hundred acres. They range from a unique yew forest restored in Killarney to an entirely new wood created at Castlearchdale, County Fermanagh, which straddles the border between the two states on the island, a symbol of the recent peace process after thirty years of bloody conflict in Northern Ireland. A school program raised awareness of the biodiversity value of indigenous woodland. A guide to growing native trees from seed was widely distributed, and a book, *Native Trees & Forests of Ireland* by David Hickie and Mike O'Toole, celebrates this aspect of our heritage.[42]

Declan Little says that the "future for the sites is mostly very good, with excellent growth." Some trees have inevitably died, and the soil conditions at one site caused problems, but a residual budget allows for "filling in" with new trees where losses occur. A politically driven promotional campaign for the People's Millennium Forest gave the unfortunate impression that not only had a tree been planted for each household in the nation, but that each family would be able to find its own personal tree in the forest, or at least the sector in which "their" tree had been planted. Wiser counsels have prevailed, and today the project is presented as offering every Irish person a shared stake in restored native forests, in a very long-term ecological process

16. **Rare native.** Irish yew in fruit. (Photograph courtesy of Kevin Collins/Forest Service.)

rather than in a particular—and inevitably mortal—tree. However, this scheme might have been merely a pious gesture had it not been followed up with the ongoing Native Woodlands Scheme, of which Brackloon was the prototype, and Gleninchaquin is a current project. The NWS offers grants of up to €5,000 per hectare to landowners who conserve and enhance native woodlands. Up to €6,470 per hectare is available to landowners who establish new woods from native seed.

Launched as part of a National Development Plan in 2001, the project started with "visionary targets," in the words of Kevin Collins, an inspector with the Forest Service. There has been an enthusiastic take-up from landowners, despite cutbacks even before the Irish Celtic Tiger boom went badly bust. Some four thousand hectares of native forests have been restored and created so far under its auspices. Collins says the NWS is regarded as innovative and effective by EU colleagues, especially for its success in building cross-sectoral partnerships between landowners, state agencies, foresters, and ecologists. Applicants have to meet tough criteria in a two-stage vetting process. This involves developing a management plan with expert advice from ecologists and foresters and consultation with other state agencies like the National Parks and Wildlife Service and Inland Fisheries Ireland (for riparian zones). Where the land is already part of a Special Area of Conservation, that seed must come from within a twenty-mile radius, and from an appropriate woodland community.

Are Irish landowners really willing to jump through so many hoops? The story of Donal Corkery suggests that some of them, at least, were more than ready for a scheme like this. "They come to us already enthused about the project," says Kevin Collins. "And we are dealing with woodlands, which are the jewel in the crown of our national biodiversity. We must get it right."[43]

In May 2011, we gathered at Ballygannon Woods, the People's Millennium Forest site closest to my home in County Wicklow. Against a gunmetal-gray sky, the sparkling green crowns of silver birch, and the deeper green canopies of sessile oak, seemed to be lit from within. Spring bluebells, wood sorrel, celandines, and wood anenomes scattered blue, yellow and white patches on the forest floor. The display put vibrant flesh on Declan Little's brief speech in celebration of the tenth anniversary of Ballygannon's restoration, evoking "thriving and vibrant young woodland areas surrounded by old woodland rich in biodiversity and color." He reminded his listeners, including many children, that "our natural heritage is an asset that can sustain jobs and livelihoods if it is managed and nurtured carefully," a welcome message at a time of deep crisis in the Irish economy and psyche. He also stressed that restoring native woodlands was quite compatible with sensitive and sustainable extraction of timber.

A walk through the woods revealed aspects of what that management entails. Many oaks were regenerating naturally on land cleared of alien plantation trees. Little pointed out that unless the deer fence was maintained, and brambles were cleared regularly until the seedlings had outgrown them, they would not prosper. John Cross, a conservation scientist with the NPWS, reminded us that the restoration of Ballygannon links directly to an older and impressive restoration of woods on the other side of the Vale of Clara, in which he and his agency have had a long and deep involvement. The link creates an expanse of native oak woodland comparable in size, if not yet in quality, to some of the famous oak woods of Killarney. Moreover, it is almost free—though some worrying patches are spreading—of the rhodendron infestations that still mar the great Kerry park despite heroic clearance efforts.

The Clara restoration had also very recently been augmented by unexpected and very welcome new residents. Several pairs of great spotted woodpeckers, absent from Ireland for several hundred years at least, have appeared spontaneously over the last few years and begun breeding there. We didn't see or hear any woodpeckers that afternoon at Ballygannon. We were probably making much too much noise. But their presence is the kind of natural endorsement all restorationists dream about. Irish woodlands came very close to suffering the total annihilation prophesied by the Gaelic bard of *Caoine Cill Cháis*. With a little help from the species that cleared most of them, they are beginning to make a fragile come-back.

8 *Future Shock: "Novel Ecosystems" and Climate Change Shake Restoration's Foundations*

Unique hazards may exist! * *Warning on a signboard greeting visitors to a secluded guesthouse on the green and prosperous campus of Stanford University in Palo Alto, California*

The past is no longer a prescriptive guide for what might happen in the future. * JAMES A. HARRIS, RICHARD J. HOBBS, ERIC HIGGS, AND JAMES ARONSON, "Ecological Restoration and Global Climate Change"

We are heading towards a situation where there are more lemons than lemonade, and we need to recognize this and determine what to do with the lemons. * RICHARD HOBBS ET AL., "Novel Ecosystems: Theoretical and Management Aspects of the New Ecological World Order"

There is the hazard of becoming more comfortable with serving as active agents in ecosystems to the extent where historical fidelity is almost entirely abandoned. . . . How smart can we be, and how much hubris is there, in presuming that we can understand and predict ecological change? * JIM HARRIS, RICHARD J. HOBBS, ERIC HIGGS, AND JAMES ARONSON, "Ecological Restoration and Global Climate Change"

But will the trees they are planting survive even a lifetime where we putting them now? Or should we start planting species higher than their normal range? Should we dare to anticipate climate change? As a restoration ecologist, I simply don't know what to tell them. * PATRICIA TOWNSEND, *on her project in the Costa Rican cloud forest*

* *

We have seen some of the challenges and dilemmas that arise when ecological restoration projects are put into practice on the ground, in the very different contexts of rural South Africa, urban North America, a cultural landscape in Italy, and Irish woodlands.

But there are also new and deep challenges facing the restoration movement in theoretical terms, which engender passionate and fast-moving debate among scientists and practitioners. Every time I thought was beginning to grasp the basics of restoration ecology, I found that the foundations were already shifting somewhere else. In the relatively short period in which I have been researching this book, growing awareness of the implications of climate change in particular has repeatedly shaken the recently established principles of the restoration movement.

So it seems a good point, in the middle of this book, to return to the midpoint of my research, when new thinking within restoration ecology began to raise disturbing questions about the feasability of the whole enterprise. First, however, it may be helpful to recapitulate, in the briefest outline, how the theoretical framework of restoration has developed. The traditional North American model of restoration—the attempt to return a degraded system to a fixed point in a "pristine" past—has been contentious for decades, though it retains a strong attraction, at least as a metaphor, in some quarters. It has largely been replaced by models that appear more practicable, whether for ecological or socioeconomic reasons, or both.

Recognizing that ecosystems are dynamic rather than static, many contemporary restoration projects do not seek to replicate a past state, though they are still based on historical reference systems. Instead, they seek to put a degraded system back on something approaching its original historical trajectory into the future within its historical range of variation—that is, to set it back, as closely as possible, on the path of development on which it *would have been traveling* had disturbance or degradation not occurred. The need to use such convoluted syntax itself reveals what a problematic task restorationists are attempting.

And recognizing that demonstrable benefits to human communities are usually needed to justify restoration's costs, restorationists are more and more inclined to add a focus on the increased flows in ecosystem goods and services that result from the restoration of natural capital to more traditional "pure" ecological concerns.

But we have also seen, way back when we met Joy Zedler and Richard Hobbs at an early stage in our journey,[1] that the difficulties encountered by some thoughtful restoration ecologists have led them to suggest that the restoration enterprise is an almost impossibly complex conundrum. Failures to achieve expected results on some projects have led to increased concern that there may be too many variables in the simplest ecosystem for science to be able to get the restoration equation even approximately right. To paraphrase former US Defense Secretary Donald Rumsfeld's notorious phrases, there are daunting ranges of "known unknowns" and even more "unknown unknowns" in every restoration project.

Throw the new raft of variables associated with accelerating climate change and other human-generated impacts on the biosphere into these equations, and some restorationists have begun to radically revise the restoration agenda. Or, in the eyes of some of their colleagues, to simply throw in the towel. I had fair warning of these complexities when I talked to Richard Hobbs in Zaragoza. But I got a much clearer sense of their full and disturbing implications when I read two papers he coauthored in 2006 and then attended a very stimulating but very disturbing session he cochaired at a SER conference in San José, California, in 2007. These papers and this discussion reopened the debate about the basic principles of restoration—principles that had apparently been resolved with the publication of the SER primer, after much discussion, in 2004.

One of the motors driving this new debate was the growing recognition that rapidly accelerating climate change changes everything. Over the first decade of the new century, each new report by the Intergovernmental Panel on Climate Change (IPCC) revealed new forces that seemed to be unscrewing every firm certainty that had guided conservation. The other, related motor was the realization that other human impacts on the biosphere, like IAPS and apparently irreversible land degradation, often interacting with climate change, create very complex and multidimensional new ecological equations. These were generating, Hobbs and a number of colleagues argued, numerous "novel," "emerging," or "no-analog" ecosystems, that is, systems whose species composition and dynamics have no precedents

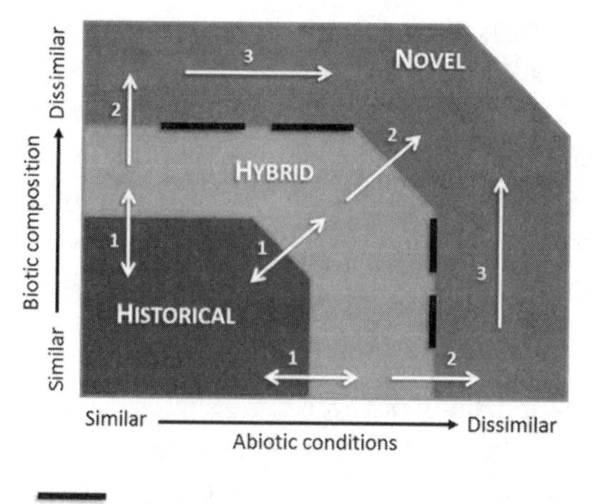

17. **Historical, hybrid, novel.** Types of ecosystems under varying levels of biotic and abiotic change, adapted by Lauren Hallett et al. 2012 from Hobbs et al. 2009. According to this view, a historic ecosystem remains within its historic range of variability; a hybrid ecosystem is biotically and/or abiotically dissimilar to its historic ecosystem but is capable of returning to the historic state; novel ecosystems are biotically and/or abiotically dissimilar to the historic state and have passed a threshold such that they cannot, in the view of some novel ecosystems theorists, be returned to the historic state. Pathways represent possible directions of change: 1) shifts from historic to hybrid ecosystems that are reversible, 2) nonreversible shifts from historic or hybrid ecosystems to novel ecosystems, 3) further biotic and abiotic shifts are possible within novel ecosystems. (Figure courtesy of Lauren Hallett.)

among historical reference systems. These developments both posed significant challenges to what is perhaps the most cherished idea underlying restoration—that a roadmap for the desireable future of a degraded ecosystem can be drawn from its past.

A 2006 paper by Hobbs,"Novel Ecosystems," provocatively subtitled "Theoretical and Management Aspects of the New Ecological World Order" has been very influential on this issue.[2] He and his coauthors define novel ecosystems as the outcome of human impacts, whether deliberate or inadvertent, which result in "new species occurring in combinations and relative abundances previously unknown within a given biome." Such changes in species composition, they continue, have the "potential" to change the system's functions.[3] These radical changes can come about through successful invasions of human-introduced alien species in relatively wild lands. But, very significantly for restoration, novel ecosystems can also occur after the abandonment of intensively cultivated systems. Hobbs's extensive experience with "old fields" in southwest Australia had shown him that the comforting and rapid spontaneous regeneration of seminatural vegetation common on abandoned European and North American farmlands often simply did not happen in other geographical regions.[4] Instead, entirely new assemblies of plants and animals might develop, usually dominated by invasive aliens. Worse, the successful removal by restorationists of one or a group of IAPs may often result not in the return of native vegetation, but in an even more successful invasion by yet another species or group of species novel to the system.[5]

Under these kinds of circumstances, where a system has passed through a process of transformation that it is very difficult, very costly, or simply impossible to reverse, the authors argue that resources should not be wasted on attempts at restoration to a historical reference system. Instead, the system should ideally be managed to produce maximum benefits in terms of ecosystem goods and services. Among the ecosystems they list in this category are pine-invaded fynbos in South Africa and salinized abandoned pastures in southwest Australia—both biodiversity hotspots—as well as many North American river systems.[6]

More Lemons than Lemonade

On the face of it, this is a sensibly realistic approach to prioritizing restoration, to focus efforts on those areas where it is most likely to be successful. The authors recognize that some colleagues will think their approach is "defeatist," but they go on to suggest that such colleagues are themselves in a state of denial about these new systems: "[reviewers'] comments . . . indicated a lack of willingness to accept such ecosystems as a legitimate target for ecological thought or management action. For instance, one reviewer commented that the examples [given in the paper] are ecological disasters, where biodiversity has been decimated and ecosystem functions are in tatters, and that 'it is hard to make lemonade out of these lemons.'" Their response is robust: "Our point is, however, that we are heading towards a situation where there are more lemons than lemonade, and we need to recognize this and determine what to do with the lemons. We suggest that the approach is simply pragmatic and provides a way for prioritizing scarce conservation and management resources."[7]

Curiously, however, while the assertion that lemons now outnumber lemonade sounds definitive, they explicitly concede only a few paragraphs later that the question of whether novel ecosystems are still the exception, or will soon be the rule, remains very open: "Are novel ecosystems on the increase? Will such ecosystems predominate at the end of the present century?"[8]

They offer no answer.

But that earlier definitive assertion that there will soon be more lemons than lemonade has engendered a serious concern that has loomed much larger as this and related debates developed within the restoration movement over the next few years. This concern sharply divides Aronson, in particular, from the other authors of the article. Is there not a tendency by Hobbs and his other coauthors to overstate the prevalence of novel ecosystems? And if there is, does that not cast a doubt over their subsequent articles, which have been read as using the novel ecosystem concept as a lever to change restoration thinking, not just in these still unusual cases, but at the level of first principles? On the other hand, are those who harbor these suspicions simply guilty of an anxious nostalgia for a past that was always complex but now looks relatively simple in comparison to these new scenarios?

If this article raised disturbing questions, the more innocently titled "Ecological Restoration and Global Climate Change," written by Hobbs's frequent collaborator, Jim Harris, with Hobbs as second signatory and Aronson still on board, was seismic. Straight off, the authors ask a key question—"How appropriate are historical ecosystem types [for restoration planning] when faced with rapidly changing bio-physical conditions?"—to which they respond with an increasingly familiar second question: "Is it appropriate to consider a temperate woodland restoration endpoint in an area likely to be flooded by rising sea level? Why establish wetland in an area likely to become semiarid?"[9]

And if sites where some species currently flourish will become deeply inhospitable to the same species in the near future, what on

Earth is to be done? This question has massive implications for conservation and restoration. The European Union's Natura 2000 Network of protected sites, for example, is very often based on their value as habitat (sometimes restored) for endangered species. What happens to these sites, and these species, if they become mutually incompatible in fifty years' time? The authors suggest that "the translocation of species is a likely technique" to deal with such eventualities. And then they add, "Active ecological restoration of appropriate sites in new locations would appear to be one answer."[10]

Yes, this does mean just what it says. As climate patterns shift around the world, the authors are arguing, the former restoration challenge of rebuilding historical ecological communities on degraded sites becomes obsolete. Worse than that, it is replaced by a challenge whose difficulty dwarfs an already daunting enterprise. The authors are saying that we have to start building ecological communities appropriate to climatic changes which have not happened yet, in places where these communities have never existed before. And we may have to take them there in trucks.

Hold hard a minute, here, one fairly wants to shout at this point. People who think seriously about restoration have long acknowledged that soil, with its vast communities of (largely still unidentified) microfauna and microflora, is fundamental to restoration.[11] It can be hard enough to replicate suitable historical soil conditions on a degraded site under restoration, because its soil composition may have changed due to any number of factors over time. But at least one can reasonably hope that many of the microbes and mychorriza appropriate to the community under restoration may still be in the vicinity. But just how are we supposed to transport them a few hundred miles, and how could we even guess when the shifting climate would make such a move appropriate?

Since Harris's own field is soil science, is he not reeling at the audacity of his own proposals here? Harris, a working-class Londoner who has retained the colorful speech and hard-nosed attitudes of his East End origins, responded succinctly to these questions one night over a beer in his local pub. "I'm not saying you shouldn't be gobsmacked thinking about this stuff," he says. "I'm gobsmacked when I think about it myself. But this is how things are, mate, we have got to start thinking like this if we are to work in the real world."[12]

Heartbreaking Understatement

Aronson was also a coauthor of this article but has since become deeply skeptical about where the thinking outlined in it is leading, as we will see in the concluding chapter. A year after the article was published, I asked him whether such relocations could possibly be timed to meet all the necessary multiple synchronicities. "Can't expect to do so," he replied tersely. "We're groping in the dark . . . it's a matter for R&D in as many places as possible."[13]

To me at least, there is a heartbreaking understatement in the way these authors briefly spell out the consequences in terms of both human emotions and future evolution. Their measured science-journal language begins to buckle, just a little, under the pressure of what they are reluctantly claiming: many familiar and much-loved natural and seminatural places are likely to change almost beyond recognition over the next century: "This [prospect of translocation] bears the burden of breaking our relations with particular places and upsetting long-duration, place-specific evolutionary processes."[14] In other words, however much you cherish your local oak woods, or your wetland, or your rain forest, if you really want to restore them you may have to do so hundreds of miles from your home, perhaps in another country. For most of us, whatever biophilia we experience is intimately linked to those species and ecosystems that we have grown up with on our own turf. The idea of restoration as reconnection with nature takes quite a drubbing in this brave new scenario.

The authors recognize that, alongside the hugely complex scientific and technical issues involved, there are "vexing moral questions." "There is the hazard of becoming more comfortable with serving as active agents in ecosystems to the extent where historical fidelity is almost entirely abandoned. . . . How smart can we be, and how much hubris is there, in presuming that we can understand and predict ecological change?"[15]

Well, indeed. This article stands out as an important and very sobering warning about the possible, or indeed, probable, impacts of climate change in some and perhaps many places. Its proposals are already being reflected in some new practices, as we will see below. But there is a clash between the confident, even hubristic, tone of their suggestions about relocating species to ecosystems that *may* be suitable in the future, and their humble recognition that our understanding of contemporary and historical ecosystems is still often rudimentary. They concede in their conclusions that the "ecosystems we manage and restore are complex systems that often have nonlinear and unpredictable behavior. Our understanding of how systems work under current conditions is often rudimentary, and we often have to learn as we go."[16] And then they add:

"To this complexity and lack of understanding, we now have to add the fact that environments are changing, and the rate of change is unprecedented. *The past is no longer a prescriptive guide for what might happen in the future*." This double recognition—that restoration was always chronically difficult and now is doubly so—makes one gasp a little at the apparent chutzpah in their early references to "translocating species" and "restoration of appropriate sites in new locations."[17]

Some SER colleagues have begun to wonder whether the impact of climate change is not being exaggerated in this paper, as a Trojan horse to undermine the restoration principle of attempting to restore to a historical reference system. One could certainly read one of their last paragraphs in this way: "There is a large component of ecological restoration that still places considerable value on past ecosystems and seeks to restore the system's characteristics to its past state. Valuing the past when the past is not an accurate indicator for the future may fulfill a nostalgic need but may ultimately be counterproductive in terms of achieving realistic and lasting restoration outcomes."[18]

In fairness, while there does seem to be an element of iconoclastic thinking for its own sake in some of the writing here, the authors end on a very cautionary note, that has not always been evident in some of their subsequent articles or presentations: "The past should serve as a guide—not a straightjacket—and also should temper our ambitions toward unfettered design and development. . . . Hence, a key question for everyone involved in restoration is the proper balance between rebuilding past systems and attempting to build resilient systems for the future . . . We must tread very carefully."[19]

Indeed, we must. The changing climate is spinning the ship of restoration, but that does not mean that we should throw away the compass.

Unfolding Sixteen Different Possible Futures

After the San José SER conference in 2007, I spent a few days at the Stanford University guesthouse. A bizarre warning sign caught my eye when I checked in with security: "Unique Hazards May Exist." As it happened, my host was Richard Hobbs, who has a long-term research project at the college and was kindly helping me develop some of the ideas in this book. I pointed out the sign and he burst into infectious laughter.

Hobbs wears his great learning—he is one of the most published and most cited scientists in his field—very lightly. He is a Scot with sparkling eyes, a bushy beard worthy of a Victorian natural scientist, and a playful, sometimes mordant, sense of humor. He tried out reading the sign as the caption to a variety of imaginary Gary Larson *Far Side* cartoons. I fantasized that it referred to the Kool-Aid laced with LSD served at Ken Kesey's mid-sixties "acid tests" in the nearby Bay Area.

In fact, as Hobbs finally explained to me, the sign refers to phenomena even stranger than Larson's transvestite cows or Kesey's psychedelic visions. It is a warning about the unpredictable consequences of complex experiments in physics in a massive concrete tunnel known as the SLAC, the Stanford Linear Accelerator Center. The scientists here investigate aspects of the deep past—questions about the origin of the universe, no less. Presumably the college authorities hope the sign will be sufficient insulation against insurance claims, should any visitors suddenly dematerialize or fall into a black hole. In any event, we both survived our stay unscathed.

But that warning sign kept haunting me. Its message seemed very appropriate to what I was learning from Hobbs's most recent papers and discussions. Indeed, unique hazards were the specific subjects of a very different set of Stanford experiments, running on the university's nearby Jasper Ridge Biological Preserve, which we had visited during the conference. These experiments, however, look not to the distant past but to the undoubtedly dangerous immediate future. At first sight, they looked like the set of a science fiction movie. On one side of the ridge, majestic valley oaks spread their placid branches over a savanna landscape; its pastoral calm is only broken by an occasional

18. **The future seen from Jasper Ridge.** SER members visit a Stanford University research site that explores variables in climate change by manipulating inputs into small patches of vegetation. (Photograph by Paddy Woodworth.)

outburst of chatter from acorn woodpeckers. On the other side, little circles of grassland, just one meter in diameter, bristle with hi-tech equipment. We are still in the countryside here, but there is a palpable air of industrial anxiety.

Spray nozzles and reflective panels subject the patches' small but complex plant communities to atmospheric variations in heat, moisture, carbon dioxide and nitrogen. Dr. Hal Mooney, one of Standford's several superstar ecologists, explained to us that this experiment "is unfolding 16 different possible futures for mediterranean-type grassland ecosystems." These little plots are about as close as we can get, it seems, to putting climate change in a laboratory.

It is hard for the layperson to grasp what such small-scale work can tell us about huge systems like the global weather, but these scraps of ground are yielding very valuable data. And this information was particularly relevant to the restorationists visiting Jasper Ridge in August 2007. Their work had always been very challenging, but they were all finding that the unique hazards posed by climate change were upping the already high odds against them in unexpected and disturbing ways.

A typical question being raised among them that afternoon could have come straight out of Harris and colleagues' "Ecological Restoration and Global Climate Change." It can be summarized like this: "What is the point of restoring mediterranean grassland to its former glory, if current mediterranean regions will only be capable of sustaining semi-desert vegetation in 50 years time? Should we instead be preparing to facilitate the transfer of mediterranean species further north, around San Francisco in the case of California, or near Paris in the case of France?" Similar questions caused constant angst at the conference itself, especially in a series of presentations organized by Jim Harris and Richard Hobbs to reconsider the principles established in the SER primer.[20] Climate change is pushing relatively stable ecosystems onto trajectories outside the limits of anything we have known before, at least in historical times, they argued. This meant that goals set for restoration in the primer may no longer be realistic.

No Blood, No Bones, but You Can Still Walk

"We are now trying," says Harris, "to hit a constantly moving target."[21] That is, we may aim to restore a degraded site, based on the trajectory of a historical reference system; but we then find that the site no longer has a climate appropriate to that system, and now favors a quite different community of species. Several of his colleagues insist that the crisis of climate change will force them to prioritize the restoration of ecosystem functions and services over the recreation of healthy communities of historically appropriate species. This is bitter medicine for some restorationists to swallow. One scientist is particularly frank about what fully embracing this new vision could mean. He likes to illustrate his point with the help of a slide showing a prosthetic leg. "These organs have none of the sensuality of the original flesh, no blood, no bones," says Young D. Choi of Purdue University in Indiana, "but they still enable you to walk."[22] In other words, when the chips are down and we need clean water, fresh air or fertile soil, we may have to use those species which do the job best under the new conditions, regardless of whether they are historically appropriate to the local ecology, or culturally or aesthetically attractive.

This turns some of the basic assumptions of restoration practice upside down. As we have seen repeatedly, a major concern of ecological restoration has long been the removal of invasive alien plants and their replacement by indigenous species. The standard theory is that native species will always combine to form the best possible ecosystem, the ideal, appropriate, and most resilient plant community, for the area concerned. But increasingly there seem to be new cases where the invasive alien can do the same ecological job—purifying water, sequestering carbon, and so forth—as well or better than the native. When new climatic conditions favor it over native species to such an extent that removing it becomes prohibitively expensive, there may be no alternative but to welcome IAPs for the services they provide. Another malign scenario: even where one could still remove IAPs, native species might not be able to do their old ecological jobs under new conditions. Might we then, under these unprecedented circumstances, consider removing native plants and replacing them with IAPs in order to restore historical ecosystem services?

The options these scientists are laying before us are certainly stark, and the baseline seems to change all the time. The following conversation from the San José conference seemed shocking at the time, but five years later it seems almost commonplace. A senior scientist with a long association with SER took me aside to impress on me that climate is changing so fast that we may now have no chance of saving some emblematic ecosystems. Not only the Arctic, which is already vanishing before our eyes, but tropical cloud forests, and South Africa's enormously rich and varied fynbos vegetation may well all be doomed, he said, and we should probably make much better use of scarce resources by abandoning efforts to save them and re-focussing our efforts elsewhere.

"We have to let some things go, like the polar bear. It is shameful that we have let this happen, but it is too late to stop it. The only polar bears that will survive," he said, "are those that are leaving the ice cap and may produce fertile hybrids with grizzlies further south. That is an evolutionary response to the crisis which we have to accept."

Such prospects then seemed too depressing to be widely articulated, and this scientist, half serious and half joking, asked for these comments not to be attached to his name. His view may have been overly pessimistic, but many of his colleagues talked about the increasing, and increasingly painful, element of "triage" in their work. Triage is a term usually used in battlefield or catastrophe situations, when doctors have to decide which of the wounded to attempt to save—and which to abandon.

The dramatic new dilemmas facing restorationists were well illustrated by Patricia Townsend's presentation on her work in Monteverde, in the Tilaran Mountains of Costa Rica. Mountain plants and animals, she says, appear to be lucky when faced with climate change, compared to their lowland equivalents. They can simply move uphill, along with their habitat, as temperatures rise. Choosing a zone just five hundred meters higher up would compensate for an increase of 3°C for mountain species. Meanwhile, a lowland species living on a plain might need to travel five hundred kilometers north to compensate for the same rise in temperature. However, that fairly reassuring picture shades into a much grimmer reality when you continue up the mountainside and consider the fate of species that already inhabit summit regions. For those that cannot fly or, in the case of plants, broadcast their seeds to surviving appropriate systems (still higher up, on neighboring mountains), there is nowhere left to go.

Townsend is already observing a significant degree of upward movement by species in her area, and is working on the creation of "corridors" so that those on the lower levels can move up unhindered. But she admits to being "very scared" at some of the decisions she is having to make. "Every year, a group of teenage volunteers comes from Vermont to plant trees for me," she says. "But will the trees they are planting survive even a lifetime where we putting them now? Or should we start planting species higher than their normal range? Should we *dare* to anticipate climate change? As a restoration ecologist, I simply don't know what to tell them."[23] This is an area where we must be extremely careful, she continues: "For we all know about biological nightmares resulting from the best of management intentions." (Massive reforestation projects from the 1940s and 1950s in the US, for example, have often actually reduced biodiversity because the species planted were inappropriate). Yet, as she wrestles with the growing uncertainty that Richard Hobbs had been warning about since the Zaragoza conference, she comes down on the side of action against inaction: "If we wait until we have all the information we need, it will be too late to do anything about it."

A radical solution to such dilemmas was offered to the conference by an expert on Arctic ecology, F. Stuart "Terry" Chapin, who proposed that we should see climate change "as an opportunity rather than a problem."[24] He pointed out that while the Arctic was now obviously vulnerable to warming, many of its species and characteristics have shown a marked resilience to wide variations in climate in the past. Many of them are well fitted to adapt through migration. Properly managed, he argued, most species will avoid extinction and may survive in areas far from their current habitats.

However, he also stressed the dramatic impact such shifts in animal populations could have on indigenous peoples and pointed out that recent changes in their lifestyles will make that impact more severe. They are often dependent on particular species for their livelihoods. In the past, when such peoples were nomads, it was relatively easy for them to follow significant prey species if they moved to other regions. Today, most indigenous Arctic peoples live in fixed housing, which will make it much more difficult for them to respond to climate change by moving their settlements closer to their shifting food sources.

Presenting a very challenging scenario for ecosystem management on a continental scale, involving all eight states with Arctic territories, Chapin proposed:

- fostering environments for the "right" (native) species;
- conserving corridors to enable the "right" species to migrate in pursuit of suitable habitat;
- considering direct human-assisted migration—that is, capture and release in suitable new habitats—where no corridors exist; and
- creating barriers against the "wrong" (invasive alien) species moving in to take advantage of changing climate.

His proposals included the conservation of habitat for species *not yet present* in particular areas, and the establishment of agreement on marine fishery reserves in Arctic regions which are *still under the ice-cap*. Such agreements, based on predicted futures rather than current realities, are unlikely to come very easily. This is especially true in the context of the rising geopolitical tensions between Russia, the United States, and other Arctic Council states over mineral wealth in this region.[25]

What to Do in a No-Analog Future?

Some voices in the restoration movement urged caution against embracing these models of a "no-analog future." Eric Higgs, a Canadian who uses the tools of anthropology and philosophy to analyse ecological restoration, had cosigned Jim Harris's climate change article. Yet he asked poignantly: "How can we come to terms with losing so much, indeed with losing sight of the very things we value most? How do we decide which species to keep only in the zoo, what plants to hold only in the seed bank?"[26] While recognizing that some Americans still need to "unlearn" the idea of returning to a pristine past, Higgs argued that restoration should continue to use historical reference systems, though "as a guide, not as a prescription." He also warned that, if restorationists abandon targets like ecosystem integrity and biodiversity, and settle just for the maintenance of basic ecosystem functions, there is a great danger that the human tendency to turn the entire natural world into commodities will be accelerated, at a great cultural and aesthetic cost, as well as at a huge biological one.

Jim Harris, however, argued that Higgs's fear was at least partly based on a misunderstanding. "Biodiversity *is* an ecosystem function," he said, "and a very important one. I am absolutely not in favor of abandoning that kind of target for restoration."[27]

I found James Aronson, who had somewhat reluctantly signed off on both Hobbs's and Harris's 2006 papers, in a remarkably pessimistic mood at this conference. "I don't often admit it, but I too think the engineers will win out for the most part, in the end, driven by business and politics . . . and by our divorce from nature," he told me.[28]

Talking to Harris, Higgs, Aronson, Hobbs, Bill Jordan, and other restorationists at the cutting edge, it is evident that these questions arise as much from internal arguments within each individual as they do from debates between different tendencies in the movement. Like nature itself, restoration ecology is in flux, and it is a disorientating if sometimes heady moment for all concerned. Aronson has since become a leading critic of the direction taken by the thinking of Hobbs and Harris over the intervening years. Higgs, once the staunchest defender of historical reference systems, has collaborated closely with Hobbs and Harris in more recent papers and workshops. He nevertheless describes himself as playing the role of "official worrier" about the new thinking he is helping to develop, and delivered a succinct account of his own concerns about it at the SER conference in Mérida, Mexico, in August 2011.[29]

The extent of the new challenges to restoration was recognized by the then chairman of SER, George Gann, concluding the San José sessions: "Unless checked, global climate change will destroy people, places, and life as we know it," he says. "Ecological restoration offers hope in two key areas: by reconnecting fragmented ecosystems [to allow] animals and plants to migrate in response to such change; and, by capturing carbon through the restoration of forests, peat-forming wetlands, and other ecosystems that act as carbon sinks."[30]

Fewer and fewer restorationists listening would disagree with these new aspirations, and all of them surely recognize just how difficult it will be to implement them. But many of them still cherish the core restoration dream of recovering, on degraded sites, communities of species familiar from the recent past. Some of them argue that it is much too early to give up on that dream, just as restoration techniques are advancing fast, and new sources of funding are opening up. They fear that some restoration scientists are embracing the concept of "novel ecosystems" rather too warmly, neglecting troublesome but very significant old puzzles for the greater excitement and higher profile that comes with identifying new ones. We will return to this debate in our conclusions.

Undoubtedly, the new complexities identified by the "novel ecosystems" trend of thought are more than a little intimidating for ecologists and environmental activists. But they are also bracing, because they indicate that ecological restoration is indeed, as one speaker at the conference put it, "stepping into reality." Cold winds blow hard in this new environment, and many hard choices—and unique hazards—certainly exist, but acknowledging new realities is always the first step to effective responses. The next chapters will take us to places where the radical tranformations of ecosystems, by invasive aliens and agricultural degradation, exacerbated by climate change and unique regional factors, have already crossed thresholds that often appear irreversible. Yet we will find that restorationists are measuring up well to these challenges in flexible ways, and often achieving quite remarkable degrees of success in novel circumstances.

9 *Dreamtime in Gondwanaland*

You will see many things here that contradict your knowledge of life on Earth. * KINGSLEY DIXON, *director of Science at Kings Park and Botanic Garden*

* *

The "Parrot," the Wetland, and the Trapdoor Spider: Paradoxes Of Restoration in an Australian City

Nothing in Western Australia is quite what it seems.

The first wild creature I saw in the region was an extravagantly plumaged bird. It had a deep blue head and a red breast, bill, and iris. Its light green back was topped with a brilliant yellow Eton collar. I took it to be a native parrot. It had landed in a tree outside my rather tacky hotel. I was having breakfast, trying to shake off the jet lag after a journey from Dublin via Hong Kong and to come to terms with the relentlessly suburban first impressions that the city of Perth had made on me. So the parrot was a welcome omen, a harbinger of exotic wildness—or so I thought. It was a bracing contrast to a familiar sense of ornithological letdown. When visiting far-flung countries settled by people from the British Isles, the first bird seen is so often not a native, but a feral house sparrow or starling. Often released by homesick nostalgics, they then become dominant species, playing their part in the invasive alien narrative.

I spent what free moments I had over the next few days exploring the network of small parks opposite my hotel. I saw many and heard many more of the "parrots"—I had learned from my bird guide that they were rainbow lorikeets. There was a stimulating variety of other birds—Australian pelicans, no less than three species of cormorant, a bewilderment of honeyeaters, even a spoonbill. The parks were hemmed in by busy motorways but were cleverly screened with ponds lined with melaleuca and eucalyptus. At one point, the trees expanded into a little woodland graced with a rather lovely waterfall.

One afternoon, I met a local birder who kindly helped me to distinguish two of the more common species of honeyeater. When a posse of lorikeets crashed noisily into a nearby tree, I commented on their beauty. "Oh, *those*," he said wearily. He told me a sadly familiar story: the lorikeets were aliens, albeit from eastern Australia. A huge population had built up since a handful had escaped captivity in Perth in the 1960s, and they were thought to be displacing several native species.

A few minutes later I was struck by another disparity between ecological appearance and reality. I was contemplating the waterfall when a middle-aged couple paused on their bikes to join me. "What a beautiful place," they said. They were lifelong Perth residents, and yet were quite surprised to have found the park here, right in the middle of a spaghetti junction of highways. "I do remember this place from when I was a kid," said one of them, and then added: "But at that time it was only a wetland. What a good job they have done in tidying it up."

Only a wetland. Tidying it up. Their phrases reminded me where I really was, in terms of ecological history: bang in the middle of the Swan River estuary, though no river was now in sight. But their choice of words also spoke of the chasm between most citizens' perceptions of landscapes, and those of people familiar with ecology and restoration. What looks tidy and pleasing to the casual observer may strike restorationists as an expression of a "world of wounds," that painful vision of omnipresent degradation that Aldo Leopold said was the price of ecological awareness. The pelicans, cormorants, honeyeaters, and spoonbill are really but a poor remnant of the avian diversity that must once have fed and roosted where the parks are today. The pretty sylvan waterfall hardly compensates for the grand fluxes of organic life and minerals that must have surged back and forth through the estuary before the settlers tamed it.

This park was an exceptionally instructive place, because the Eliza escarpment, which overlooks it, had itself gone through a series of transformations since white settlers arrived here in the 1830s. Thanks to a complex urban ecological restoration project, its current state, if not better than its first, is certainly a great improvement on everything in the intervening period. Before settlement, the escarpment had served the indigenous Noongar people as a convenient point to drive kangaroos, panicking them over its cliffs and harvesting them for meat. The Noongar found a precious source of fresh water in the springs at its base. They believed that Waugal, their variant of the Rainbow Serpent, a creator figure common to much Aboriginal mythology, had shaped the Swan River that flowed beneath it. They respected parts of the escarpment as one of his sacred dwelling places.

White settlers imposed more drastic uses on these steep slopes. First they quarried limestone, undermining the escarpment's fragile stability. Then they unwittingly continued this degradation by planting alien pines and agaves. The deep tap roots of the pines penetrated and parted the friable limestone, while the shallow roots of the agaves spread across the topsoil, and eventually their great phallic flower spears fell over, pulling roots and soil with them, after their brief (if long-awaited) blossoming. In the 1990s, drivers on the motorways into Perth's booming business district, which run below the escarpment, were alarmed to find rocks rolling down in front of their Porsches and BMWs.

Since the escarpment forms part of the border zone of Kings Park and Botanic Garden, home to a dynamic restoration ecology section in the science department directed by Kingsley Dixon, the crisis represented by the crumbling escarpment became an opportunity to showcase ecological restoration in the interests of Perth's citizens. The job was never going to be easy, but it became very complicated when a research team funded by the Western Mining Corporation found that the escarpment was home to four species of trapdoor spider endemic to the region, one of them threatened with extinction. "The trick," says Stephen Hopper, who was then director of the garden, "was to get the exotic plants off this landscape, hold the soil in place, and then get native shrubs and herbs back into that landscape—without losing the spiders."[1] This in turn was made acutely problematic by the tight time frame imposed by the spiders' life cycle. They cannot build their traps after they are two years old, so a long period of disturbance would wipe out entire generations, perhaps a whole population.

Work continues today on this restoration; 1,300 pines have been removed, and 300,000 native seedlings have been planted, while the spider populations remain healthy. With the caveat that must be entered for most ecological restoration projects—success can't be fully assessed in the short term—the Eliza escarpment looks like a shop window for state-of-the-art restoration in an urban context. There was some initial opposition to the loss of the beautiful trees that had long been enjoyed by walkers and runners on a highly attractive recreational trail. But it was alleviated through a good public information campaign that, combined with pragmatism, has forestalled the kind of bitter controversy that has arisen over the removal of well-loved trees by restorationists in Chicago. "A few pines have been retained for 'European heritage reasons,'" according to another garden staff member, "but we keep a tight rein on them."[2]

"When you engage [citizens] with science, you have to be completely transparent. You get durability then in terms of public and political credibility," says Kingsley Dixon. "You have to be honest and win the debate based on a better future. If you are not transparent and you hide it, ah, it's a bit sneaky."[3]

The motorways are not going to be removed in any foreseeable future in order to restore the wetlands of the Swan River estuary, nor is it reasonable to argue that they should be. And it is likely that rainbow lorikeets will continue to displace native bird species. But restoration seems set to continue to slowly but significantly alter the city's relationship to the natural context that attracted the settlers here in the first place. There is a plan to restore a salt marsh on waste "reclaimed" land under a section of the escarpment. Perhaps, in ten years' time, the cycling couple may return and find pleasure in exploring a new wetland that needs no "only" to qualify its importance as a reflection, however pale, of the estuary's ancient ecosystem. Elsewhere, some 270 hectares of native bushland are being restored in Kings Park and Botanic Garden properties right across this brash mining boomtown capital.

In the meantime, let's move beyond the city to look at two stories of large-scale restoration in Perth's vast hinterland. The first involves the reconstruction of a complex and very biodiverse forest from ground zero. The second is about the attempt to reconnect eight ecosystems across a thousand kilometers. There is nothing modest about ecological restoration in Western Australia.

Putting A Whole Forest Back Together: How a Mining Giant Restores What It Destroys

This is the most sophisticated [ecological] restoration operation on the planet. * KINGSLEY DIXON

To enter the Huntly bauxite mines in the jarrah forest that stretches over the Darling Range, southeast of Perth, is to reencounter Leopold's "world of wounds" on a grand and devastating scale. The power of Big Industry—in this case Alcoa World Alumina—to simply erase nature from the face of the Earth is bleakly evident here: if you stand in a newly mined pit, four meters below where the topsoil used to be, a salmon-pink but utterly sterile flatland stretches so far away that you cannot see the tops of the big eucalypts that surround it.

That vanished topsoil, and some elements in the subsoil and overburden underneath it, had supported the rich variety of plant and animal life that is manifest where the forest still stands. Jarrah and marri eucalypts dominate a relatively open canopy that shares sunlight with 800 plant species in the mid- and understories. Twenty-nine different kinds of mammals, 150 bird species, and 45 types of reptiles find nourishment and shelter here.[4] To compensate for the chronic infertility of the soil, the plants have developed a vast repertoire of root associations. Soil and litter invertebrates are thought to play exceptionally varied and crucial roles in maintaining this complex and interdependent community. The forest is not only valuable for its biodiversity; it plays a vital role in the hydrology of the region. It is the main source of water for Perth, by far the largest city in Western Australia, and for the irrigation of the wheatbelt to the west of the range. The forest is, as Alcoa itself declares, one of the most species-rich in plants in the world. But the company has turned it into the site of the world's biggest bauxite mine.

This whole system appears at first sight to be falling down before the advance of giant bulldozers, to satisfy the world's hunger for aluminum. Not a single trace of fertile life remains where you are standing. Most of the forest will eventually to suffer the same fate, and when the company's lease comes up for renewal in 2045, most of the aluminum will be gone. Most of the forest too, you might well think, as you stand in this lunar wasteland. But then ecologists employed by the company, accompanied by very reputable independent scientists, may guide you back into the surrounding forest, where no mining has yet taken place. Stopping just a couple of kilometers down the road, they will walk you into a patch of forest where the trees are obviously younger than the ones you have just passed, but where they can identify an almost equally rich variety of species beneath the canopy. You are standing, they will tell you, in jarrah forest that has been rebuilt from ground zero by the company after mining.

If you knock down a forest and strip the minerals beneath it to a depth of several meters, can you really put the forest back again afterward, with all its constituent species?

Common sense suggests that such a task must be nearly impossible. But Alcoa's scientists and outside assessors say they are doing it all the time in this jarrah forest, right down to the last obscure species of sedge or sundew, many of which are endemic to the region. The company even claims that results from 2008 show 8 percent *more* biodiversity on restored sites than in equivalent areas of unmined forest.[5] The restoration has been as successful with mammals as with plants, with all recorded species coming back to the restored areas. Alcoa has still to demonstrate 100 percent comeback for birds, and especially for the very diverse range of reptiles present in the Darling Scarp, but they are getting close. The problem here is linked to the need of some species, like the rufus tree creeper, for old fallen timber to provide nesting holes and refuges. Efforts are now being made to bring in old logs from other sites, in advance of natural ageing here, so that these species too can be accommodated. Success rates with invertebrate, microfaunal, and microfloral communities, all of which may be crucial to the long-term ecological health of the forest, are less demonstrable so far. As usual, these are the areas where current knowledge is most limited. But Alcoa scientists are working on this.

Recent data also indicate something even more surprising: that restored areas are significantly more resistant to dieback, a devastating disease caused by an introduced soilborne germ (*Phytophthora cinnamomi*), than unmined parts of the forest. "It seems that the new site conditions favor the plants, not the pathogen," says one of the Alcoa restoration staff.

In 2007, *Restoration Ecology* published a special supplement that indicated that the restoration of the jarrah forest has been extraordinarily successful.[6] In 2009, a conference of the Society of Ecological Restoration International in Perth, Western Australia, heard another series of papers delivered that confirmed these findings. Few restoration projects run by environmentalist organizations, let alone by commercial corporations, have been so extensively monitored and have ticked so many of the right boxes. Awards have come from the government of Western Australia, the World Environmental Center, SER, and even the United Nations.

Kingsley Dixon, whose institution has done significant research to assist the jarrah forest restoration, does not stint superlatives about it: "This is the most sophisticated restoration operation on the planet."[7] I balk at that a little, and ask him if he does not mean "the most sophisticated *post-mining* restoration operation on the planet." "No, it is *the* most sophisticated, period, simple as that."

Still, it's hard not to be just a little skeptical. "Greenwashing"—putting a positive spin on stories of deep environmental degradation—is a notorious technique of corporate PR in mining and other industries. Many companies will claim to have "rehabilitated" a mining site, but if you visit it you will find this means that a monocultural plantation of alien trees or a golf course has replaced the original and much more biodiverse native vegetation. And indeed, had you visited "rehabilitated" sites in the Darling Range jarrah forest in the 1960s, you would have seen endless rows of pine species, which are inappropriate to this ecosystem in every sense and whose roots easily lost their grip in the sandy soil, causing further erosion in degraded zones. "Our early rehabilitation would now be regarded as a failure," says John Koch, senior research scientist with the company. "Pre-1988 we put in the trees the government asked us to. It is still an open question whether we have to return, to rehab the areas we rehabilitated then to today's standards." But given the company's record here of doing more than is required by law, it seems likely that it will. "We've got the money to restore the rehab," says Koch simply.[8]

A combination of public protests and a change of heart by local management appears to have propelled Alcoa, in this jarrah forest at least, to the forefront of both ecological restoration and restoration ecology. There were a number of reasons for this remarkable progression: the Darling Range supplies Perth's water, and has great recreational value to its citizens, including mining executives and workers.

These were the powerful lenses that put the mining juggernaut's operations under close scrutiny. It is much easier to behave irresponsibly in the far reaches of the outback, or in developing countries low on the international radar, than it is when you are mining in the middle of the watershed of the capital of a democratic industrialized state. John Koch spells this out:

> Alcoa in [Western Australia] took the stance in the 1970s that we are going to be a long term operator here for fifty to a hundred years, with a large financial investment, three refineries at multibillion dollars each, and we must maintain access to the bauxite lease. We are close to Perth and operate in a sensitive area in a drinking water catchment in a unique forest. Therefore our large environmental commitment is an investment in the future. We need to do an exemplary job and stay well ahead of government and public expectations.[9]

Dissident voices about the success of the jarrah forest restoration can still be heard, but they are hard to find and not very convincing. One attendee at the Perth SER conference in 2009 told me that senior scientific researchers fear to speak openly about flaws in the work done because of the generous funding their universities receive from the company. But his refusal to follow up this conversation, even on an entirely off-the-record basis, makes it difficult to assess how well-founded this criticism might be. In my own experience, I found two researchers indirectly funded by Alcoa, Richard Hobbs and Dixon, to be rigorous in their scrutiny of the jarrah forest restoration and outspoken in their criticism of the company's other operations. "On the one hand, Alcoa are doing this marvelous restoration work, on the other they are behaving like any other corporate bastards," Hobbs told me in reference to the company's locally notorious failure to eliminate toxic runoff from its Western Australian refineries.[10] These are not situated in the jarrah forest but cause considerable ongoing concern and anger. Alcoa's environmental record elsewhere in the world is "appalling," Dixon said bluntly.

Alcoa representatives at the Perth SER conference did accept that there were very big variations in their environmental standards internationally. They used the disingenuous defense that this was because their ventures in developing countries are often partnered by governments with little concern for environmental regulation; they have also been heavily criticized on their record in the United States. The reality is that corporations are generally only likely to invest significantly in environmental protection and restoration when governments insist that they do so; in Western Australia there has been a happy coincidence between (fairly) strong government regulation, and an enlightened local management.

For all his admiration of Alcoa's achievements here, Dixon does not hesitate to put them in the context of its enormous wealth. In our interview, I expressed amazement at the degree of this company's commitment to restoration. "And? *And*? You know what? For a company pulling a profit out of this state in excess of a billion dollars, for six hundred hectares of pristine Jarrah forest knocked over, I think what they do is admirable, it's laudable, it is beyond their compliance requirements, but it is a minute speck in terms of the amount of profit they make."[11] Australians have a refreshing habit of straight talking.

Dixon's point is well taken, but it would still be churlish—and less than pragmatic—to fail to acknowledge that Alcoa is investing far more in restoration in southwest Australia than other very wealthy mining companies here, and that the company has gone to meticulous pains to earn its status as a world leader in the field. In any case, I found researchers directly employed by the company, like John Koch, very willing to acknowledge areas where there is still room for improvement in the forest itself.

Over the last thirty years, a team of ecologists, led first by John Gardiner and now by Ian Colquhoun, has pioneered a series of meticulous techniques to achieve their current results. These include "double stripping" of soil layers prior to mining. The first five centimeters, rich in seeds, mycorrhiza, and microbes is lifted off first. Then almost immediately—within days—this topsoil is laid down on an adjacent area undergoing restoration. A deeper layer of subsoil and overburden can be stored longer without ill effects. Mining for aluminum involves the removal of up to four meters of bauxite below the overburden. Before restoration, the site is recontoured by heavy machinery to mimic local topography, then deep ripped to break up compaction and give the plants access to water. The recontouring is the most expensive part of the operation, Koch says, but the biological bits don't come cheap either.

Seeds of every species previously found in the area are then planted where possible. But some plants are "recalcitrant"—they won't propagate when planted by seed. So Alcoa has built an AU$3 million laboratory to germinate them by tissue culture, after which they are planted individually. In Dixon's view, it is this kind of extraordinary—and very expensive—attention to each species that makes the Alcoa restoration unique. "No other group has the resources to afford to pay $8 for a [single] little sedge plant that has been brought through a tissue culture led through synthetic seed technology," he says.

While the company has been remarkably successful—100 percent (or more, in some cases)—in restoring plant species richness, John Koch agrees that its failure to replicate the original abundance of some species remains "a concern," but he believes that this will be resolved over time. A layperson must wonder, however, if the removal of the bauxite itself will have no negative effects in the long term, even though there is no evidence that any of the plants need it for nutrition. "The trees seem to respond to all the same factors as an unmined forest," responds Koch. "Call me in 2020 and I will let you know how the 50-year old jarrah is going!"[12]

"That is the good thing about Alcoa, we have twenty years of good data now," says Richard Hobbs, but then he adds, "Even that is a tenth of the time it takes to get anything like a mature forest going, so there is still a long way to go."[13] Like all restorations, Alcoa's jarrah forest is a work in progress, and it is too early to declare it an unqualified success. But it remains a most impressive achievement, a paradigm for other projects. Steve Hopper was the director at Kings Park and Botanic Garden when crucial research for Alcoa was undertaken there. He says, "Thirty years on, Alcoa has flipped over from a quick-and-dirty-just-get-trees-into-the-landscape model [of restoration] to a point where they aspire to 100 percent plant diversity, and they are almost there. So that is an example of what can be done. One of the reasons it has worked is that the economic drivers have remained the same, and the political process has enabled a long-haul research component to be built into it: all ducks in a row, a great outcome."[14]

19. **Restoring from Ground Zero.** The recovery of a segment of jarrah forest in southwest Australia in 2001, following bauxite mining in 1980. The Dandalup Dam, which supplies water to Perth, is in the background. (Photographs courtesy of Alcoa.)

Putting Charlie Darwin Back In The Landscape: The Campaign To Reconnect Eight Ecosystems

The Gondwana Link is the most exciting thing that is happening [in restoration] in Australia, if not in the world. Working at big scale but with the underlying philosophy that everything you do should make sense in its own right. Even if the big vision falls apart you are still doing good stuff on a local scale. * RICHARD HOBBS, *professor of restoration ecology, University of Western Australia*

The Gondwana Link is audacious—and bloody expensive. We are reinventing landscapes here. It's an almost agricultural approach to ecological restoration, plus a precision not seen in restoration here before. * ROBERT LAMBECK, *director of corporate affairs, Greening Australia*

"Let's get Charlie Darwin back into this landscape!"[15] Keith Bradby has just pulled over to the edge of a red-dirt road, to show me a plant which has some characteristics of two species of *Grevillea*, a widespread Australian botanical family with many members. One species is found a few dozen kilometers to the west of where we have parked, the other is found a few dozen to the east. But the one Bradby is holding does not quite fit either of them, and seems to be developing into something completely new. "This is evolution in action," he continues enthusiastically, "but if the plant communities are no longer connected across the country, evolution can't happen any more."

Bradby is the coordinator for the Gondwana Link, a massive environmental project that aims to reconnect and restore wet forests, mountain ranges, flat scrubland, and arid woodlands across eight distinct ecosystems and a thousand kilometers in southwest Australia. The Link gets its name from Gondwanaland, the ancient composite continent that included Australia, South America, southern Africa, and India before it began to drift apart—about a hundred and twenty million years ago. The link project, about as young as this century, is an attempt to prevent one of Gondwanaland's richest biological legacies from fragmenting in a different sense, this time into shrinking islands of declining biodiversity.

This enterprise starts with some massive assets and as many daunting challenges. The area it maps out forms a necklace of natural jewels across Australia's only "biodiversity hotspot." And southwest Australia is special, even within that select global club. Like South Africa's Greater Cape Floristic Region, it is one of the few exceptions to the rule of thumb that species richness increases from the poles to the equator. This is a distinction with surprising origins, which have critical implications for restoration here, as we shall see. As you would expect for a hotspot, this region has very exceptional numbers of endemic species—plants or animals unique to the area. Above all, it includes some huge tracts which are as close to pristine wilderness as you can still find anywhere on the planet. Aboriginal people walked very lightly here. Europeans arrived as settlers only 180 years ago, and while they have transformed vast tracts of land, other areas remain almost unaffected by their presence.

The challenges to the Gondwana Link are as dramatic as its assets: they include immense forest and bush clearances for farming (with catastrophic ecological consequences) and a gung-ho mining industry sitting on some of the world's richest mineral deposits of gold, copper, nickel, and possibly uranium. A young but cheeky wine industry is causing some fresh environmental headaches today, just at the point where I start my journey across the link. Margaret River is a fashionable surfing and holiday town at the center of this new boutique wine industry. It is near the coast, between Cape Leeuwin and Cape Naturaliste, at the continent's southwestern tip. Members of the Cape to Cape Catchments Group, an impressive local conservation organization, tell me that the diversion of water to vineyards—and even to ornamental "lakes" to enhance the views at wine-tasting restaurants—is seriously affecting streamflow in the area. With big companies moving in fast, the industry is expected to double in volume in the next decade. I arrived just after a state government funding cut had slashed the group's already minimal staffing. Their mood was understandably low, though they took a little comfort from the fact that the Gondwana Link project would be helping them to draw up a conservation action plan over the next few months.

However, while much of the cape's area is undoubtedly in ecological trouble, it retains some magnificent stands of karri and marri wet eucalyptus forest. Driving the four hundred kilometers southeast along Route 10 to Albany, the next substantial town, these stands become more frequent. Soon they start to merge into impressive unbroken belts of rich old growth. The landscape begins to exude a broad impression of robust ecological health, though rolling pastures open up the view periodically, reminders of the huge clearances just a little to the north and east. Karris are majestic trees, with old growth individuals reaching almost ninety meters. They thrive in the lush, wet climate of the extreme southwest, supporting a remarkable range of plants beneath their canopies. This understory in turn is rich in small marsupials, birds, and reptiles. These forests were threatened by logging until recently, but are now extensively protected in national parks. Perhaps the most beautiful of them lies along the lazily meandering Warren River, where you can lose yourself in cathedral groves of karris that rival California's redwoods. Splendid fairy wrens, tiny bundles of blindingly vivid purples and cobalts, perform courtship displays on the forest floor. They move with a nonchalance that suggests little human disturbance, but perhaps they just belong to that category some bird guides list as "confiding species."

At the Walpole-Nornalup National Park, a little further on, even the karris are dwarfed by towering red tingles, another eucalyptus. This magnificent forest still accommodates some of the strange and seldom seen small marsupials that have suffered catastrophic losses from predation by introduced foxes and feral cats. Driving at night through these phantasmal trees on my way out to Margaret River, I repeatedly see little bundles of fur scurry away from the edges of my headlamps' beams. A quokka? A brush-tailed phascogale? I never come close to an accurate guess at their identities, but the sense of abundant nocturnal life is palpable.

The contrasting sparseness of the human population of Western Australia is evident on the empty roads, even by day. Fewer than three million people are spread across an area one-third the size of the United States, and most of them are in Perth. Until I reach the suburbs of Albany, I only encounter seven other cars in eight hours of driving. Here I hook up with Keith Bradby, and we set off to find the areas where the Gondwana Link has been most active so far, between the Stirling Range and the Fitzgerald River National Park. Our final goal is the mining town of Kalgoorlie, deep in the Great Western Woodlands, where the Link is just beginning to make an impact.

Immediately after we leave Albany, the climate becomes visibly drier. Rivers shrink rapidly from lazily looping broad floods like the

Warren, now far behind us, to the struggling start-stop trickles of the Pallinup and the Brewer as we move eastward. The challenges facing the Link again become harshly apparent. First we pass monocultural plantations of blue gums, eastern Australian eucalypts farmed for wood chips or planted for carbon credits. You can see clearly here that carbon trading is not always the friend of biodiversity: the tightly packed ranks of trees admit no undergrowth. Then we enter the wheatbelt region through vast paddocks (you can't say "field" in Australian) cleared for crops and grazing. Almost at once it is apparent that something is very wrong here, and it is not just the absence of native vegetation. Occasionally, a paddock is choked with yellow sand. Here the removal of soil has led to erosion and "blowouts," unleashing the buried energy of the ancient dunes on these failed facsimiles of English pastureland. Much more frequently, the green paddocks are pockmarked with great gray stains, spotted with vegetation more typical of salt marsh than of meadows. A number of the farms, though established barely fifty years ago, are already dying from salinity.

The farmers who came to clear them found rich native vegetation here, both bush and forest. They drew the understandable but mistaken conclusion that the soil beneath it would be highly productive for agriculture. They failed to understand the consequence of a key geological fact: this ancient part of the earth has not been refreshed and enriched by the rock-grinding, soil-building processes of either glaciation or tectonic mountain-building for roughly 250 million—yes, million—years. Only native plants can grow really well here, unless you pump industrial quantities of fertilizer into what little soil there is. The farmers are not the only ones to be deceived by appearances. Ecologists, too, often missed just how distinctive this landscape is, and how restoration needs to follow a very different path here from the broad paradigm dominant in most other parts of the world. The Gondwana Link project is therefore particularly instructive in that it is as much a process of learning by doing as it is the execution of a grand master plan. Even the core concept of natural corridors, implicit in the link part of the project's title, is having to be reworked and reconsidered as new characteristics of these ecosystems are revealed by successes—and failures—in restoration practice, and parallel advances in scientific knowledge.

The first white settlers treated their new paddocks as if they were carpeted with the deep, rich soils of Kansas. They ignored warnings from as early as 1929 that thick layers of salt, deposited by rainfall from the ocean over many millennia, were only kept below the surface by the structure of native plant communities. Clear those plants, and the salt may rise like an ecological avenger to poison the best attempts at growing wheat or raising cattle.

Bradby, now in his fifties, is from an urban working-class background in eastern Australia, but his childhood was rich in visits to rural relatives, and there is little of the city boy about him today. The tough impression given by a badly broken nose (it dates back to his school days) is immediately contradicted by sparkling eyes and a ready smile. But his lived-in face has an expression of sadness to it that reflects the old 1930s antifascist motto, "pessimism of the intellect, optimism of the will."[16] Bradby started out as a radical and still dares to dream of a better future, but he knows in his bones how heavily the odds are stacked against success for restoration here. He cut his teeth as an environmentalist on this ground in the 1980s while earning a precarious living as a beekeeper and a seed collector. The Australian government was then trying to revive a hubristic campaign slogan to put "a million acres a year under the plough." Bradby was one of the leaders of the local opposition. He and his colleagues eventually saved countless precious hectares of bush and woodland in this region from inappropriate development. Many farmers regarded him as the devil incarnate, hell-bent on stopping decent people from making a buck. What could be wrong with turning a wilderness into a national breadbasket? But the spreading plague of salt has added bite to his arguments, and hard-headed locals are much more willing to listen to him now.

"But in any case," he says, "my attitude has always been that you can't cluster, like in a city, and just go to the coffee shop with all your green mates. This comes from being a rural person where you are always meeting neighbors of different sectors and interests. When you go into a shop, it's got miners in it, it's got farmers in it, and you discover they are human beings. I've never subscribed to this footie [soccer] team view of world history where there are the goodies and the baddies. A farmer does not drive around on his tractor thinking, 'how can I rape this environment a bit more?' I always tried to stay on speaking terms with everyone; we are all human beings struggling to survive under difficult conditions."

One of Bradby's strengths as a consensus builder is his willingness to recognize his own mistakes. "If you had asked me about sand blowouts and salinization in the mid '80s, I would have said a definite 'Yes, farming is dying here,' and I would have been proved wrong, very wrong." Salination has afflicted many farmers, but not all of them, and the farms are so big that there is usually enough halfways good land left to keep turning a profit. And there have been dramatic technological advances, including the use of hydroponics. "They have also switched from quite destructive plowing techniques to minimum tillage techniques, much kinder to the soil structure," says Bradby. They use a lot of herbicides, and they are getting pretty good yields. Plenty of money is still being pulled out of farming in the Fitz-Stirling—generally by the guys who ignore all the basic sustainability tenets and just get on and crop big areas. But if you look at their hundred-year trendlines, the cost of production relating to returns on produce, are ever narrower, and crossing." But he is not predicting now that even long-term economics will force the farmers to abandon the region: "Double the price of cereals with climate change, and these guys—or their descendants—will be there a lot longer. But it is often not the economics that takes them off the land. It's the stress, the social isolation, and exposure to chemicals."[17]

So, looking toward that kind of future, Bradby says he conceived a vision for the Gondwana Link where environmentalists would say "yes" instead of "no," where they would work with farmers and miners and not against them. Spend some time with him talking to a member of Parliament representing miners in a Kalgoorlie bar, or at a flower show with the ladies of the tiny town of Ravensthorpe, and you quickly see that he actually *likes* rural working people. They respond with confidence to his good-humored and let's-cut-the-bullshit approach to dialogue. He is very passionate about his cause, but he has learned that zealotry is a form of bigotry and leads only to righteous isolation. Many people believe that the Gondwana Link project has only become viable through Bradby's remarkable personal qualities. Not a few people question whether it could survive and thrive without him, a question we will return to. He is entirely self-taught in science, but extraordinarily well-informed; he loves a little story about the difference between botantists and seed collectors so much that he tells it to me twice in three days: "A botanist studies rare plants, Paddy. A seed collector *finds* them." And he makes no secret of a view which disturbs many scientists: emotion matters when you are dealing with nature. "I've earned my scientific credentials," he told a hall full of scientists at the Perth conference, "but it is feeling for the land that really counts."

"I'm not sure if our science has yet been that rigorous, but our common sense has," he tells me at one point.[18] He is forceful when he needs to be and is not shy about expressing robust opinions. But listening is his strong point, and he can make himself invisible in group settings. He may carry the Gondwana Link on his broad shoulders, but his charisma is of the quiet kind, a world away from the eco-guru who knows it all and demands unquestioning loyalty from his followers. He takes gentle pleasure in self-deprecating stories told in a clipped, witty style; then he draws strong conclusions from them.

"I was invited up to Shark Bay, when it had just been declared a World Heritage Centre. I was sent to meet the horrible evil salt miners at a place called, believe it or not, Useless Loop. They shut the works down so I could give a talk in their community hall. Good reception, I felt good about that." But his sense that things had gone well was shaken the next morning when he was roughly accosted while standing alone outside the manager's office: "This bloody great haul truck pulls up, this six-foot-ten lanky Australian burnt brown by the sun, in his shorts and singlet, gets out and walks up to me and I think, 'Uh-oh.' Well, he towers over me and says, 'You said in your talk that there are wedge-tailed shearwaters here, the dark-breasted form, and that they are endangered and blah blah.' And I said, 'Yes, that's my information.' And he said,'Well I reckon you're an arsehole, I reckon you're *wrong*. I reckon it's actually the white-breasted form of that shearwater that nests here'—and he just happens to have a bird book in his pocket—and he threw me in his truck and we drove to the place where the miners had left a special place for the birds to keep nesting, and he was right, they *were* all the white-breasted form of the species.[19] Well I reckon that in every community, and especially in every rural community you scratch, you find someone of good intent, and I think the conservation sector has wasted a lot of those people. They are left dormant. So we are seeking them out and working with them in the Gondwana Link."

So ecological restoration is social restoration is human restoration? "Totally entwined," he says. "Totally entwined." He talks about how a school at Shark Bay got involved in the restoration of native mammals, and then the whole community got involved. In the remote regions of Western Australia, this kind of enterprise gives people "pride in where they live, instead of seeing it as a derelict outpost." Restoration, it seems, can have a similar social impact across the world, whether the setting is rural South Africa, backwoods Australia, or suburban New Zealand.

The monocultural wheatbelt now separates two islands of biodiversity, and we are approaching the first of them as he speaks. The abruptly soaring outcrops of the Stirling Range, a spiritual center for the local Aboriginal Noongar people, boasts 82 endemic plant species. The most imposing peak, known as Bluff Knoll in English, is called Bular Meila by the Noongar. This translates as "the place of many eyes," where they believe their ancestors still gather. In one sense at least, they do: this is the site with the most ancient traces of animal life (1.2 billion years old) found anywhere in the world. The Stirlings now form a national park, but the rich texture of the green mantle that covers the hills is deceptive. Swathes of this vegetation are now severely threatened by dieback, the introduced fungal pest we first met in the jarrah forest. *Phytophthora cinnamomi* has become the scourge of Western Australia's bush and forest. Ironically, it has penetrated the Stirlings largely through the building of park roads to facilitate human access to the natural wonders of the range. The fungus hitchhikes on the soil used for road construction. Meanwhile, the cleared farmland has cut the Stirlings off both from the karri forests to the west, and the Fitzgerald River, the other biodiversity island, and another jewel on the link, to the east.

"The Fitzgerald River National Park is . . . the most important mediterranean ecosystem reserve in the world. It stands out for its scientific, conservation and educational values in the same way that the Galapagos Islands do," UNESCO'S former director of ecological sciences, Bernd von Droste, wrote after a visit. Bradby was one of those who campaigned to have the Fitzgerald region actively managed as a UNESCO biosphere reserve, where people and environment work together. Driving along the edge of the park toward sunset, he allows himself to take some pleasure in that victory. Ubiquitous western gray kangaroos, with a few of the much rarer black-gloved wallabies among them, bound away from us through the bush. It is close to here that we find the *Grevillea* that, he argues, needs a reconnected landscape to continue its evolution into a new species. Bradby now believes that isolated protected areas alone will not stem the loss of biodiversity, in Western Australia or anywhere else. He approvingly uses the phrase "ecological apartheid" to describe the tendency of traditional conservationists to divide the planet into supposedly pristine preserves without permanent human presence or impact, and vast spans of human-dominated landscapes, whether urban or intensively-farmed countryside. So the Gondwana Link's most visible efforts so far have been in the farming areas between the Stirlings and the Fitzgerald River. There is a five-point agenda behind the Link's policy in this area:

- to preserve such pockets of native biodiversity as remain;
- to restore biodiversity to many places where it has been lost;
- to create an unbroken corridor, at least two kilometers wide, of preserved and restored areas, reconnecting the Stirlings and the Fitzgerald River;
- to do this alongside and in the midst of continuing sustainable farming practices; and
- to draw the indigenous Noongar back to their ancestral lands.

The first three items obviously fit the overarching Link policy of reconnecting ecosystems, and the last two with the policy of maintaining and restoring healthy human connections to this landscape while minimizing negative human impact. We will return a little later to the surprisingly vexed question of how appropriate "corridors" are in these ecosystems; first, let us explore some of the organizational supports and ecological practices which are making the Gondwana Link a reality on the ground in the Fitz-Stirling area.

It is important to understand that Gondwana Link is more a vision than an organization and works through encouraging—stimulating, cajoling, and sometimes goading—a broad coalition of other groups to achieve its goals. Bradby always stresses the vast range of activities that find shelter under the Link's umbrella. There are the "big bold strokes," like reconnecting eight ecosystems or treating the vast Great Western Woodlands as "one chunk of bush." But there also many "little actions"—anything, in fact, that promotes biodiversity and restoration in the region. He sees the role of the Link at this level not as a command-and-control authority—it has none—but as a facilitator, recognizing that each little action is worth doing for its own sake but attempting to network it with other actions so that its effect will be multiplied. The Link has no state or federal funding, and philanthropy in Australia is not exactly munificent. So Bradby has found substantial funding from US-based foundations like The Nature Conservancy and the Pew Charitable Trust. "As an Australian, I'm embarrassed to take it, but I take it," he says.[20] There has also been some local funding, especially from Wesfarmers, the agricultural cooperative turned business corporation. But it is the US funding that has empowered the Link to make a high-profile impact here. This funding has enabled Australian NGOs like Greening Australia and Bush Heritage Australia to purchase five large farms between the two parks. The focus of the former in the Fitz-Stirling has been restoring farmland to native bushland, while the latter works to preserve the significant pockets of undisturbed bush some of these farms still retain.

I had been introduced to the story of the Fitz-Stirling farms in the care of the Link during a vibrant presentation at the Perth conference by Justin Jonson. He is a wiry and dynamic young man who clearly brought the high energy of his New York State background to his job as Greening Australia's restoration manager for the Link. In fifteen minutes, he zapped through a maze of detail on his research, showing how he has matched native plant communities to soil types on the cleared land. His brilliantly colored maps of the farms are like mandalas, fragmented and randomly redisplayed on kaleidoscopes. But they are based on sound science. A few days later, and shortly before my trip through the region with Bradby, I had had the opportunity to look at Jonson's maps on the ground, as it were, and see how well the brilliant colors matched reality. I did this by hitchhiking on a short field trip with an international group of researchers led by Richard Hobbs and Jim Harris to Peniup, the Link farm closest to the Fitzgerald River National Park. The first impression of an early restoration like this is of a certain artificiality. If the Gondwana Link is indeed one of the first examples in Australia of restoration at an agro-industrial scale, as some of its advocates claim, then it should be no surprise that the rather tidy rows of young trees and shrubs at Peniup should look like what they are—a plantation. Closer observation, however, yields clues that suggest that this is a plantation with a difference. The rows are not in serried military ranks, but weave a little drunkenly from left to right and back again as they extend toward the horizon. Nor is the gap between rows uniform. Focus in on the plants themselves, and you find shifting assemblages of species that few if any local farmers would have planted.

Swathes of local plants—from genera including *Eucalyptus, Acacia, Grevillea, Banksia, Hakea, Allocasuarina, Melaleuca,* and *Dryandra*—have been replanted on old wheatfields and pasture here over the last five years. The lurching rows and the sward of varying width that separates them are strategies that attempt to mimic the irregularity of natural distribution of seed, so that the rows will ultimately merge seamlessly as a "natural" bush community. Jonson and his colleagues have made on-the-job mechanical adaptations to industrial-scale seeding machinery to achieve these unconventional intervals between the plants. The plantings incorporate several ecological and economic strategies. Some are linked to carbon credits, one of the ways in which the Link hopes to finance itself in future. But they include a built-in biodiversity that most carbon farms eschew. When we are discussing this on the farm, Jim Harris adds the observation that the carbon stored by the root systems and by the microbial life which sustains native communities, may well exceed the above-ground carbon. The latter, however, is the only measure recognized in the still very tentative interface between science and business that is carbon budgeting. Other plantations will facilitate the commercial exploitation of native trees like sandalwood; but they are also experiments, aiming to show that a crop can be produced as efficiently—if indeed not more so—when the trees are interplanted with a range of other native species. These will remain as bush when the sandalwood is extracted. If these practices could be transferred to commercial farms in the area, the biodiversity benefits would be very significant, a cumulative benefit that both extends the work of the Link to private land, and makes its advances on its own ground more secure.

Jonson has certainly done his homework. His study of plant communities appropriate to myriad soil types on the farm has been supplemented by tapping into the memory and experience of the last man to farm this land. For the purposes of restoration he has identified 9 soil-specific seed mixes, including 120 of the 240 species he and his team have found as natives on the property. All the seed used has been sourced on the farm itself, which conveniently also includes some 900 hectares of intact plant communities, now managed by Bush Heritage. Despite his meticulous research, however, Jonson says he is not attempting to produce any precise replica of what was here before. "In this situation, I don't use [historical] reference ecosystems, per se. Since we are operating in a biodiversity hotspot, with so many different communities and so much local endemism, I don't believe we could reproduce them all, I just try to make sure I have a representative selection," he adds.[21] Representative enough, that is, for nature to complete the task of restoring these communities to something approaching their former richness. He has even allowed himself the liberty of creating a "novel ecosystem," a "light Yate eucalyptus community," which he has never actually encountered but believes will work well on the land where he has planted its putative component parts.

Ten days later in Fremantle, Hobbs singles out Jonson as an ideal restoration manager from the point of view of restoration ecologists. "What we try to do as scientists is get involved with the management of projects and ask them to do their management in a way that is amenable to research at the same time. Justin is one of the first managers we found who was willing to pause, stand back, take a breath and actually discuss the options. Most managers just want to get their bit done, they have a lump of funding and need to rush out and spend it quickly. And there are managers who don't want researchers, who think they know the answers anyway, and nine times out of ten maybe they do. But there is no learning in the process, because you can't actually poke them and say 'what did you do here, what made the difference'? They know in their heads but they can't articulate it or measure it, so that is the real challenge."

Hobbs is evidently very impressed by the work at Peniup, but he enters a caveat which reflects his controversial view, shared by Harris, that the best restorationists can hope for, in many of the ecological circumstances that obtain today, is restoring ecosystem function rather than the ecological communities that existed prior to disturbance.[22] "What you are looking at is not perfectly restoring an ecoystem but you are trying to restore some sort of functionality. . . . What Justin is doing is probably as good as it gets, but restoring an ecosystem in terms of its soil structure, et cetera . . . well, we are so far away from that, it could [only] be an aspirational goal. And yes, the time factor is important, there won't be meaningful results at Peniup for five, ten years." This of course means that monitoring will, as ever, be crucial to extracting the lessons that are the real value of experiments like Peniup. Bradby and the Link are committed to maintaining very high standards of postrestoration reporting. One must hope that the Link's precarious and unpredictable funding in the current difficult climate will not put this principle at risk. One instance of the instability of that funding, and perhaps inherent in the Link's "broad front" strategy, is the fact that Jonson's contract as Greening Australia's restoration manager lapsed shortly before the Perth conference and was not renewed. Bradby found funds to keep him on in the short term, and hopes to find a long-term solution in linking him to a research institution, but he does not hide his disappointment at the NGO's declining commitment to the Link at such a critical moment. In his view, the Australian conservation sector has still not risen fully to the exceptional but crucial challenges posed by the Link's vision. "I'm still nervous about the Fitz-Stirling area. For a number of years it was immensely fragile [in terms of organization]. We are three or four years behind where we should be, due to the cultures of the organizations involved."

I ask him if his main concern is the Link's failure to acquire more land—they need to triple the strategically situated acreage bought so far, if the dream of "connecting" the Stirling Range with the Fitzgerald River reserve is to materialize. He nods, but goes on to say that he sees the core problem "more in terms of the solidity of the groups who work in that area":

> The Fitz-Stirling is very *lightly* staffed, shall we say, more lightly than it was eighteen months ago. That may reflect a classic organizational dynamic: the first four years are the best, the honeymoon, with a head of steam. Possibly we moved too fast for some of the groups involved, they want to pull back and consolidate. Greening Australia felt overreached, and yet, once they got their teeth in, they wanted

> to extend such work nationally, and have maybe taken their eyes off work here. I would have built a bit more sequentially than that. Bush Heritage was not as enthusiastic as Greening Australia in the early years, but now it is bringing more resources to the table. But there is no question but that the Fitz-Stirling has stalled for a bit.
>
> Let's try and tease out numerous fronts of stuff here. One of the juggles we have is . . . You have got to help make sure things happen, but make sure the groups involved have the exposure and the territory to feel confident that it is their turf. And when you try to get them to do things collaboratively with others, that is a juggle in itself. There is a tendency to keep to the focused detailed work, rather than keeping the big vision open, but there is also a responsibility to keep working to the big goal, while giving the groups and the steps they are achieving along the way as high a profile as possible. We do find it a juggle. . . .
>
> Some of the people working here come to identify more with the Gondwana Link than with the organization that pays them. Their organizations sometimes get stroppy about that, but it is not something we do, it is something that happens. I suspect that our institutional structures are working against our geographic strengths. We have a lot to learn there. The strength of what we are doing is the geography to some extent, and yet we still work to these organizational silos. Part of the juggle," he continues, "is that the [NGOs] have not wanted the Gondwana Link to be an organization in itself. Some of that for was for very good reasons. But sometimes it has worked against us strongly. And now we are going to be a group, with a broad membership base made up of the groups working with us, and a board made up of serious players, an ongoing institutional basis within which people can come and go.

Bradby does not hide behind false modesty in recognizing that his role in the whole project has been essential to its success to date, though he stresses the huge contribution made by the Link's only other full-time employee, Amanda Keesing, and the groundwork done by many individuals and groups over many years before the Link was established. But he believes the time is approaching when he should move on: "I think that in three years, which will be ten for me in total, I will go. If it's not consolidated to such a point that someone with

20. **Moment of reflection.** Keith Bradby in the eucalyptus woods at Nowanup, midway between the Stirling Range and Fitzgerald River national parks, which features a project to restore indigenous Noongar people to their ancestral landscape. (Photograph courtesy of Mark Godfrey.)

different energies and perspective can take over, there is something wrong . . . I think it would be too identified with me by that stage." Bradby's remarkable frankness is one of the traits that draw some restoration ecologists toward the whole project. "Keith is good in that he discusses the bumps in the road, a lot of these things fail because people won't discuss them," says Hobbs. "The reason that Gondwana Link works is because Bradby is at the helm and he has a very quirky way of doing things which just happens to work, it is pragmatic and flexible, which is all the things which [government] agency things aren't, a lot of the other big schemes are government initiatives. He is good at working with other NGOs, but he does have big problems working with agencies . . . in the Fitz-Stirling area he has only recently developed a meaningful dialogue with the Department of Environment and Conservation."[23]

Kingsley Dixon is also a firm supporter of Bradby, but he says that this support is not universal among the scientists who originally supported the establishment of the Link. His account of their critique is a kind of template for the tensions between the theorists and the practitioners of restoration that occur on so many projects.[24]

A couple of very prominent people said to me—they must remain anonymous—that Keith is a disappointment because he gives lip service to the idea of science, but doesn't make it part of the actual equation. So mistakes are made, but mistakes should then be guiding the science to come up with the solutions. . . .

Keith's view is that he encounters university scientists, who are often the worst people because they live on a formula of grant funding, whatever the outcomes and outputs are, [that is] only about publications. In this institution [Kings Park and Botanic Garden], we have to focus on a very different paradigm. That's why I enjoy being here, although I am a university academic. We do the good science but all the science has to have an impact on the practice of restoration. The scientists need to engage back with the Keiths of the world, go back and say we'd like to work with you on some aspects. Like I'm talking to Keith about enabling techniques for broadacre multispecies plantings. They [on the Link] are doing a little bit, but we really need to get a lot more species back in, this is an area that we are very interested in because the mining companies are prepared to fund us to come up with solutions. So I want to see broad landscape impacts from that. But I still have to build a confidence bridge with him because if he goes and gets ten million dollars from a sponsor and I say to him you need to spend a million of that on research, he says to me, "What are my outputs and outcomes?" And I say, "Well, research is risky, that is why you do it, that is what is actually tax deductible, if it was a certainty you wouldn't be doing it." We are confident that we can minimize the risks to maximize the outcomes as much as possible.[25]

Bradby himself makes it clear that he values can-do organizational drive over scientific debate on the detail of restoration: "I'm a little bit impatient with a few processes, but I have noticed that the ice bergs are melting, so I think we should all be impatient.

"Do we have it right?

"Does it matter if we've got it wrong?

"Should we be doing all this restoration work where we are doing it?"

He lets the questions hang in the air for a moment, and then continues. "I argue a bit with a lot of the biologists who love prioritizing, because prioritizing assumes a case of money this big"—he sketches a small box with his hands—"and we have to work out how to spend it. In fact, part of what we are doing is saying, 'If the case is this big, let's make it a lot bigger.' Very little if any of the money that we have spent has been diverted from other stuff, it has become available because the Gondwana Link came into existence. I'm not sure about how we meet our ecological targets, will we change the ecological history of the world, et cetera. But I think we are significantly influencing the organizational and institutional cultures and the amount of bucks available for this work in Australia. That I think is of importance today." So you might say that Bradby is restoring funding to restoration.

The Link's wealth can be expressed in social as well as dollar terms. Bradby always gives credit to those farmers who were working for conservation in the region before environmental organizations were heard of. He talks warmly about Ken Newby, a farmer who botanized in his spare time, seemed to find a new species on every outing, and tripled the known flora of the area he lived in. A number of farmers are now working with the Link, putting in belts of native plants on unproductive land. This is part of the Reconnections project sponsored, rather controversially—and to the dismay of the some of the participant NGOs—by Shell. Between the Link's own farms and patches of private contribution to the effort, looking at a map running from the Stirling Range to the Fitzgerald River, there are still more gaps than connections here, but the fragile dream of the Link is taking definite shape.

An even more fragile restoration of the Noongar people is also forming tentatively on one of the purchased farms. Like many Australian environmental projects, the Gondwana Link is informed by a sharp awareness of the destruction of the millennial Aboriginal relationship with what they simply call *country* by white settlement and dispossession. At Nowanup Farm,[26] Simon Smale of Greening Australia introduces me to Eugene Eades, a charismatic Noongar elder and former boxing champion who has taken a few hard knocks in his life. Eades is inviting his community back to the land from which they were driven not too long ago, first to "reserves" and ultimately, all too often, to urban squalor. On these visits, they find Stone Age artifacts almost everywhere they walk. These are not from some remote prehistory: when a Noongar woman picks up an ax head or grinding stone, she knows that it was probably used by relatives only a few generations back. This is a reminder of the devastatingly accelerated cultural shocks to which Australia's Aborigines have been subjected through rapid colonization by an industrialized society.

An imaginative magistrate asked Eades to lead a pilot project rehabilitating Noongar teenagers who have fallen foul of the law. (Aboriginals make up a grossly disproportionate share of the prison population here.) He substitutes lessons in bushcraft and Noongar culture for detention cells. A sparsely furnished barn, with wind and dust howling around it on the day I visit, is their new home, but the four teenagers currently enrolled there seem to love it. It is festooned with vivid paintings and posters in Aboriginal style. One set illustrates the six seasons into which the Noongar divide the year. Others proclaim the slogans of Eades's program: "Healing the Land, Healing Our People's Spirits"; "Life's Messy but There Is a Way Through." The renewed pride and dignity in these boys' faces was touchingly evident when I visited, but they will face heavily stacked odds when they return to their "normal" lives.

Bradby is passionate about the need to help Aborigines reconnect to *country*, but he is unromantic about their current levels of ecological knowledge. "There is little traditional ecological knowledge left in Noongar culture," he says bluntly, "and little for us to learn from what there is. So why do we want to help them come back out here? Because we are decent people, and because this is their land, for God's sake." At one point on our return journey we spend a congenial hour with a group of white citizens, mainly women of a certain age, at Ravensthorpe. The group maintains a fine citizens' herbarium. They also run a very popular annual wildflower show, which is in progress when we visit, displaying great pride in the rich natural heritage on their doorsteps. Outside, a large plaster relief catches my eye. It is a deliberately naïve piece of art that portrays happy white citizens under the benevolent smile of an Aboriginal figure, situated at a convenient distance above the clouds. I suggest to Keith that this is a rather patronizing representation of the indigenous people. "You don't know the half of it," he tells me, and recounts the story of white settlement here, which occurred as recently as the second half of the nineteenth century. The newly arrived white men repeatedly raped Aboriginal women. Finally, in retaliation, one of the native men speared a white man to death. The response of the white community was brutally effective. They poisoned every water point within a hundred-mile radius of the settlement.

Earlier, driving from the northern Fitz-Stirling toward Kalgoorlie, we encounter the famous Rabbit-Proof Fence, familiar to many moviegoers from Philip Noyce's eponymous 2002 film on the cruel fate of the "stolen generation" of Aboriginal children. The fence is also a stark reminder that biosecurity measures can be costly failures: European rabbits swarmed through and under the fence long before its last post was hammered home in 1907. Today, ironically, it serves as a quite effective barrier against a native bird that is sometimes farmed and sometimes regarded as a pest by farmers. Huge numbers of emu try to migrate to the wheatbelt from the interior in times of drought. They pile up against the fence a dozen deep, apparently unaware that they could find gaps in it. They are shot as vermin by the thousands and left to rot. Bradby says the smell is appalling, but the country is so big and so empty that nobody complains about it.

The fence is still the boundary, visible from space, between the cleared agricultural land we have just driven through and a vast area so little known that it has no name on Australian maps. Nor have the local Aboriginal peoples ever given the region a single designation, though their cultural mapping named many of its constituent parts. For the moment, the Gondwana Link simply calls it the Great Western Woodlands. Rather bigger than Switzerland, it is the largest area of (more or less) intact semiarid woodland in the world. A dazzling variety of eucalypts, some 350 species, ranging in color from the carnal pinks of the bark of salmon gum to the argent foliage of silver gimlets, dominates the region. But there are also grasslands, scrublands, salt lakes, and granite outcrops.

The Gondwana Link is just beginning to operate here, mainly through the Australian Wilderness Society. The woodlands might appear entirely intact on a quick drive-through, but in fact they form part of Australia's legendary Goldfields region. Many mining companies operate here, and many more will in the future. Rather than trying to oppose them, the Link is seeking advance agreement on an overall biodiversity plan that will ensure the total protection of the most valuable areas, plus best practices during mining, including restoration afterward. The quid pro quo for the mines would be a fast-tracking of environmental permits outside the totally protected zones, where they are often blocked for long periods by environmental impact surveys at present. This is one of the Link's "bold and audacious strokes," a conservation strategy to manage this vast area as "one big chunk of bush" as Bradby puts it. But does the laudable desire to achieve that kind of momentum risk sacrificing some species to extinction including, given the endemism rates here, some species not yet known to science?

Bradby replies carefully. "That is not an accurate interpretation, exactly. There are already laws on species, which the mines have to observe. We can make the environmental costs of mining much more cost-effective to the mines by looking at the big picture: how does the whole landscape work? What are the ferals doing there? And it *is* sometimes the case that a mine is being held up because of a species on site which is thought to be rare, but which is actually quite common elsewhere. By not taking the traditional environmentalist stance of going head-to-toe with the mining companies we could be accused of selling out. In fact, just by talking to mining companies, you can cop that. But that to me is quite silly, just plain silly." Bradby uses expletives as freely as any Australian, but when he really wants to emphasize a point he uses none, he just speaks very quietly. "The mines are major forces in our landscape, and we should be building a constructive dialogue with them," he insists. So far, the companies are responding positively, and Bradby has driven up for an encounter with all the stakeholders, ranging from Aborigines to industrialists, in the region. We enjoy a remarkably amicable field trip and dinner together, and the next day they convene again for a private meeting.

While they gather, I drive out to the granite outcrop of Gnarlbine Rock, once a vital watering point for the Aborigines and then for white prospectors. Betty Logan, one of the Aboriginal representatives on the previous day's field trip, had talked to me about the significance of these massive rocks for indigenous people. They had been valued as precious sources of scarce water, which gathered in their hollows, and are still a favored place for family gatherings and picnics. There were also beliefs associated with them about which she was reluctant to speak. At one rock she held back from climbing with us because, she said, her father had died very recently and the place held too many memories. From the high point of a second rock, she pointed out a third one, where her mother's grandfather had brought

21. **Generation restoration.** Schoolchildren on a day of biodiversity and cultural education at Nowanup. They are learning to radio-track mallee fowl, whose chicks are very vulnerable to introduced foxes and cats. (Photograph courtesy of Amanda Keesing.)

the first white explorers. He had shown them, as a courtesy to strangers that he would rarely find reciprocated by the Europeans, where they could find water.

The rust-colored dome of the Gnarlbine Rock is ribbed like the carapace of a gigantic tortoise. It is stippled with small carpets of purple flowers. It barely rises a hundred meters, if that, from the surrounding flatlands. But its summit offers a breathtaking view over the bush, stretching out in every direction much further than the eye can see. It is no doubt presumptuous for any white person to imagine the Aboriginal Dreamtime. But it is tempting to imagine everything that lies beyond the woodlands to the west—the Fitzgerald River, a wheatbelt with a corridor of restored bush, the Stirling Range,—all the way to the magnificent karri forests near Margaret River. And to imagine them once again organically connected, linking the heart of Australia once again to the ocean. Maybe it's a dream, maybe it's a vision, but I can't help thinking that every seed planted, every square meter of bush protected along those thousand kilometers, brings it closer to fruition.

It's a seductive moment, but sooner or later one always has to come back down to Earth. In this case, to listen to a new debate about reconnection in the extraordinary landscapes that makes southwest Australia such a special place. Because, with the best will in the world, there are scientists who argue that reconnecting all this native vegetation should not be the focus of restoration at all in this context. Taken too literally, they say, it would run counter to the significant degree of natural fragmentation that is one of the very factors that makes the area so rich in endemics. Worse, it could provide vectors that would allow alien invasive plants and feral animals to penetrate and damage those areas that still remain more or less intact. To grasp this argument, we need to reconsider some concepts, like natural succession. Such concepts are helpful to understanding ecosystems in those parts of the world—North America and Europe—where ecology cut its scientific teeth, but they may tend to distort our vision of what is actually happening on the ground in this part of Gondwanaland. We need to go on a field trip with Kingsley Dixon and to listen to Steve Hopper.

Why An OCBIL Is Not Like A YODFEL: Contradicting Our Knowledge Of Life On Earth

I just think we should just ease off the pedal on the idea that we are going to create the grand corridor that connects one side of Australia to the other because while that might help a few organisms. . . . It could place at risk a considerable number of others. * STEPHEN HOPPER, *director of Kew Gardens, London (2006–12); former director of Kings Park and Botanic Garden, Perth*

There is a bigger gain, which we don't understand, but hopefully we can be part of understanding in future. This is the gene flow across a landscape. There are patterns across many plant species that show interactions between so-called isolated populations 100–150 kilometers away. We know that pollen moves back and forward where it can. Enabling that to happen again is an insurance policy against bigger change, such as is happening with our climate. * KEITH BRADBY, *coordinator of the Gondwana Link*

First impressions from abroad suggest to me that Australian approaches to restoration should be similar to those in North America. After all, this country has a broadly similar modern history. European settler occupation involving the dispossession and near-extermination of indigenous peoples has led to ever-intensifying agricultural and industrial development. This has been accompanied by rapid ecological degradation. First impressions, however, can be very wrong. Settlement history, dramatic though it is, remains only one small part of a country's story. Australian ecological history is so different from America's that restoration is following a very distinct trajectory Down Under. This is especially true of the continent's only biodiversity hotspot, on its southwestern tip, where the Gondwana Link cuts its ambitious arc.

During a field trip prior to the 2009 SER conference in Perth, Western Australia, the conference director, Kingsley Dixon, showed a group of visiting restorationists some of these contrasts in dramatic close-ups. Dixon's bush classes were as entertaining as they were erudite. "You will see many things here today which contradict your knowledge of life on earth," was his mantra every morning, and he was as good as his word. Distinguished ecologists from other lands felt they were back in botany 101. They struggled to identify plants that often, courtesy of convergent evolution, looked deceptively familiar at first sight, only to find that they belonged to genera, families, or even orders they had never encountered before. And then they found that each plant seemed to have four or five confusing cousins, and that these were only a stone's throw away.

We tiptoed through bush whose densely-packed biodiversity (up to 110 plant species per 100 square meters) richly deserved Dixon's sobriquets of "knee-high rain forests" and "coral reefs out of water." Even more striking, from an ecological point of view, were the astounding levels of plant endemism—49 percent of the region's species are found nowhere else on Earth. This high diversity and endemism can be partly explained by something that seems crazily counterintuitive to the layperson: the *infertility* of the soil, at least inland. This is because, as we saw earlier, this ancient landscape has not known great soil-enriching geological phenomena—such as glaciation or big tectonic shifts—for 250 million years. It's hard to get your head around this idea at first, but scientists here now believe that infertility can be a driver of diversity, because plant species will evolve many different and ingenious ways to source nourishment when it is very scarce. The exceptional numbers of species of fly-trapping sundews (*Drosera* spp.) that we found on every excursion was just one example of this. If you can't find nutrients in the soil, eat flying meat.

The most puzzling aspect of these ecosystems, and ultimately the most revealing, is the radical variation in species between apparently identical sites. The maximum similarity in species composition between plots only five hundred meters apart, and on the same soils and landforms, rarely exceeds 50 percent in the kwongan uplands we were visiting. Or, as Dixon put it, "many of our plant species like to live where mother lives."[27] For most of his guests, this ran counter to the norm, which is that plant seeds have high mobility, as they are borne considerable distances by wind, water or animal host. But, in southwest Australia, the apple really does not fall far from the tree; indeed it falls and sprouts right beside it. So seed dispersal can never be taken for granted in this landscape, and the implications for restoration science and practice here are radical. Without seed dispersal, natural succession may not be very natural here: it either happens excruciatingly slowly, or it does not happen at all.

These implications were spelled out at the Perth conference by Stephen Hopper.[28] In his plenary address he described a new paradigm for what he calls OCBILs. You might think that this is a newly discovered species of Australian marsupial, but no, it is an acronym he has coined for "old, climatically buffered, infertile landscapes." He pointed out that ancient places, geologically speaking, like the Southwest Australian Floristic Region, South Africa's Greater Cape Region,[29] and Venezuela's Pantepui Highlands have "special biological traits, combined with unparalleled species richness and a lottery-style recruitment and community assembly [that] render restoration efforts complex, slow, uncertain, costly, or impossible."[30]

Those of us from the northern hemisphere (and from much of the southern one too) learned from Hopper that we live in YODFELs—young, often-disturbed fertile landscapes. No one ever thought restoration was easy in any context. But it is certainly more straightforward where you can rely on seed dispersal and natural succession to do a lot of the heavy lifting once abiotic structural restoration and selective planting have kick-started the process. In southwest Australia, spontaneous restoration is conspicuous by its absence.

Speaking to me two months later in his office in Kew, beside the Thames, Hopper can look out his window at a landscape in which "there are places which are dead easy to restore, wetland margins and coastal foredunes and so on. If you get the biophysical stuff right, and allow enough space, then regeneration will just happen. Or take the classic case, the farm paddock in the UK, ringed by oak trees and seedlings. If you stop the disturbance [farming], the seedlings are in there the next year, in five years you've got a copse and in ten years you've got a woodland." Restorationists in the United Kingdom might find this reading of their situation just a little rosy, but they would certainly recognize the contrast with what Hopper describes in his home region: "I was encountering abandoned farmland eighty or ninety years old with just one or two eucalypts and then an acacia, in a sea of [alien invasive] weeds."[31]

But while he saw that natural succession was not happening on old fields, and that invasive aliens were literally making hay in these habitats, he discovered that even very small patches of native vegetation—up to particular points—were exceptionally resilient. As a young researcher documenting rare plants, he found that "very localized places including road verges five meters across, on laterite, had relatively intact, weed-free [native] vegetation."[32] Farm paddocks, rigorously cleared of native plants decades before, ran right up against these marginal strips, which had been subjected to the "usual uses and abuses of road verges. Yet they looked like they would if they were in a nature reserve." However, he also noticed that a single major disturbance event could wipe out these precious remnants. "On [open] sandy ground, with a mix of native plants and weeds, after just one storm covered a road verge in sand, the native species were gone overnight." Like his colleagues in the Midwestern United States, he found that such relict plant communities could also be found in cemeteries: "At Gingin Cemetry, a sand plain on a hill, the shire had been slashing and burning the native plant growth every two years or so for decades. You would have expected that, in a sandy environment, this would have flipped the place into weed city. But it didn't, there was just a slow incursion of veld grass. But in the mid 1990s the shire decided they would . . . level out the soil. Next year the native plants were all gone, and it was wall-to-wall weeds. It just struck me that something is going on with this thin layer of soil, that if you don't disturb it with bulldozers or something like that, [it supports] very resilient communities of plants."

In a nutshell, Hopper's theory says that native plants in an OCBIL landscape will resist invasive aliens much better than natives in a YODFEL landscape,[33] but that they will fall over and die if the fragile soil in which they sit is disturbed by digging, or if they are smothered in sand. What's more, unlike YODFEL natives, they have little or no capacity to recover spontaneously after such disturbances. Given that many of the plant communities in the southwest Australian landscape are naturally fragmented, and that so many plants here are endemic to very small areas, the prospects for restoring anything remotely approaching pre-settlement biodiversity to the huge cleared areas within the Gondwana Link seem precarious indeed in the light of his theory.

Taking this argument further, Hopper questions the very idea of "connectivity" that underpins much of the thinking behind the Gondwana Link project. He asks whether the currently dominant view that "corridors" are always a "good thing" in conservation is really useful in restoring an OCBIL landscape. As if that were not enough, he argues that such corridors as might be established could make things worse than they are now: "What OCBIL theory challenges is the notion that the naturally fragmented parts of the landscape will equally benefit from creating corridors. So if you have in the Fitz-Stirling area very subdued ridgelines of granite outcrops or laterite soils, then the notion that a strip of vegetation from one ridgeline across the valley floor and up to the other ridgeline is going to be beneficial to the ridgeline species is challengeable. In fact you could argue it is creating portals for weed invasion, and for foxes and other feral animals to move across. Most significantly in southwest Australia, you could argue that it is a portal for dieback to reach those ridgelines where they are really packed with those species and families that are most susceptible to the disease."

Despite articulating this grim scenario, Hopper is much more optimistic about the Link project than you might expect—largely because he believes he has seen its leading figures learning on their feet over the last few years, in a classic case of adaptive management, interpreting the original concept of a Link corridor in much more nuanced ways. "So I would argue that—and I see the people in Gondwana Link respond in this way—you get as many of the ridgelines as possible cared for and managed in the ways that they have evolved: that is, as relatively disjunctive and isolated parts of the landscape. How can we then target our conservation corridors along the routes they would once have taken? Where the Link corridor idea is spot on is where it creates corridors in those parts of the landscape where wild vegetation was continuous on soil types—along rivers, for example."

He insists however, that at least from a botanical point of view, the core of restoration on the Gondwana Link should be preservation first, and then the very slow, painstaking—but still very approximate—replication of the preserved areas on appropriate soils on degraded land. "I would say the most significant thing that Gondwana Link has done is buy up the remaining patches of wild vegetation. Those patches are just irreplaceable. They will be the nuclei of future restoration." He recognizes that where the Link is actively restoring native plants on farmland, the organizers have begun to appreciate the scale of the problems the OCBIL theory poses for them. Meticulously detailed soil mapping prior to planting vegetation appropriate to each soil type, as Jonson has done at Peniup Farm, will yield dividends, though it will be a long and often painful and testing investment. But while Hopper thinks the Gondwana Link is in good hands, he is concerned that its current very high profile in Australian restoration is generating too many copycat projects where the science has not been so rigorous. "I just think we should just ease off the pedal on the idea that we are going to create the grand corridor that connects one side of Australia to the other. While that might help a few organisms . . . it could place at risk a considerable number of others. I think a better story is that in creating this green belt we are in fact caring for the incredibly complex and diverse vegetation types that still remain. We are working at the right scale to try and put back into this very complex soil environment the right species for the place and the landscape."

But since the soil on Western Australia farmland, though often only cleared for as little as fifty years, has been both repeatedly disturbed and radically altered by fertilizer, what chance is there for such sensitively soil-specific plant communities to make a really significant comeback on the wide open spaces of degraded and salinated farms? Can relatively small-scale and very complex work like that being done at Peniup be replicated on a sufficient scale to represent true restoration across the wheatbelt without draining scarce resources from more feasible projects?

"That's the big debate now. Do you cut your losses and accept that there are some parts of the landscape where the geochemical processes are so badly disturbed that it is not worth even contemplating doing it? My argument would be that this is the big debate that restoration ecology invites the whole world to think about. What I would say, is that *if* it has been decided in some places—and the Gondwana Link is leading this—that people have already taken the decision that we want to try to reverse this degradation and put the native biodiversity back, it's bloody hard, but we *can* get the resources and we *will* learn as we go along, and we accept that this is going to take a very long time, it's going to involve local custodians over decades to deliver the outcome. I remain quietly optimistic that that's perfectly within the remit of people to do, with the right science, with the right combination of skills. I accept it seems a huge and daunting task. At present, restoration ecology is such a young science, and the practitioners of ecological restoration likewise are learning."

The Link project, Hopper concludes, is a good demonstration that "where the initial assumptions are wrong, where science has shown that they were wrong, the process now is different from what it was when they started. So if you have that degree of humility—which means to me, 'let's invest in science,' which is about challenging the status quo in testing ideas—you can find the best way forward."

Bradby shares Hopper's concern that grandiose proposals for links across the whole continent are not founded on good science and may be unhelpful to effective conservation. But he believes that much of Hopper's critique of the Link itself is based on misconceptions about its original intentions, and he also believes that Hopper puts undue stress on the extent—and implications—of the natural fragmentation of plant communities. "The vision has not changed since we started in 2002," he says bluntly, "and it was always about a lot more than a 'GL corridor.' Perhaps Steve's impressions or understanding of what our initial assumptions were has changed?" As for the fragmentation of plant communities, "this is at least a tad debateable. Even though the communities are on separate parts of the soil mosaic . . .

at the very basic levels the pollen moves across the landscape between patches." He also believes that Hopper's intense botanical focus distorts his overall vision of what is happening on the Gondwana Link patch: "The southwest is made up of more than plant communities; it also has bird, mammal, reptile, and invertebrate communities too, and they get around and have sex all over the place." He resists any implication that the Link is just about "buying and replanting. We are also doing our damnedest to introduce good bush management, which includes fox control, dieback quarantine." He believes these measures can counter the dangers that Hopper sees in the reconnection of disjunct patches.

But the Gondwana Link literature does put great stress on a vision of connectivity. So what kind of corridor is the Gondwana Link? Does Bradby see it as similar to the corridor concept as used in conservation in the northern hemisphere? His answer is nuanced, even ambiguous, which may not be a bad thing, given the complexity of the issues involved.

> Yes. It is that kind of corridor, *sort of*. Not that many or any species are going to go the whole length of it. It is not [like] elk going from inland forest to the tundra on an annual migration, because we don't have that. But, to persist, our species do need to travel around for food, and to have sex outside their immediate [family] relations. Twenty to 30 years ago, people were doing heaps of papers and research demonstrating that bush in small isolated fragments loses species. So here we are, trying to join things up again. But the Link is more about bringing wholeness and health back.
>
> Without question Fitz-Stirling is a good example. We know that landscape reasonably well now. We know that there are wallabies and other mammals and bird populations on the remnants out there that will be extinct within thirty to forty years unless the bits are joined up again. The tammars [a wallaby] of the Stirlings and the tammars of the Fitzgerald need to cohabit, and will be able to if we are successful. It is not impossible that the quokkas of the Stirling range, which are a precarious little population, will wander eastward. And of course every bit of bush you save, every bit you connect, every bit you buffer in this landscape is valuable in itself, given our local rates of endemism.

He concludes by returning to one of his favorite themes, *putting Darwin back in the landscape*:

> But there is a bigger gain which we don't understand, but hopefully we can be part of understanding in future. This is the geneflow across a landscape. There are patterns across many plant species that show interactions between so-called isolated populations 100–150 km away. We know that pollen moves back and forward where it can. Enabling that to happen again is an insurance policy against bigger change, such as is happening with our climate. Corridor is such a limiting word, though. This is the sense in which [Gondwana Link] is *not* a corridor. "Corridor" implies that species will scuttle back and forth along a protected corridor *within a hostile landscape*. I think what we are starting to achieve in the Fitz-Stirling, and certainly can do in the Great Western Woodland, is to show that the landscape does not need to be hostile to wildlife. We can have a landscape that birds and mammals and plants can move across, and where you can also have your farming and your living and so on.
>
> We have gone through a very totalitarian farming model—ecological apartheid, to which conservation and national parks have also contributed. So that on one side of the fence you have this "tip-toeing" sort of zone, and on the other side it is "burn at will." I try not to use the word balance too often, but in that mad rush to clear the land our farmers did move the boundary much too far toward the production line. Ours is not a landscape like Europe's where wildlife moves reasonably freely through agricultural areas.

The landscape, reflecting the relationship between agriculture and environment in this unique part of the world, he concludes, "needs to be designed much better."[34]

I'm giving Bradby the last word here, not because I am sure that he is right,[35] but because of the depth and breadth of his experience of this unique landscape. Time may prove him wrong, as it has before, in many details, and perhaps even in the big pictures. If it does, though, there is reason to believe that lessons will be learned and applied, again and again. However, so much institutional memory should not be delegated to one person in the long term. The scientific debates about how to restore the Gondwana Link are vital, significant, and lively. But shoring up the vulnerability of its structures must be its most urgent priority in the immediate future. Like every other major human enterprise, ecological restoration ultimately depends on individuals, organization, and social contexts.

10 *Restoration on a Grand Scale: Finding a Home for 350,000 Species*

The question is not whether a tropical forest can be restored, but rather whether society will allow it to occur. * DANIEL JANZEN

The wild is at humanity's mercy. Humanity now owns life on Earth. . . . Part of the problem is in the name. Stop labeling the wild as the wild. There are simply many varieties of gardens. There is no footprint-free world. . . . Restoration is fencing, planting, fertilizing, tilling, and weeding the wildland garden. * DANIEL JANZEN

What restoration needs above anything is permission of one sort or another by society to do it. We can worry about doing it "better" later. * DANIEL JANZEN AND WINNIE HALLWACHS

* *

Commonplaces about conservation whirl through the brilliant mind of Daniel Janzen and emerge, transmuted, into challenging paradoxes. These paradoxes manifest themselves repeatedly in the bold, large-scale restoration experiments that he has undertaken in what has become the Área de Conservación Guanacaste (ÁCG) in northwestern Costa Rica since the mid-1980s.[1]

Or perhaps the paradoxes spin out of the equally brilliant mind of his wife, Winnie Hallwachs. Maybe Janzen simply expresses her insights in the pithy, provocative, no-bullshit sentences that are the hallmark of his essays and his conversations. He consistently uses the first person plural in talking about these experiments, and his mantra about their relationship, surely one of the great partnerships of contemporary science, is, "She thinks, I talk."[2] Either way—or, more likely, both ways—his analysis is always bracing, often very insightful, and occasionally irascible; the manifestations on the ground are inspiring to some, disturbing and controversial to others.

Today the ÁCG stretches from the Pacific Ocean through dry forest, rain forest, and cloud forest up to the peaks of several volcanoes and pushes down their eastern (Caribbean) flanks. It is a measure of the couple's extraordinary contribution to the expansion of the original Parque Nacional Santa Rosa (fifteenfold in twenty-five years) that a guest at his dry forest home should ask him, as if he could wave a magic wand, why the ÁCG does not extend all the way across the country to the Caribbean Sea. "That's another hundred and twenty kilometers and would cost as much as the entire expansion of the ÁCG has cost to date, $53 million. I don't have it in me to do that again," he replied flatly.[3] In fairness to the guest, it is the only time I have heard the man say there was something he could not do.

But let's go back to his analysis: the restored secondary tropical dry forest in the Santa Rosa sector of the ÁCG looks very impressive. Its canopy already approaches the height of those few remnants of the old-growth forest that were spared clearance for the cattle ranching that has dominated Guanacaste's agriculture for many decades, and clearances for hardwood extraction (especially mahogany) for centuries before that. Janzen claims that visiting biologists entering this world for the first time are often convinced they are looking at old-growth forest, when the land they are standing on was degraded pasture only eighty—or even just forty—years earlier.[4]

That's understandable. The new forest's biodiversity is already exceptionally rich by almost any standards except, of course, the local ones. (This small corner of a small country contains more species—350,000 according to Janzen's latest ballpark estimate—than that of the United States and Canada combined, in a space smaller than Rhode Island.[5]) But if you walk through a restored area that is contiguous with one of the old-growth remnants and then cross that boundary, the impact of the transition is palpable. The first thing you notice changing is the temperature, which, in the dry season heat, drops a most agreeable five degrees almost as soon as you pass from one zone to the other. It may be—and probably is—an illusion, but it gives a feeling that you are breathing purer air. The temptation to call this elder environment "pristine" is powerful. The next thing you notice is visual, the signs of great age. The trees may be not much taller than their counterparts in the restoring forest but, solidly established and comfortable citizens that they are, their girth is often three times that of the stripling youths in the surrounding areas. Others, of course, are dead, their slowly decomposing trunks enriching the system in ways that the restored sectors will not enjoy fully for many, many years. And there are other signs that this part of the forest is not the product of any recent restoration.[6] The trees themselves will mostly be semideciduous, shedding their leaves for such short periods that they are often described as "evergreen," whereas the secondary forest is largely dominated by dry-season deciduous pioneer species. The understory in the secondary forest has become the midstory here; the epiphytes, bromeliads, and mosses that thrive on the newer growth have largely vanished.

How reassuring and appropriate, you might think: here is the "historical reference system" forest side by side with the restored one, and no doubt the latter will approximate to the condition of the former in five hundred years' time, give or take a few decades. That is Janzen's ballpark estimate for the period by which the quality of the restored forest will, as he puts it, "fool the alert biologist into being viewed as "undisturbed" by European-style agriculture."[7] The alert reader will note that Janzen does not rate the alertness of many biologists who visit the park today.

All well and good. But Janzen is the first to point out that the old-growth remnants, these apparently stable repositories of mature biological wealth, will undoubtedly be changed, and not necessarily for the better, in the short- and medium term. And the agent of change will be, ironically, the rapid progress of the restoration projects that surround them: "As successional changes occur through forest restoration, the changes can be both startling and distressing," he writes. "This is especially true if the restoration is part of a conservation movement intended to save familiar nature."

These old-growth patches are very valuable seed sources for adjoining restored sectors, he continues, but as time goes on this relationship becomes reciprocal, and that changes the older system into something else. It happens like this: "In the early stages of landscape-level restoration, the small patches of remaining semideciduous old-growth forest are also sites of concentration for vertebrates in the dry season. They offer shade, moist soil in dry spells, nesting trees and tree holes, and distinctive species of seeds and fruits. This in turn generates massive animal-dispersed seed flow from deciduous species in the ocean of surrounding secondary succession into the more evergreen forest. This in turn alters the species composition and competition regime of natural succession in tree falls . . . which in its turn alters the nature of the old-growth forest."

And he continues, though surely not without irony: "This has led to the suggestion that they may do better surrounded by rice fields than by oceans of secondary succession (as in so-called "buffer zones"), if they are to retain their old growth structure."[8]

Hence the paradox: if you are moved to restore a landscape to the condition of a remnant ecological "jewel," you may significantly increase biodiversity even in the short term; but you must accept that, in your lifetime and perhaps for several more human generations, the jewel that inspired your project in the first place will itself shift into another condition, indeed, into several other conditions over time. There will be a dialectical dance between historical reference system and restored area, producing new syntheses in each. In short, some of the very characteristics that made you find the jewel attractive, a model for restoration, were produced precisely by the isolation imposed on it by deforestation.

"Red in Tooth and Claw"? *C'est la vie.*

Janzen takes a ruthlessly unsentimental view of *Homo sapiens*, and of our role in a Darwinian nature, forever "red in tooth and claw." Every creature is always beating up on or duking it out with some other creature in the ecosystems he studies.[9] He is not one of those conservationists who spend precious time lamenting humanity's record of creating degradation and causing extinctions. Rather, he takes it for granted: in his view the human genome is irrepressibly urged to control the whole planet. "Obviously the human animal is hard-wired to harvest what is within grasp and push aside that which interferes," he and Hallwachs responded brusquely to my e-mail query on the subject. "What animal is not?" Did this not, I had followed up, perhaps somewhat naïvely, amount to an acceptance of the biblical injunction that we rightly have dominion over all living things? "I do not accept or reject it. It simply is, like you have two eyes," he replied, adding one of his favorite phrases, "*C'est la vie*."[10]

If this is indeed the way we are, his argument continues, then the golden road to conserving biodiversity lies in a major paradigm shift on our part. Conservationists must stop presenting nature as something separate to be preserved in splendid isolation for its own sake. Instead, the human genome needs to absorb and designate what he calls "wildland nature" as an essential part of its own needs; otherwise we will inevitably destroy it. In a vividly written 1998 polemic for *Science*, Janzen argues paradoxically but cogently that only by relabeling wild nature as a garden can we save the wildness of species and systems:

> Part of the problem is in the name. Stop labeling the wild as the wild. There are simply many varieties of gardens. There is no footprint-free world. Every block of the world's wildlands is already severely impacted. Not only are they internally impacted through macro-events such as the megafaunal extinctions and selective extraction of old-growth timber, but the very frameworks of their existence—global warming, acid rain, drained wetlands, green revolutions, wildland shrinkage, introduced pests, and many more—are set by Homo sapiens. The question is not whether we must manage nature, but rather how shall we manage it—by accident, haphazardly, or with the calculated goal of its survival forever?[11]

And yet, in a further double paradox, there is little doubt that the Janzen-Hallwachs wildland garden will be one with minimal human management, though with, in the very special sense in which he uses the term, maximum human presence, rigorously controlled. This presence will include ecotourism, education and research, but he also stresses a less tangible but crucial development of a sense of a social ownership.

So if you are going to meet Janzen and Hallwachs, or even simply read them, you might want to fasten your intellectual seatbelt, because there is turbulence ahead, though it will be an exhilarating ride. Hallwachs is soft-spoken and sweetly courteous, gently reconducting Janzen's wilder declarations onto the rails of rational discourse when she deems it necessary. But Janzen has a sharp tongue in his head, and does not suffer fools with the remotest semblance of gladness—unless the fool might be willing to part with her money or lend his influence to support his project. Bill Allen, in his meticulously researched and very readable account of the first two decades of the ÁCG enterprise, *Green Phoenix,* sums up our Janzen as "at once an admirable, crusading genius and an abrasive bully."[12]

That phrase, however, was descriptive of the man during the 1980s and 1990s. Janzen was then negotiating the rapid expansion of the ÁCG through an obstacle course that included the Nicaraguan Contra rebels and their US sponsors,[13] Costa Rican bureaucracy and nationalistic resentment of *gringos,* the diverse and contradictory agendas of international environmental NGOs, and a host of contentious local and regional issues. He had plenty of excuses for being tetchy and impatient. The park, expanded from Santa Rosa's 10,800 hectares to some 163,000 today, is now consolidated,[14] and while Janzen continues to be embroiled in controversies and to seek more land for the ÁCG, things are calmer there now than they were in the last century. Perhaps his senior years have also mellowed him; certainly several of his Costa Rican ÁCG colleagues commented to me, as one might of a favorite but irritable uncle, on a certain sweetening of his temper these days. His response to my introductory e-mail, as always cosigned by Hallwachs, had been generous; he robustly and copiously engaged with my questions while making it abundantly clear that he had no time to waste "feathering the academic conservation nest."[15] I hastened to dissociate myself from university research. The fact that I was writing for a general readership may have been decisive in their giving me hours of interview time, social time, and a daylong field trip out of their notoriously hyperbusy schedule.

Restoration or Conservation?

I was puzzled, however, by the way he balked at my use of the word "restoration" in the same e-mail. I had used it in relation to his work with forty thousand hectares of tropical dry forest in the ÁCG, generally regarded as one of the biggest and most audacious restoration projects in the world. Moreover, his numerous essays on Guanacaste use the word repeatedly, and sometimes in their titles. But here is his bald rebuttal, stitched into my e-mail after my first reference to the R-word:

> I DON'T DO WORK ON RESTORATION. I DO WHATEVER SEEMS NECESSARY ON THE SHORT AND LONG TERM (BACK BOARD CHESS) TO MAXIMIZE THE SURVIVAL OF THIS LARGE BLOCK OF TROPICAL BIODIVERSITY FOR THE NEXT 1000 YEARS. (IF THAT LONG, MEANS SOMETHING SORT OF PERMANENT IS IN PLACE). IT IS PERHAPS AS MUCH AS 2.5% PERCENT OF THE WORLD'S BIODIVERSITY, WORTH DOING. AND A REALISTIC TARGET. ALL THE REST IS BYPRODUCT, SPIN OFF.

So I approached our first meeting with some qualms as I took the short walk from the Santa Rosa sector headquarters to the unconventional home that he and Hallwachs share for the annual semester when he is not teaching at the University of Pennsylvania. It is a house without doors and, it sometimes seems, without walls, open to the creatures with whom, he says repeatedly, they feel more kinship than with most of their own species. Bats hang from the rafters. Agoutis scavenge around the edges. Snakes and scorpions and tarantulas come and go at will, along with his beloved moths. Janzen has listed one of his life's goals as the completion of the inventory of the estimated fifteen-thousand-plus moth and butterfly species found in the ÁCG. He has completed basic work on upwards of five thousand of them, with the help of his unique corps of locally recruited and trained "parataxonomists."[16] Despite his inexorably advancing years—he was approaching his seventy-first birthday when I interviewed him in early 2010—he is not conceding defeat.

He appears to be one of those people who regard a tidy desk as the sign of warped mind. Manila files, external hard drives, and myriad transparent plastic bags containing all sorts of biological specimens are strewn in apparent chaos around the house. In the midst of all this, he sits at his computer, totally focused, lean and muscular body stripped to the waist, long silver hair flowing like an Old Testament prophet's. He hammers out another paragraph while I hover in anticipation. Hallwachs glides out from the shadows to say hello. Her Quakerish dress and diffident manner mask a powerful personality—a very

22. **At home in nature, nature at home.** Winnie Hallwachs with Espinita, a prehensile-tailed porcupine she and Dan Janzen adopted and reared after they found it orphaned in 1995. The couple also keeps open house for bats, a boa constrictor, and tarantulas, none of them adopted, in their home in the restored dry tropical forest of the Área de Conservación Guanacaste in Costa Rica (Photograph courtesy of Erick Greene.)

useful gambit, they both tell me later, for blending smoothly into the gender-conservative Costa Rican establishment.

A friend has advised me to bring the gift of a large watermelon. I almost discounted his counsel, because it seems like bringing snow to Eskimos in this land of endless fruit stalls. But it is received by both of them with pleasure and laughter, and becomes the ceremonial centerpiece for a sunset picnic deep in the park a couple of days later. And then, without further ado, we are down to brass tacks:

What's the problem, I ask, with the word "restoration"?

"You don't set out to *restore*," he tells me. "You set out saying, 'I'm going to *conserve* the things that still live here, keep them alive and on the table.' And in the act of doing so, because you are no longer burning or cutting or shooting or hunting, then they themselves go to work and reconstruct some sort of wild area, which maybe, if there were no climate change, a thousand years ago from now might look pretty similar to what was here before the Europeans arrived. Now you can label that 'restoration,' and I probably have, on many occasions in the past, because whoever was paying the bill, whether a government or private donor, wanted to label it that way, because it was the fashionable catchword that came on . . . but the goal was conservation in the first place."

I still find it hard to pin down the precise source of Janzen's reluctance to accept the restoration label today, at least as a subset of conservation.[17] But it may have its roots in two related aspects of his long experience in Costa Rica. One is the hostility to restoration he experienced from conservationists and conservation funders in the 1980s; the other is the resilience and recuperative capacity of dry tropical forest, the ecosystem that has been, at least until quite recently, the focus of his work, and which, in the special conditions of Santa Rosa, made restoration seem like little more than natural succession. Both aspects shed some light on key restoration issues.

Janzen experienced the once-widespread hostility to restoration when he made an early funding pitch at a conference in Washington in the mid-1980s: "In all innocence, I went in and said that we want to buy all this trashed forest and nearly dead ecosystem, and we can grow it back. I expected general agreement and happiness, and I got dead silence and a lot of glum-looking people. I had no idea what I had stepped into. I was at right angles to their mantra. The mantra of organizations like The Nature Conservancy at the time was: 'what is cut down is lost and gone forever, so you had better give us the money to preserve it now.' And here is Janzen, this famous tropical ecologist, saying, 'Oh no, you can grow it back.'"

After a year of intense discussion, Janzen persuaded The Nature Conservancy that the public could deal with a more complex message, which included both strategies, but the experience may have made him wary of the restoration label. Additionally, his formative experience of Guanacaste was very different from that of most restorationists, because his priority target, bringing back dry tropical forest on degraded pasture, does not involve many of the features typical of restoration in many other contexts. Dry tropical forest[18] is the poor relation of tropical rain forest, not only in the popular imagination but also in research and conservation priorities. Perhaps this is because it was easier to clear for pasture and therefore had almost completely disappeared before the emerging environmental movement could focus on it. This means that the forty thousand hectares restored in the ÁCG represent a significant proportion of the remaining dry tropical forest in the world, and most of what remains—in a healthy condition—in Costa Rica. It may also have been neglected by biologists because it is (slightly) less diverse than rain forest, but by comparison with temperate zones, dry tropical forest remains a biological treasure trove.

If Janzen's early focus had been on rain forest, he might never have become a conservationist, because rain forest, though obviously under great pressure today, was not likely to disappear in his lifetime. But as soon as he began to produce a series of acclaimed papers on the drier ecosystem's species and their interrelationships in the 1960s, he saw almost immediately that it was shrinking at an exceptional rate. He can remember when almost unbroken tropical dry forest ran from north of the Mexico–Guatemala border all the way to the Panama Canal, and nowhere in more profusion than in Costa Rica. At that stage, he lived in a typical academic silo, divorced from conservationists by the catchphrase: "I studied it. They saved it."[19] The walls of that silo started to collapse when he found that half the trees he had chosen for a multiannual study along a hundred-mile stretch of road had vanished in a single year. What he has described as the "Africanization of Central America"—deforestation for pasture, plus the introduction of highly invasive African grasses to support beef production—was sweeping the region. He realized that if he were to have anything left to study he would have to work exclusively in conserved areas. He moved the focus of his studies to the newly established Parque Nacional Santa Rosa in 1972, though at the time it was mostly degraded pasture, with only occasional islands of remnant forest.[20] He still had no intention of becoming a conservation crusader. "It was a purely selfish move to protect myself," he says.[21]

He quickly observed that supposed conservation measures in this then small park, especially the removal of cattle from abandoned pastures, did not assist the restoration of such forest as was left. Quite the contrary. Without the presence of these "biotic mowing machines," the African grasses, especially jaragua, quickly accumulated unprecedented fuel loads. This resulted in far hotter fires that did much more damage to the remaining forest than the old rancher-set burns on grazed pastures had ever done. Once again, the object of his studies was vanishing before his eyes—and well-intentioned "preservation"—called conservation—was actually speeding up the process. And yet, over the next thirty years, that seemingly unstoppable tide of degradation would be rolled back as the forest was restored throughout the greatly expanded ÁCG. It is important not to portray Janzen and Hallwachs as the "onlie begetter[s]"[22] of this heroic reversal of ecological fortune, though the international media have often done so. They make no such claim themselves. The first plans for both expansion and restoration in the Santa Rosa sector came from Costa Rican park officials.[23] And an incident as late as 1984 reveals that Janzen was still reluctant to get with the conservation program. When he received the very prestigious Crafoord Prize for Biosciences in Sweden that year, a journalist asked him "what he was doing to protect tropical forests."

"Great Data, Pile of Ashes"

Remarkably, this apparently predictable question seems to have surprised and even shaken him. He still retreated behind the "I study/they save" excuse. He replied "I don't do that, but I know people who do, and if you raise money for it, I will get it to them." The Swedish reporter did just that. From then on, Janzen began to think about becoming actively involved in conservation. He donated half of the prize money to bringing telecommunications and power lines into Santa Rosa, and the other half to buying the first land for expansion. Within a year it would be the two Americans who most clearly articulated the ideas behind the park's recovery to an international audience. It was Janzen in particular who then took on, with extraordinary aplomb, the roles of international fundraiser and regional wheeler-dealer in the protracted and complex negotiations to buy up tens of thousands of hectares and secure the park's position as the nascent ÁCG. On top of this, he took on a lead role as innovator and fixer regarding the endless conservation challenges that arise in its management. Today, he likes to describe his role in the park as "coach and cheerleader."[24] While Janzen's academic output remains impressive by any standards, he is caustic today about those scientists who still think they can focus exclusively on their studies, and leave the saving to someone else:

"[An] AWFUL lot of conservation biology is taking the temperature of the house while it burns. Great data. Pile of ashes," he wrote to the restoration ecologist and RNC activist James Aronson in 2009.[25]

At the heart of the recovery of Guanacaste's dry tropical forest lay an unremitting focus, by all concerned, on a simple and very convenient ecological fact about the ecosystem as it is in Santa Rosa: if you can stop the fires, tropical dry forest will very rapidly reinvade degraded pasture, shade out the grasses, and reestablish the threatened system. And this indeed may be one of the reasons Janzen resists the word "restoration," which to him involves lots of active human management and manipulation, "such as planting trees and trying to arrange species to generate a particular end-product."[26] What happens with the recovery of Costa Rican tropical dry forest, and even rain forest, he has insisted again and again, is simply natural succession playing out its age-old drama, just as it does after a volcanic explosion, a hurricane, or a landslide. However, experience has shown that the rapid regeneration at Santa Rosa is not necessarily the rule for the dry tropical forest system everywhere. As is so often the case, soil is a key factor, and the volcanic substrates in Santa Rosa transform quite easily from grassy hardpans to rich loam if fire is eliminated. But spontaneous soil restoration on the serpentine rock base in the neighboring Santa Elena sector has proved to be a much slower process, with consequent long delays in forest restoration.[27] In either case, though, Janzen's main point stands: "The key management practice was to stop the assault—fire, hunting, logging, farming—and let the biota re-invade the ÁCG."[28] (Rain forests in the ÁCG require a much more complex approach).

Janzen credits two experiences, both in 1985, with forming a core element of his and Hallwachs's new perspective on conservation in Costa Rica. The first was the result of a request from the director of the then–national parks system, Álvaro Ugalde, a former director of Santa Rosa. He asked Janzen to write an environmental impact statement on the damage being done in Corcovado National Park by an "invasion" of 1,500 gold miners. In a lightbulb moment, Janzen switched his focus, only days after arrival there, from biology to sociology.

Why, he asked himself, did the miners feel they had a right to minerals on national park land, and what would persuade them to leave? He talked to them and learned that they believed that the land was "empty" and "unoccupied"; unlike a cornfield or a pasture, it was not being put to any obvious human use. The point stuck in his mind: it is human presence, not human absence, which is critical to the long-term success of a park: "Ownership needs to be psychologically and sociologically visible if a conserved wildland to remain conserved."[29] Other facts were impressed on him during this research: all the food for park staff was flown in from the capital. This meant that local merchants did not benefit from having a conservation area on their doorstep, but the miners kept their cash registers ringing. "The park was being subsidized from the state, and this made the park guards even more outsiders. They were never in the general store, shooting the breeze with everyone. They never met the miners except as the enemy. There was this incredible disconnect between the park as entity and the community around it."[30]

Finally, his amicable debates with the miners reminded him of something he knew already. Costa Rica has a very different culture to other Central American countries. It has a long democratic and quasi-pacifist tradition—it has no national army. "A Mexican or Guatemalan miner would have either shot at me or run away in these circumstances," he says.[31] But all Costa Ricans have a bit of the lawyer in them, and dialogue comes naturally. That had significant implications for pushing a radical conservation agenda.

Later the same year, he and Hallwachs were invited by the Australian government to advise on "how to create an Australian presence in [an] enormous expanse of tropical dry forest."[32] They advocated a mix of science and ecotourism, research, conservation, low-yield long-term forestry, and watershed management, managed by resident Australians. And then, on the flight home, they realized that they had never asked themselves how they could create a similar Costa Rican presence in Santa Rosa. They began to answer that question before they got off the plane.

Creating a Human Presence in the Park

"In the first two weeks of September 1985, Winnie and I generated an unsolicited strategic plan for the long term survival of Santa Rosa's dry forest through creating for it the psychological and sociological presence of owners, the 'owners' being at once both its direct custodians and society near and far."[33] So, if the key management practice was to "stop the assault," the key sociological practice was "to gain social acceptability for the project locally, nationally and internationally."[34] Their ideas, though often heterodox, found sufficient resonance with local and national conservation managers to ride out the ever-present accusations that they were either crazy *gringos*, environmental imperialists, or both.

Creating a sense of ownership of the park took many forms. Park staff began to be locally recruited, contrary to Costa Rican (and US) practice. As the park expanded onto adjacent failing and former ranches, their *sabaneros* (cowboys) were often hired to manage for conservation land that they had grown up managing for livestock. That transition was eased by the short-term reintroduction of cattle, to reduce fire risk by grazing, until the reinvasion of the forest had shaded out the jaragua. The focus on fire control, directed by a dedicated team but with emergency duties shared by the entire staff, including all office workers, forged a new corporate solidarity. Upward mobility was encouraged; Róger Blanco, the current and very high-powered director of research at the ÁCG, started as a park guard with horse, uniform, rifle, and little more. There has been a big emphasis on engaging all the schoolchildren of the region, through myriad innovative education programs. Janzen has called this "biocultural restoration."[35] And he and Hallwachs have also recruited dozens of local people as the aforementioned "parataxonomists" who assist on the massive inventory of the ÁCG's plants, moths and butterflies, and on other similar programs, without formal scientific training.

Two of Janzen's anecdotes give a flavor of his intimate, hands-on, approach. After signing off on the buyout of a group of sixteen squatter families, on land he was purchasing for the park, he recalls that he "woke up in the night and *boom*! I realized I was paying them too little. They say squatters get their land for free, but I know they had paid with their backbone labor. I realized they could not buy a replacement for what we were taking from them." He reconvened the meeting the next day and told them that he was increasing the offer. Jaws dropped; hearts opened. On another occasion he was with Julio Quiros, a federal forest warden on loan to the ÁCG when they heard hunting dogs within the park's recently expanded boundaries: "And we go to get them. But I stop, and say to Julio, 'We are not going to arrest this guy; we are going to sit on a log with him. And you are going to tell him: "Inside the park no hunting. Outside the park, do what you like. We are not going to enforce the law outside the park."' This was totally against Julio's training. We found the guy; he was the bottom of the world, clothes in tatters, five snarly dogs, beat-up rifle. Julio did what I said." Months later, Janzen was addressing a village meeting in the same sector about the benefits of conserving biodiversity by not hunting in the park. The response was less than enthusiastic, until a voice from the back of the hall shouted: "He's right! I believe him." The speaker was, of course, the erstwhile poacher turned, if not gamekeeper, at least defender of wildlife within the park. After intense discussion, the meeting swung in Janzen's favor.

It was probably no accident, then, that when I applied for permission to research restoration in the park, the itinerary I was given prioritized a number of field trips and interviews with Costa Rican staff before an appointment with Janzen and Hallwachs was agreed on. I was being vetted, quite rightly, by the locals.

My first port of call is the Horizontes Forest Experiment Station. The material conditions here come as a shock, a salutary reminder that Costa Rica remains a very poor country. National parks are grossly underfunded, contrary to the impression given by so many articles in glossy magazines, which paint pictures of a tropical eco-paradise. Costa Rica is indeed an exceptional country in many ways. Its lively democracy has escaped much of the history of dictatorships, human rights abuses, and civil wars that have so deeply scarred its Central American neighbors. Its approach to conservation—and to ecotourism—is often courageous, radical and innovative, putting many more developed countries to shame. But conservation is still grimly constrained by lack of resources and poor infrastructure. And radical and innovative ideas often have to struggle to survive here.

In any case, the sleeping quarters at Horizontes are bare concrete bunkhouses. The bathrooms have no hot water. I am advised to keep my door shut for "biological security," as the patio outside is favored by snakes, spiders, and scorpions at night. It is kindly meant advice, but then I find that my door does not shut properly. Sure enough, I find a tarantula near the patio on my first evening stroll. And when I make a midnight exit to the toilets, a small serpent with field marks worryingly similar to those of the venomous coral snake falls off the top of my door. I gingerly tap it away along the concrete with the tip of my flashlight. When I return, I find the same snake, or one very like it, now *inside* my room. It had come through a hole in the masonry that had escaped my attention earlier. I shake out my bedclothes with more care than usual but otherwise decided to entrust my biological security to fate. Somehow, I sleep better than I often do at home.

Breakfast the next morning is rice and beans. So is lunch. So is dinner. Same each day, every day, but sometimes a little egg or chicken would be added.[36] I think at first that this austerity was due to the fact that Horizontes was a remote station within the ÁCG. But when I go to Santa Rosa headquarters, conditions are similar. I am given a room next to a very senior park official. I have more space than at Horizontes, but the furnishings remain, shall we say, radically minimal. There is no hot water, and I have a whiplash scorpion for night company. Clearly, no one works in Costa Rican national parks for the perks.[37]

If these material circumstances at Horizontes are tougher than I had expected, the human hospitality could not have been warmer. The commitment of my hosts, Milena Gutiérrez, Felix Carmona, and Ronald Castro, to the work of the ÁCG was passionate, and their local knowledge encyclopedic. Sector Horizontes is in the lowland dry tropical forest zone, just south of, and very narrowly connected to, the original park nucleus of Sector Santa Rosa. It had long been exploited for cattle ranching. On my first morning, Ronald Castro leads me past immobile thick-knees standing sentry near their nests outside the station office to a watchtower for fires. I follow him up vertical ladders to see the lay of the land. "When we came here about twenty years ago, you could see every rock to the edge of the station," he shouts above the wind from the Pacific. Now a variety of plantations carpet most of the area, ranging from mahogany to spiky pochote. The station is not being restored to dry forest at this stage, but is being used to develop different models of agroforestry to maximize biodiversity as a public service to Costa Rica. Castro talks about "escaping from traditional thinking." Some of the plantations shock conventional foresters: monocultures are generally avoided, and there are unusual mixtures of tree species, all natives. Castro and his colleagues monitor the impact of these plantings on speed of growth and timber quality, and they are registering positive results so far. The aim is to export these ideas to commercial plantations, thus fostering biodiversity enrichment in forestry on a national scale. It is all part of the grand plan to make the ÁCG "visible" to broader sectors of Costa Rican society, to embed its activities in the economy. The point is, you might say, to ensure that what happens in Guanacaste does *not* stay in Guanacaste. Meanwhile, the pasture grasses are shaded out under their canopies, and the limited understory that the plantation structures permit extends the habitat for many of the animals in neighboring Santa Rosa.

Later, while walking me through one of these plantations, Castro and Carmona comment on the spread of *chán*, a native shrub that assists the shading out of the alien pasture grasses. They explain that its seeds are also greatly valued for flavoring a soft drink of the same name. It is not produced on a commercial scale but is popular with local people in summer, and is a source of income as a cottage industry. They are concerned that the plant's ecologically welcome expansion in the park is being limited by illegal harvesting. Just as they are saying this, a dozen local people of all ages pass us by on bicycles, carrying empty sacks. The staff knows them as harvesters, and the harvesters know that they are so known, but the encounter is amicable. The exuberant and idiosyncratically Costa Rican greeting, *"Pura Vida!"* (literally, "pure life" but variously translatable as "hello," "all's well," or even "thank you"), rings out across the open woodland.

How to Nurse a Rain Forest

Before daybreak one morning, Gutiérrez and Carmona drive me out of Horizontes, inland across some still-functioning ranches, and north along the Pan-American Highway. Then we turn off and head into another section of the ÁCG, on the steep slopes of the actively volcanic Rincón de la Vieja. Within a few kilometers we move up from the dry forest system to rain forest, and by the time we reach the isolated San Cristóbal Station in Sector San Cristóbal, the dawning light has dimmed again beneath a drenching mist. They take me across a muddy field to a barn that serves as a classroom and formally introduce me to a restoration project very different from the park's work along the coast.

While Janzen and his colleagues had initially focused on restoring the dry forest on Santa Rosa's pastures, they knew that the rain forest above them was not only enormously valuable in itself; it was a vital annual refuge for many migratory dry forest species. It became a target for purchase and expansion by the park. You cannot successfully restore the biodiversity of one system without restoring the biodiversity of its neighbor; as ever, scale and defragmentation are crucial to success. More recently, as mitigating climate change shot up the conservation agenda, they also realized that restoring a contiguous gradation between both systems would also allow the boundaries of the ecosystems themselves to move under shifting conditions with, hopefully, minimal loss of species.

However, much of the rain forest had also been cleared for pasture. And when the park expanded to include these zones, they soon found that this was a system that would not restore itself rapidly and spontaneously on degraded land, even when adjacent to healthy rain forest. "Stop the assault and let the biota reinvade" would not work here, or rather, it would work much too slowly to be practicable. The reasons for this are complex, but include the adjacent rain forest's shortage of mycorrhizal fungi spores, and the paucity of wind-dispersed seeds. Even those that do succeed in germinating in pasture grow very slowly in the open landscape. Add in the reluctance of rain forest vertebrates, which disperse most of its seeds, to move through the cleared pastures, and it becomes evident that rain forest will be a long time spreading of its own accord.[38] Janzen came up with a characteristically unorthodox proposal to solve the problem. He proposed using a fast-growing alien tree species, *Gmelina arborea*, to shade out the grasses. He argued that once the grasses had died out and a canopy had formed, the wind and the vertebrates would do the rest. Typically, he came up with a further plan to extract the gmelina for commercial use when it matured, though this has faltered for lack of a market.[39] The eyebrows of more than one conservationist rose sharply at the idea of using an alien tree, especially one notorious for creating monocultural deserts in plantations across the tropics, as a nurse tree for restoration. But Janzen got his way, and then he left Carmona and Gutiérrez to get on with the job. All being well, this restoration will eventually not only fill in the vast gaps in the rain forests on Rincón; even more ambitiously, it is beginning to recreate an unbroken biological relationship between these forests and those on its sister volcano, Cacao, currently divorced ecologically by a broad swathe of old pasture.

The budget for the gmelina project, raised from the Wege Foundation of Grand Rapids, Michigan, falls far short of financing its planting throughout all the degraded areas. So Carmona and Gutiérrez have been experimenting with planting gmelina in "islands" and "strips" of various sizes in the hope that the rain forest will reach such a critical mass in the zone that spontaneous regeneration will indeed begin to occur, or that, at the very least, the patches of regenerated forest will be close enough to each other to facilitate the free movement of animals—and therefore seeds—between them. They have also improvised by building the islands or strips around "nucleus" rain forest trees still standing in the pasture, to accelerate the recovery. As though to underline Janzen's point that all restoration is place-specific, they found out that fire, their old enemy from the dry forest, could be their ally here. They had been having great problems plowing through the grass "mattress" to plant the gmelina. Given their training, they were reluctant to use fire, but an unintended blaze showed them that it was easier to control small fires in this climate zone. "We found the use of fire by accident," says Carmona. "It was like magic, a very valuable mistake."[40]

The results they show me after ten short years are most impressive. With the extraordinary growth characteristic of the tropics, rain forest species have spurted up among the areas cleared by the gmelina and are now often outcompeting it in the canopy. We try to make our way through an island to reach one of the well-labeled nucleus trees, clutching battered photographs from 1999 that show them standing isolated in grassland. Again and again, Carmona has to use his machete to hack through dense and varied rain forest growth. Muscular green shoots burst vigorously through brown seed husks in the mud wherever an open patch remains within the islands and strips. The fantasy that the forest is growing before my eyes, like the accelerated images in David Attenborough's *Secret Life of Plants*, is seductive.

Very conveniently, the mature gmelina themselves are already beginning to die off naturally in some plots. Their spectral shapes could give the impression of a die-off due to acid rain or chemical pollution, but this is the natural mortality of a fast-growing, short-lived tree. You may well be wondering, though, whether the gmelina is not regenerating, and thus compromising the native integrity of the rain forest. Most conveniently again, although they have produced copious seed in these restoration plantations, the gmelina seedlings do not prosper without direct sunlight—the opposite of most of the rain forest species nearby. So the new understory is entirely free of a new generation of the species that had helped bring it to birth.

Understandably boosted in confidence by the success of their project, Carmona and Gutiérrez have come up with an innovation of their own. As casual conversations throughout the day reveal, they are very aware of the complex politics and financing of the ÁCG. Not only have the original funds for gmelina been exhausted, but falling interest rates on the ÁCG's endowment—its establishment was one of Janzen's key innovations—mean that things are tighter all round. So they now propose using all or any of four native rain forest trees, which they can easily produce without external costs, as nurse plants for new islands and strips in the corridor. Janzen is totally opposed to the plan, insisting that these trees have been tested experimentally for this purpose, and been found wanting. And he has made his opinion known in no uncertain terms. He tells me later that the idea is "baloney." My guides are completely unfazed by the opposition of the man they always respectfully refer to as *el doctor*. "I know he has his doubts about our plan," says Carmona. "But this is rain forest, and *el doctor* does not know the rain forest very well." Whether he likes it or not—and he probably does—one of the world's leading ecologists clearly has a fight on his hands. The disagreement, passionate on both sides but without rancor on either, suggests that, whoever is right in this instance, the policy of putting the future of the park in the hands of local people is bearing fruit.

"*El doctor* always says you learn through doing, and that this park is one big classroom," he continues. "We can differ with him. He knows we have our own ideas. We are Costa Ricans. There is no boss here."

"Our work is our boss," chips in Gutiérrez, "and I love it."

Carmona's confidence is no doubt related to the unconventional way he was recruited into the park service, though he is without formal biological training. He was headhunted for the park, he says, because management were looking for local people with leadership skills. He had organized the youth of his nearby village into a successful football team, and that mobilizing talent was recognized as the right quality for a park job. He jumped at the opportunity and has never looked back. With his green fatigues, mustache, mass of curly dark hair, and commanding physical presence, he bears more than a passing resemblance to the Hollywood stereotype of a 1960s Latin American revolutionary. But his fight is a broader one, rooted in his own experience and the complex challenges of this century.

"I was a peasant, born in the dry tropical forest. I loved the trees and had always wanted to know more about them," he says. "The park management gave me that opportunity. Then they let me move to the rain forest because I wanted to learn more, and I am delighted with the work. I feel totally at ease with it, there is joy in it. It's hard, not easy—you can spend a week without sleep fighting a fire, always in the frontline. Or waiting up all night to catch a hunter who never comes. A professor of botany from San José could not stand the conditions here. But I will carry on. When people talk about climate change, I can say, 'Well, I have *done* something about that.'"

Such declarations come more easily in Spanish, a language in which people have generally yet to become shy of sounding "worthy." And what lies behind them is a quite exceptional degree of commitment. I have had the privilege of meeting many people dedicated to very demanding work for the environment while researching this book. But in no organization have I felt that commitment so universally expressed in practice as at the ÁCG. Carmona's local origins and human qualities are clearly factors in the success of the project in Rincón-Cacao. There was much opposition to it in the town of Dos Ríos, which stands right in the middle of the proposed biological corridor. Its people had grown up with open spaces and ranching, and they did not welcome the return of the jungle to their doorsteps. To many of them, that meant domestic visits from poisonous snakes and insects, even threats (largely imaginary) to their children from jaguars. Many of the men, used to supplementing their family diets by hunting, did not welcome the arrival of park guards, with new conservation laws. Things got better when the ÁCG gave the town six hectares on which to build the town's first high school, a creative way of spending conservation money. But I suspect that it was the almost daily interactions with Carmona that won more people over. Extrapolating from his skills as a football coach, he talks about how he involved the town's teenagers directly in the project. And, when there was a medical emergency, he used an ÁCG Toyota to race people to hospital. As for the hunters, he asked them to show him seed sources in the forest, and they found that this unconventional *guardaparques* could be a good companion. When we visit a local shop for food, he is greeted like the neighbor he has become.

Changing Paradigms: Pyromaniacs Turn Firefighters

The next day, I drive up to Santa Rosa to meet Róger Blanco, director of research at the ÁCG and a close associate of Janzen. He is a coiled spring of energy, quietly determined, tough and wiry, yet the kind of man who somehow remains scrupulously neat while hiking through thick forest. He has a dramatic story to tell me about a most unconventional restoration project further north. But first he gives me the obligatory tour of the Parcela del Príncipe. On one side of the road, there is a dense field of jaragua a couple of meters high; on the other side stands an equivalent area of thriving dry tropical forest. The only difference between the two is that the jaragua is deliberately burned every year, while the forest zone has been kept strictly fire-free, and regenerated, entirely of its own accord, on old pasture exactly similar to the jaragua field. The contrast in vegetation could hardly be greater. This is a textbook demonstration of the success of the "stop the assault and let the biota reinvade" strategy for this ecosystem at Santa Rosa. It is strategically situated for propagating this achievement to the public, as it flanks the main entry route to the park.

Blanco moves on briskly back onto the Pan-American Highway, and as we head north the ÁCG is now on both sides of the road. As I had discovered to my cost, this grandly named continental thoroughfare is actually, when you are lucky, a two-lane blacktop pitted with potholes and clogged with clapped-out trucks. There are plans to widen it, and this poses a big problem for the free and safe movement of animals within the ÁCG. Animal behavior experts generally reckon that a two-lane road is no great obstacle: every creature from frogs to jaguars regards such a road as a new kind of dry riverbed. There will be casualties, but they will try to cross it, and they will mostly succeed. Make that road a four-lane dual carriageway, however, and the obstruction becomes critical, changing behaviors, distorting migration patterns and populations, limiting genetic interchanges. Blanco and other ÁCG staff have persuaded the authorities to retain the ÁCG's stretch of road as a two-laner, a calmer and more relaxing interlude on long journeys. They will even allow the returning forest to touch canopies over the road at times, forming a green tunnel, reminding travelers they are in a very special place.

As we travel, Blanco bombards me with observations about the park, moving seamlessly from its ecology to its politics. Apropos, he makes an interesting distinction, in Spanish, between *ilegal, legal, and alegal,* one that will deepen in resonance as our field trip progresses. He says that most civil servants—and by implication, most employees in other national parks—operate strictly on the basis of fulfilling what is *legal,* what is required of them, and no more. While he firmly defends the legality of the ÁCG operation, he acknowledges its maverick reputation, and describes it as being *alegal* when circumstances demand flexibility and innovation: "We don't just do the minimum the law requires, we do what is not required as well."[41]

Fire and its prevention are the main theme again as we turn east off the road, for a meeting with the park's firefighting team at a station known, like the ranch that was there before it, as Pocosol. I thought this meant "little sun," and wondered at the irony, since the ranch buildings are exposed and blisteringly hot. But I learned that it means "dimwit," as in "not much lightbulb," in Costa Rican Spanish, though no one now remembers why the ranch got this odd name.

Julio Díaz is the veteran leader of the fire squad, which he describes as the *punto de lanza,* the sharp edge of park operations. For twenty-five years, the squad has been the vanguard unit in the battle to restore the dry forest. They work every day of the year to keep fire out of the jaragua, and out of the still fire-vulnerable regenerating forest. The painfully slow regeneration on the steep slopes and poor soils of the Santa Elena peninsula means that the period of vulnerability is greatly extended there. "And on a hot day, when there has been no rain, a malicious look can start it blazing," he says wryly.[42] When the grass ignites, the great problem is getting his team into remote areas quickly, and then keeping them supplied, sometimes over a period of several days. Good maintenance of forest roads is critical to his work, both for access and as firebreaks. He lays out the tools of his trade on the ground for me. Most of them are low-tech—axes, spades, the inevitable machetes, and lots of straw brooms, which, soaked in water, are used to beat back grass fires. "Fire was normal in ranch management, in our culture here," he says. "So people know how to put out fires because they know how to start them. We *Guanacastecos* love fires, we are all pyromaniacs at heart," he says, concluding his short presentation, and everybody laughs. Then he adds that the idea of full-time firefighters was a culture shock, and at first, "our yellow uniforms were definitely seen as the badge of clowns. But now they command respect."

Díaz has a major secondary role in training all park staff as auxiliary firefighters, and also in raising awareness of the dangers of setting fires among the general population. Yet again, the park policy is not to make any blanket prohibition, which would be anathema to local people, but to teach people how to limit their fires when they need to burn pasture. Janzen later summarizes the spirit of the policy: "We still have fires here every year. If you applied my anal north German organizational skills in a hardcore way, we could knock all fires out entirely. But it would disrupt the sociology here, disrupt what other people are doing, so badly, that the price would be much higher, more damaging to the ÁCG than being a little tolerant. The same applies to every aspect of our policies."

As we were leaving Pocosol, a surreal image caught my eye in the undergrowth near the station buildings. Ten miles from the sea, but apparently beached in a forest clearing, stands a very fancy motor boat, the kind that can accommodate a dozen sports fishing enthusiasts in considerable comfort. Ah, explains Blanco wryly, that was a gift to the park from the new Four Seasons Resort on the nearby Peninsula Papagayo. The hotel group wanted to make a contribution to patrolling the no-take zone in the park's newly established marine reserve.[43] However, their gift consumes so much gas that the park cannot afford to use it. And because it is a gift to what remains, in the last analysis, a state institution, it cannot be sold for cash revenue. There are some bureaucratic knots even the freethinking management of the ÁCG cannot untie. And we were about to encounter another.

The Great Orange Pulp Restoration Project

Blanco's next destination is the parking lot of a fruit-processing factory, at first sight an unlikely venue for a presentation on the park's most radical restoration project to date. Our route there takes us off the highway, and inland around the gentle lower slopes of Orosí, the northernmost of the park's trio of volcanoes. This is still a dry forest zone, though once those slopes steepen, it shades into rain forest, as usual invisible under its blanket of mist and cloud. There is an agroscape again alongside the road, but here it is not pasture but citrus plantation. Tightly packed rows of bottle-green orange trees flicker by as we drive eastwards. This territory, bordering the ÁCG from the north, was bought by the fruit juice company Del Oro in 1992.[44] It is in the shadow of its gleaming installations that Blanco gives me his version of the park's biggest recent drama, a daring attempt to marry big business interests and restoration initiatives. It is a tale of short-lived triumph followed by bitter intrigue and heart-breaking, though hopefully instructive, failure.

Initially, the park regarded the arrival of its new neighbors with mixed feelings. On the one hand, a monocultural plantation was never going to accelerate flows of biodiversity in and out of this part of the ÁCG. But Del Oro did offer a degree of security and predictability for its boundaries that had not been the case with the patchwork of ranchers and squatters that the company and the ÁCG had just bought out. Besides, the company's purchase included a tongue of rain forest stretching up into ÁCG territory that Janzen had long been trying to incorporate in the park. As was his custom, he made social contact with the newcomers and offered the managing director, Norman Warren, a straight swap: a piece of the ÁCG's degraded pasture, more conveniently situated for orange production and already cleared, in exchange for the tongue of forest. Over one of several amicable dinners to discuss mutual interests, the Del Oro executive commented that he was delighted to find that his trees were exceptionally free of pests. "Ah," riposted Janzen, never one to miss a trick, "that is because there is so much biodiversity in our park. Our birds and bugs are eating your pests. We are supplying you with ecosystem services. What can you give us in return?"

On another occasion, Warren confided to Janzen that he had a big problem. He could use most of an orange, extracting juice from the fruit, and essences for soap and perfume from the peel. But he was still left with thousands of tons of pulp to dispose of every year. He had offered to sell it as fertilizer to local farmers, but they weren't even nibbling at the deal. "And then," says Janzen, "the lightbulb went off. And you want an idea? That's an idea. They have a problem, how do you get rid of all these orange peels? Hey, I got 350,000 creatures, some of them are going to want to eat orange pulp."[45]

So they decided to do an experiment, dumping the pulp and spreading it on the most degraded pasture to be found nearby in the ÁCG. As a restoration method, this was a pretty wild shot, but then Janzen insists that he did not see it as restoration. He saw it as helping a neighbor solve a problem, so that he would help the park in return: his modus operandi whether he was dealing with squatters or industrialists. But the pulp turned out to have biological properties that no one expected. "The first hundred truckloads were a true scientist's experiment," he told me later. "I did not have a clue what was going to happen, other than that I bet that something was going to come and eat the pulp. We put it on the shittiest piece of real estate you can imagine, the crappiest degraded pasture, and waited to see what would happen."[46]

What happened next was a revelation: "The pulp killed all the grass, drowned the roots, which was gorgeous. And it created this great black loam soil. But it had another very striking effect. The broadleaf seed bank was sufficiently resistant to this one year of anaerobic conditions that it then germinated, and you had one seed here and one seed there . . . the most gorgeous seed crops you could imagine, it seeded the whole place. Then the cecropias went nuts afterwards." These extraordinary results led to what promised to be a landmark restoration deal between Del Oro and the ÁCG in 1998. The park agreed to absorb a thousand truckloads of pulp annually over twenty years and provide Del Oro with other specified ecosystem services like high-quality water and bug control. In exchange, Del Oro would give the park the fourteen hundred hectares of quality rain forest that Janzen had been coveting for the ÁCG for many years.[47] The agreement, signed in the presence of the Costa Rican president, gained a high media profile because it was one of the first in which a corporation agreed explicitly to pay for ecosystem services. It looked like a win-win-win moment: Del Oro solved its pulp problem while its already high reputation as a "green" producer soared; Costa Rica got another big international credit as a pioneer in environmental protection; and the ÁCG got more prime biodiverse territory, plus a rapid accelerator for the restoration of its degraded lands.

Now, twelve years later, Blanco and Gutiérrez walk me up to "Plot No. 2," the site where the thousand truckloads of pulp had been dumped in the first year of the agreement, handily within ten kilometers of the Del Oro factory. The plot is bursting with vegetation. Cecropias tower ten meters above our heads, and the ground cover is sumptuous. Blanco hunkers down and grubs out a fistful of earth: the pungent smell of humus rises. Across the path, a control plot of similar size tells its own story. Forest plants are still struggling to rise above the level of the jaragua. As long as there is no fire, they will eventually win, now assisted by the seed rain from the plot next door. But the process will be slow and the species-richness will probably continue to be sparse compared to the experimental area. And on the control site Blanco's proffered soil looks gritty and exhausted, with no discernible smell. The progress of the second pulp-mulched plot had borne out and exceeded the hopes raised by the first experiment. Starting as a jaragua field with a paltry twenty-three other plant species, within eighteen months there was no jaragua left and eighty-one other species; three years later the plant species count had reached a hundred and twenty-three. Animals were thriving as well, with twenty pairs of thick-knees moving in to nest in the early stages as the jaragua thinned and before the canopy began to close. Janzen calculated that the pulp-on-pasture technique could reduce the timescale for bringing back mature dry tropical forest by seventy years.[48]

Janzen began to develop a grand strategy for the extensive area of degraded pasture still dominating this northern sector of the park. "What was coming up [in the pulp-treated plots] was not burnable. This was a very fire-prone area, but on Plot No. 2 I'll give you a pack of matches but you are not going to be able to burn it up. We were going to build a fire lane, chunk by chunk, a marching stipe that would cut in half the burnable area."[49] Hold on a moment, I want to say, because this was very early days for this project. Might it not have been prudent to test the outcomes over longer periods on small plots before committing to spreading pulp over hundreds or thousands of hectares? No sooner do I think this than Hallwachs cuts in, sounding a note of caution, insisting that drainage would have had to be carefully controlled and monitored. In the early stages and middle stages of its decomposition, the pulp fairly seethes with bacterial life, and one would want to be sure that no toxins were entering the general water supply. Janzen dismisses this danger as exaggerated by their opponents, but Blanco concedes that they had not taken into account the potential impacts of events like Hurricane Mitch in the fall of 1998, and that this had led to some problems with water down the catchment from the site.

In any case, in terms of biodiversity, one would surely want to know more about the differences between restored tropical dry forest accelerated by pulp fertilization, and the tried-and-tested spontaneous restoration method, before applying it wholesale in the park?[50] Shouldn't its impacts be tested with multiple replicated experiments, with control patches?

Well, yes, those are points I would like to have discussed with Blanco, Janzen, and Hallwachs in a lot more detail. Unfortunately, a stormy media and judicial campaign against the project has made them moot for the time being. Thanks to an action taken by TicoFrut, a Costa Rican orange producer not enamored by the arrival of Del Oro on their turf, the first consignment of pulp delivered after the agreement was also the last. TicoFrut accused their rivals of polluting a national park. They gained the support of Alexander Bonilla, a high-profile Costa Rican environmentalist.[51] "We were naïve about the world of business," laments Blanco. And it does seem likely that the ÁCG had inadvertently fallen victim to commercial rivalry. But what underlies the supreme court judgment in favor of TicoFrut's suit is a literal interpretation of the law. This favors the traditional concept of a national park—a place where nature is left strictly to its own devices—against the proactive conservation innovations fostered by the ÁCG's management. Spreading orange pulp is clearly a significant alteration in the ecology of the park, and that is against Costa Rican law. Blanco recalls one of his superiors from San José saying to him disapprovingly: "You did what did not correspond to you." He had, in other words, acted in a manner that he considered *alegal,* but that the courts, and some of his senior colleagues, regarded as *ilegal.*

He draws some comfort from the fact that one of the judges commented that, while the law as it stood had to be upheld, history might judge the court as having being as blinkered as the theologians who had tried Galileo. And he hopes that a new law that would mandate that the parks foster biodiversity may create an opening for the legalization of this and similar innovative initiatives. Sadly, even if it does—and Janzen says he is very skeptical about this avenue of appeal—then orange pulp may not be easy to come by. Forced to withdraw from the deal, Del Oro has since built a very expensive pulp processing plant and has succeeded, after all, in selling the refined output to farmers as fertilizer. And of course, as Blanco ruefully observes, the company continues to benefit from the ecosystem services the park supplies to it, but that the company no longer pays for in any shape or form. This point reveals a key weakness in the argument for payment for such services, at least under our dominant market economy paradigms. They do not flow out of a faucet that can simply be turned off—and of course, even if they did, conservationists would be acting against their own interests by refusing to supply them. Unless payment for ecosystem goods and services becomes mandatory, conservationists have no hard bargaining power.

For Janzen, the episode ultimately simply confirms his conviction that the ÁCG must work unceasingly to insert its own ethos into Costa Rican social mores. "This project became a very revealing political controversy," he has written. "It exposed as-yet-to-be-resolved weaknesses in the ÁCG's societal underpinning. . . . [I]t is clear that a centralized, biodiversity-naïve and ecosystem-naïve urban national process has not yet come to be comfortable with a conservation area conducting its own management decisions in accordance with the needs of its wildlands; especially when those decisions smack of facts or ideas unfamiliar with whatever classical environmental awareness the urban center carries."[52]

In our interview, he brought the whole story down a much more homely image: "As my mother always said, if you are going to run stop signs, you have got to look in both directions." Or, as Bob Dylan once put it, "To live outside the law you must be honest."[53] The maverick approach of the Janzen-Hallwachs tandem, coupled with what they call the "outcast" culture of the ÁCG management—equivalent to Blanco's *alegales*—requires extraordinary integrity, blended with the instinctive fancy footwork of the ace jaywalker.

The exceptional set of circumstances that have brought the ÁCG team together cannot be exported wholesale as a template to other restoration projects. As Janzen warned in our interview, "a one-shirt-fits-all model does not exist in conservation." But the perspectives and principles that inform and support ÁCG operations are remarkably fertile and flexible, and could well germinate green shoots in very disparate soils. To begin to grasp them, there is no substitute for reading Janzen's articles, and following their muscular arguments. In the hope of encouraging you to do so, I have taken the liberty of attempting to boil them down to some of their essential points on key restoration topics, with occasional interjections from my conversations with Janzen and Hallwachs and from their informal communications with other scientists, interspersed with some comments that I hope will contextualize and link these remarkable ideas.[54]

Preserving the Mother Fishes

We should not leave the ÁCG, however, without a very brief trip to the seaside, in the company of children. Cuajiniquil is next to a small coastal fragment of the park, north of the Santa Elena peninsula. It recently became the home of María Marta Chavarría, a veteran ÁCG biologist and member of the first group of parataxonomists set up in the 1980s. I meet her in her small home, which is tucked in among fishermen's dwellings and has a deck that sits directly above the sea with sweeping views of a serene bay lined with mangroves. A manta

23. **Bio-literate.** María Fernández, a schoolgirl who participates in the Área de Conservación Guanacaste's bio-cultural restoration program, shows off fish species she has found in the maritime zone of the ÁCG. (Photograph by Paddy Woodworth.)

ray glides by in the translucent sea, as common a sight from her deck as a woodpigeon is from mine. It's early on a Saturday morning, but she is already busy with her new job: educating the children from the village and visiting schools about maritime biology in general and the marine sector of the ÁCG in particular.

She claims modestly she has absolutely no expertise in her new field of work, but her home is alive with children, there just for the fun of it, when I arrive. It turns out that she has taught most of the children in the village to swim; as in so many fishing cultures, from Ireland to Vietnam, most local fishermen do not swim and so cannot teach their offspring. A star biology pupil, María Fernández, quickly slips into the role of teacher herself, talking me through a series of laminated cards that pictorially identify the rich variety of fish that can be found in the vicinity—sometimes just by standing on Chavarría's deck and looking down. She is not yet a teenager, but she has strong views about the world. Some of them are sound, some of them perhaps requiring the benefit of a little more experience.

"I thought the sea was bad, I have discovered it is marvelous. Our parents' generation did not understand it, they did not know that by wiping out the mangroves they were wiping out the nurseries for the fish. Our generation must be more aware of these things. Everything is made from Nature," she continues. "I'm proud to live in Costa Rica, where we value it, and where there are lots of trees. I'd hate to live in the US, where there are only buildings."[55]

In a rare quiet moment, I ask Chavarría what she thinks she can achieve here.[56] She pauses, and then tells a story. One of her informal pupils came to tell her about a conversation she had had with her father, a fisherman. "Her father, she said, had told her that I was a very bad woman who did not want the village to eat, because I did not want him to fish. I ask her how she had replied. "'I told him he was wrong,' the little girl said. 'I told him you wanted to preserve the mother fishes, and not dirty the water, so that there would always be fish here.'"[57] The vision of bio-cultural restoration, articulated by Janzen in the last century and developed by his Costa Rican colleagues in this one, seems to be taking root on the Pacific edge.[58]

11 *Killing for Conservation: The Grim Precondition for Restoration in New Zealand*

Manaaki Whenua, Manaaki Tangata, Haere whakamua.
(Care for the land, care for the people, go forward.)
MAORI *proverb*

Though one usually thinks of New Zealand as an island or archipelago, it is one of the world's smallest continents. . . . The biotas of New Zealand and Madagascar are the closest we shall ever come to observing the products of continental evolution in island-like isolation, unless we discover higher life on another planet. . . .

New Zealand is distinctive in the two-stage destructive impacts it received from human colonists, and in the innovativeness with which its biologists are now seeking to mitigate those impacts . . . all these features make New Zealand one of the world's biological prizes. * JARED M. DIAMOND, *biologist and author*

So what is the biggest conservation tool New Zealand has got? I would have to say 1080 [a poison]. It eradicates the possums, and it knocks back stoats, ferrets, and rats . . . mice are the thing that now strikes fear into our hearts. * DAVID WALLACE, *farmer and former chairperson of the Maungatautari Ecological Island Trust*

The problem is that people want to see the outcome, the iconic bird in the forest. They forget [the process of restoration], what you need in terms of learning for the future that could be applied elsewhere. * BRUCE CLARKSON, *dean of Science and Engineering, University of Waikato, and restoration advocate*

* *

Imagine you are walking along a ridge trail in a major national park. You are surrounded by temperate rain forest in apparently pristine condition. Tree ferns proliferate, creating an exquisite variety of curvilinear architecture in the understory; the robust girth of venerable beeches and pines bears witness to centuries of almost undisturbed growth. The thick canopy refracts the dawn light in a dazzling exhibition of kinetic beams. Rare breaks in the foliage allow you to see across to a receding series of ridges, where a similarly rich green tapestry stretches, and stretches again, and then yet again, as far as the most distant horizon. Birds, most of them endemic, one of them close to extinction and several of them also shadowed by that fate, are making the forest ring with a diverse chorus. Here at least, if in few other places, these species all seem to be thriving; some are even abundant.

A single steady column of early morning sunlight suddenly breaks through the canopy, giving form and texture to a long dark patch on a tree you are approaching. Drawing nearer, the patch resolves into the body of a mammal, rather bigger than a domestic cat. It is hanging from the tree, right on the public way–marked path. It turns out to be a brush-tailed possum; the jaws of a formidable iron trap have broken its neck. The carcass is at some point—you probably won't want to look too closely—between decomposition and desiccation.

If you are a typical citizen of New Zealand, you will probably pass on unperturbed, with your heart perhaps even a little lighter for the dawn light's gruesome disclosure. Three hundred meters further along and thirty meters higher up—the gradient is steep here—an oblong metal cage will catch your eye, again right beside the trail. The fluffy russet fur of a stoat's tail is still pressed through the roof in rigor mortis; its owner's neck has also been snapped, by a rather more delicate trap, just as it tried to reach an egg laid out for bait in the center of the cage. If you are a foreigner on a first visit, you will surely be wondering why sights more appropriate to the macabre displays mounted by nineteenth-century gamekeepers are so common in a national park. Once again, however, the typical New Zealander will likely continue on his or her way untroubled; this is because he or she believes that there is a direct and necessary connection between the eradication of mammal pests and the recovery of their many rare bird, reptile, and insect species, and the restoration of healthy native landscapes.

Killing and restoration are intimately linked on this island nation,

24. **Restoration trap.** One outcome of the possum eradication program in Te Urewera's "mainland island." Possums, introduced from Australia, are a plague in New Zealand, threatening many native species with extinction. (Photograph by Paddy Woodworth.)

which is distinctive in so many biological, historical, and sociological ways. On the morning I hiked this trail in Te Urewera National Park, accompanied by Greg Moorcroft of the New Zealand Department of Conservation (DOC), we had dropped off Kobey Bremner, an eighteen-year-old on a starter job with the department, for a three-day hike in the next valley. He took with him only a backpack and a high-powered rifle. His mission was to shoot as many deer and feral pigs as he could find. At the remote hut where we spent the next two nights, Asher Morley, a young student entomologist doing holiday work at the park, spoke with enthusiasm about New Zealand's endemic wetas. These are a group of giant, cricket-like insects straight out of fantasy movies—indeed, the New Zealand design studio that serviced both *Lord of the Rings* and *Chronicles of Narnia* is called the Weta Workshop. But Morley spoke with equal zest about his initiation into hunting as a teenager, when he managed to kill fifteen feral goats on his second outing. And he described approvingly how council workers in his hometown were

25. **A canopy of restored rarities.** The Te Urewera forest at dawn rings with the songs of threatened endemic bird species, thriving again thanks to the trapping and poisoning of introduced predators like the possum. (Photograph by Paddy Woodworth.)

squirting Vaseline into the nests of rooks. Why? So that the adults' flight feathers would become disabled, thus eliminating, primarily by starvation, two generations of the birds at once.

The conservation map of Te Urewera looks like a war zone with multiple enemies in play. In an attempt to create "islands" where native species can prosper, possums, stoats, and rats are trapped along hundreds of kilometers of bait lines; the intensity of the trapping increases toward "core areas" of up to 2,500 hectares, where dogs, cats, and deer are also controlled. The department can point to remarkable successes as a result of this tough policy. Te Urewera has uniquely rich remnant populations of endemic species on the verge of extinction, but they were collapsing under pressure from these predators and competitors in the 1990s. The number of pairs of kokako, a charismatic forest songbird, had sunk to eight in the core area of Otamatuna in 1993. Thirteen years after predator trapping started in earnest, that figure had soared to 112. Similar success stories are documented for species ranging from mistletoes to kiwi, New Zealand's national bird.[1]

Trapping—and poisoning and shooting—are taken-for-granted parts of the tool kit of conservation here. Public opinion is generally very supportive, though there is some robust opposition to the use of poison, particularly the controversial "1080" (a compound of sodium fluoroacetate). In Te Urewera, traps and guns are the weapons of choice. The broad acceptance that good conservation practice involves killing animals, especially in a country still largely dominated, despite a recent Maori renaissance, by Anglo-American cultural values, will surprise anyone familiar with the animal rights lobbies in the United Kingdom or United States.

Desperate Times Need Desperate Measures

What is so particular about New Zealand that its citizens are willing to endorse the wholesale slaughter of mammals as part of large-scale, nationwide programs of ecological restoration? Three related facts stand out. First, these large but relatively isolated islands evolved without the presence of any land mammals, with the exception of three species of bat. Its birds, reptiles, amphibians, and insects developed without any experience of mammalian predators or competitors. As one would expect, on fertile islands with climates ranging from subtropical to alpine, there are high levels of endemism. A number of birds became flightless, and many more became or remained ground-nesting and/or ground-roosting. To this day, a New Zealand bird will tend to look to the sky if it senses danger, ready to avoid a swooping raptor, but oblivous to the cat or stoat coming at it from ground level.

Second, the most dangerous mammal of all, *Homo sapiens*, arrived on the scene exceptionally late in the country's ecological history, probably only about eight centuries ago. No other significant landmass—a "continental" one in the view of Jared Diamond—with the exception of the polar zones, was colonized by humans so recently. But in this relatively short period, human hunting and agriculture, coupled with the introduction of mammals ranging from mice to deer, have had a devastating impact on both native fauna and flora.

It is these human impacts that have created New Zealand's third distinctive biological feature—it has become one of the world capitals of extinction, along with Hawaii and Guam. It has lost fifty-one

PANEL 3 • Restoration by Any Means Necessary

All the good work here with kiwi can be undone by a pack of feral dogs in one month. * GREG MOORCROFT, *DOC ranger, Te Urewera Mainland Island, where a hundred pairs of kiwi are protected in core areas by a network of traps*

I'm the biggest sap when it comes to killing, generally. But I make no apology for killing pets. They should not be here. * *DOC employee at Pukaha Mount Bruce Wildlife Centre*

Whack-a-Rat and Save a Native Bird! Waste the Pests and Save the Rest! * *Slogans in a media and schools campaign organized by Pukaha Mount Bruce*

Yes, it's in the culture, in our children's books, there are lots of stories about topics like the good kakapo versus the nasty stoat. * COLIN MISKELLY, *former DOC researcher*

Wire your traps open for a couple of nights so the cat gets used to entering it, then set it to catch. Bait the traps with fish, fresh meat or cat food. Once trapped, a feral cat should be disposed of humanely. Check with the SPCA for approved methods of destruction in your area. * *Advice to citizens on controlling feral cats from Waikato District Council*

From July 2006 to June 2009, there were a total of 1416 goats, 114 pigs, 26 red deer and one fallow deer shot in Boundary Stream Mainland Island [BSMI] and the BSMI buffer . . . . Cat kills increased from previous years to an all-time high [of 128] in 2008–9. * *Boundary Stream Mainland Island 2006–9 Report, 10–11, http://www.doc.govt.nz/documents/conservation/land-and-freshwater/land/boundary-stream/boundary-stream-06-09-report.pdf*

I used to save ferrets in Portland, now I kill them here. * *Confused but willing American volunteer at Boundary Stream*

Early results [of poison use on mainland islands] are promising, although high costs, accumulation of toxins, nontarget game mammal poisoning, short-term declines in rare species, and public wariness of poisons are concerns. * JOHN CRAIG ET AL., "Conservation Issues in New Zealand," 68.

The Opposition

1080 not the way, shame, shame, shame * *Graffito on a bridge in the Taupo district, North Island*

Influential groups opposed to the general principle of killing alien predators to protect native species and restore native landscapes in New Zealand are rarer than takahe. But there is significant opposition to the use of poison, especially 1080. This is attributed by most conservation scientists and activists in New Zealand to the hunting lobby, who fear that prey species like deer will either be wiped out or rendered toxic and therefore inedible by these poisons. There are, however, some scientists who urge more caution with poisons, and some environmentalist groups who campaign actively against them. For a cool assessment, see www.stop1080poison.com/files/Download/Sean%20Weaver.pdf, and for a more partisan position, powerfully presented, see http://www.enufisenuf.co.nz/.

species of bird, one bat, and an unknown number of insects since the first human contact.[2] The number of bird extinctions accounts for 10 percent of the total estimated extinctions on the planet over the last twelve thousand years,[3] and close to half the species lost globally in the period since European colonial expansion. Many more birds and reptiles were lost to the two large islands (North Island and South Island, often referred to collectively as the "mainland"), and were reduced to remnant populations on offshore islands that the introduced predators had not reached. The most recent extinction event—a negative triple whammy involving a wren, a snipe, and a bat after rats reached Big South Cape Island and two adjoining islets in 1964—had a major impact on public opinion. Up until then, influential biologists had continued to argue that habitat fragmentation, and not alien mammal predators, was the decisive factor in New Zealand extinctions. This catastrophe demonstrated beyond reasonable doubt that the predators were the main culprits.[4] The late Don Merton, one of the early leaders of New Zealand's attempt to reverse this tide of extinction, wrote of this moment: "[The] threat of rats had been brought painfully home to each of the team members who had held the last of a species in their hands—then lost it—and were determined never again to witness this ultimate event."[5]

Looking back at this moment today, Colin Miskelly, an inheritor of the eradication and translocation movement that followed it, reflects that "this happened when it shouldn't have happened, in modern times."[6] New Zealanders took the approach that desperate times required desperate measures, as a number of other species, especially birds, seemed destined for the same fate, and in short order. The 1960s and 1970s were heroic decades in conservation here. The population of

Kakapo, a large flightless nocturnal parrot, had collapsed to eighteen individuals—and all of those were males—in the remote Fiordland region of the South Island in those decades. While a further two hundred birds were found on the nearby Stewart Island in 1977, this group was under terminal threat from feral cats. Only translocation of all known surviving birds to several small, predator-free islands has for the time being brought the kakapo back from oblivion. But it was—and remains—a close, close call. The total population had dropped to fifty-one by 1995, but had more than doubled again by 2009. This gives real, if slim, grounds for optimism, but the birds remain "biological refugees in their own land" as Don Merton put it.[7]

The story of the South Island takahe, an outsize and flightless version of the globally widespread purple gallinule (or swamp hen) is even more dramatic. It was thought to be extinct by the beginning of the last century. Then a relict population of less than three hundred birds was found in tussock grass country high up in Fiordland in 1948, where its continued survival was threatened by competition for food by introduced deer. Despite concerted captive breeding programs, translocation to four islands and, most recently, a second (and very heavily protected) mainland area, its overall numbers are barely increasing.

But the poster-bird for survival in New Zealand is undoubtedly the black robin, which is endemic to the small Chatham Islands. It has never had a big population, but by 1980 there were only 5 left. Against all odds, intensive management saved the species, at least for now. Techniques included translocations—involving the restoration of a small forest to provide suitable habitat on one of its new homes—and cross-fostering of eggs and chicks with the tomtit. While great credit must go the recovery team, led by the indefatigable Don Merton, perhaps the real hero is "Old Blue," at one point the only remaining viable female. She survived three times the average robin's lifespan and her many offspring played a crucial role in raising the population to 254 birds by 1999. It is much too early for the conservationists to rest on their laurels, however, as future accidental introductions of mammal predators on its sanctuary islands could pose new threats. Meanwhile, robin numbers had declined to 180 by 2007.

This exceptional and ongoing extinction crisis has conditioned New Zealand's distinctive approaches to restoration. Projects here, until recently, have tended to be species-led rather than driven by a desire to restore entire ecosystems. Of course, the restoration of ecosystems is often a basic requirement for restoring particular species, but the New Zealand context leads to a shift in the main focus from vegetation in general to particular birds, mammals, reptiles or even insects. (Weta restoration has a surprisingly high profile, given the less-than-charismatic charm of most bugs to most humans). Inevitably, the immediate threat from mammal predators pushes active pest eradication to the top of the agenda in most projects. Finally, the knowledge that small islands can provide "sanctuary arks" (because it is relatively easy to rid them of all predators) has led to New Zealand's most remarkable conceptual contribution to restoration theory and practice: the "mainland island," and, in a recent twist, the "island mainland." Could ways be found—through fencing or through creating *cordons sanitaires* of traps and bait stations—to protect areas of the mainland from predators, as the sea protects islands? Where this was not possible, could quintessentially mainland ecosystems be re-created on islands?

But before we revisit mainland islands in more depth—the trail in Te Urewera lies at the heart of one of them—let's pull back the focus and look at the overall state of New Zealand's land, the ultimate context in which restoration must make its mark. If your image of New Zealand is formed by the lushly verdant or starkly volcanic landscapes that feature in so many movies, a drive from Auckland to Hamilton on the North Island is likely to be disappointing; but it is a good place to start in order to grasp what has happened here, and what may develop in future. This trip is rather like driving through a British rural landscape—but without the historical depth and breadth of cultural and agricultural practices that millennia of settlement have imprinted upon Britain's countryside.

A "Skinned" Land Full of Aliens

The drive south from Auckland rapidly enters the watershed of the Waikato River, which gives its name to a region that is mostly lowland, flattish, or gently undulating. The Waikato was dominated by tall native forest, except for some extensive peat bogs and deep swamps, and occasional outcrops of steep hills,[8] at the time of the first Maori settlement here, probably some eight hundred years ago.[9] These first human colonists soon opened up broad swathes of the land for cultivation, using fire as their primary clearance tool. They intensified their tillage in the nineteenth century as new crops became available from European traders and settlers. In the latter decades of that century, white farmers cleared most of the remaining forest at breakneck speed, introduced patches of exotic trees such as pine and willow, and drained most of the bogs and swamps, converting them to pasture.

Today, dairy and stock farming dominate most of the region, and large green paddocks[10] are the ubiquitous unit of landscape. Green, that is, with patches of brown showing through, especially on the hillsides. Overgrazing has taken its toll almost everywhere, and the farmed land of New Zealand has an extraordinarily tight-cropped and bare look to it. "We did not just clear this land," a rueful New Zealand farmer was quoted as saying, "we *skinned* it." Cattle trails cut deep gullies into this stressed surface, and these become watercourses in heavy rain, eroding more with each downpour. Trees planted as windbreaks or simply as reminders of "home," are scattered in the scalped landscape, looking oddly out of place. Rows of Lombardy poplar seem especially incongruous, along with big untidy huddles of Monterey cypress, so rare in its native Californian habitat but so common in so many other places in the world. Plantations of Monterey pine, too, and other exotics occur from time to time. Willow species flourish wherever water runs. But the chances of spotting a native tree on this drive are close to zero. Now here is a scary set of statistics: New Zealand has 2,300 native vascular plant species, 86 percent of them endemic, but this impressive total is topped by the number of introduced species established in the wild, which currently stands at 2,400 and counting.

The typical birds you may see on this trip into the interior are also indicative of the state of things ecological here. If you are a European birder, you will feel a flurry of excitement as the road crosses tidal flats on the margins of Manukau, southeast of Auckland's airport. Glimpses of stilts, oyster-catchers, plovers, and sandpipers all promise a cornucopia of endemic shorebirds to come. But your heart may sink as you move into the farmland, where flocks of European starlings, sparrows, and finches, with an occasional European blackbird or song thrush, offer the only avian interest. Or almost. One elegant raptor species will inevitably catch your eye, again and again, quartering the fields. You will also spot it with alarming frequency, inelegantly grounded as roadkill. The Australasian harrier is one of a minority of native species that has benefited enormously from human clearances, and from the mammals humans have brought with them. It becomes a casualty so often on the highways because they are littered with mammal roadkill; the harrier has not yet learned that cars have right of way on this bountiful new ecosystem.

The process of conversion from native landscapes to a cosmopolitan "New Europe" reaches a blandly successful climax in our first destination, Hamilton City. There is plenty of suburban greenery here and miles of verdant river walks along the Waikato, which winds gently through the downtown area. But almost every bird you will see is still an import. As for trees and shrubs, the city is close to the top of the New Zealand league for its eradication of indigenous vegetation. Less than twenty hectares (0.02 percent) of the city itself, and only 1.6 percent of the ecological district that includes it, retain patches dominated by native plants.[11] Even the nectar-feeding tui, or parson bird, one of the few endemic birds still relatively common, cannot support itself in this environment. It visits the city so rarely that a local restoration society, Tui 2000 is using it as its flagship.[12]

One of the many people endeavouring to change this situation is Bruce Clarkson, chair of biological sciences at the University of Waikato. A tall, rangy, restless, and decisive man, he is a scientist who tries to practice what he preaches.

"When are you coming to New Zealand?" was his greeting when he approached me after I spoke about my work on this book at the SER conference in Perth in 2009. When I told him I had run out of funds for further trips—and confessed that his country had not been high on my radar in any case—he promised there and then to find sponsorship for my visit. He was, with an absolute minimum of fuss, as good as his word. His rapid-fire exposition of the cutting-edge nature of New Zealand's "mainland island" restoration schemes was his lure, and he persuaded me of their importance even before I began to read about them. But today he has decided to start off my visit with something apparently more mundane—urban ecological restoration in his home city. We start at Waiwhakareke National Heritage Centre, a grandly named project that might become a mainland island in the future. At the time of my visit in February 2010, however, it was still at an early stage in the category of restoration-from-the-ground-up. Waiwhakareke (roughly translated from Maori as "pole in the water") was originally a restiad bog. This peatland ecosystem, created by decomposing rushes (restiads) rather than the Sphagnum mosses that form most of its northern hemisphere equivalents, occupied 30 percent of the Waikato region prior to human settlement. Drainage, especially by Europeans in the nineteenth century, and more recent mining for horticultural peat, has now reduced it to a few remnants. The sixty-hectare Waiwhakareke site centers on Horseshoe Lake, and is surrounded by pasture. Before restoration began in 2004, the vegetation around the lake was dominated by invasive willows. The water had become super-trophic from fertilizer leaching in from its shallow catchment, and from cattle excrement delivered directly to the pool.

A partnership between the city council, which owns the land, the university plus a second college, the local Maori community, and other citizen groups has now initiated an ambitious planting program. It includes five distinct vegetation types, appropriate to the local soils and topography. These range from potentially majestic stands of kauri trees on the ridge crest down to manuka-flax vegetation beside the lake. Most ambitious and long term of all is a small patch of cane rush and wire rush that, all going well, should produce a mature restiad peat bog—in a thousand years' time. When fully planted, the park will be linked directly to the local zoo so that the public can move from an exhibition of (mainly) exotic animals to a botanical environment which, while native, has become much rarer than exotic-dominated landscapes in the region. Future plans include the erection of a predator-proof fence—a topic to which we shall shortly return—which might permit the reintroduction of very rare endemic birds and reptiles to the park. The planting scheme thus has amenity and educational value to the local community, and research value to restorationists as well. It will be rigorously monitored and creates a unique opportunity to test methodology for restoration in the Waikato across a range of systems. It will also triple the amount of native vegetation cover in the city, at a stroke.

Waiwhakareke is on the outskirts of town, and as we drive back to the center, Clarkson pulls up on a suburban street that, at first sight, has much in common with similar suburbs in Florida or Sussex. But below street level there is something very distinctive. Descending a walkway, we enter a small world that is part of a remarkable plan. If all its dots are joined up, this scheme could end up having a much bigger impact than the heritage park on the ecology of the whole city. Mangaiti Gully is part of a system of small ravines, created by a network of streams flowing into the Waikato River some fifteen thousand years ago. Too steep for development, they wander like veins of green blood through the city's body, and make up 8 percent of its area. In most cases they were neglected until recent years and, as you would expect, they were very effectively invaded by alien plants from the outlying farmland, and by exotics from neighboring gardens. In the 1960s, a citizen named A. J. Seeley began to convert his gully to native vegetation, and now there is significant public and private momentum to make gully restoration universal in the city. Some of the older restorations have already achieved fine native canopies free of alien species. Mangaiti is being developed as a model for council-resident-community group restoration. Willow is being steadily replaced on the stream banks with flax, tree ferns and small flowering native trees, all the way up to kauri on the slopes. Across the road, Hukanai Gully is being restored as an educational resource and play space by the local school.

Clarkson is always pragmatic and agrees that there should be no rush to remove a noble Monterey pine from this gully, nor a stand of eucalypts from Hukanai. Nor does he sweat over a neighbor's reluctance to cut down some well-loved trees in the gully he is personally restoring under his own Hamilton home. Time and a native understory will do the job in the long run. Unlike the Chicago North Branch Restoration Project, there is no drive here toward the fundamentalist purity of an imagined past. The dynamic is rather toward building consensus to create a future which can never precisely mimic the past, but which should be much more distinctively reflective of this country's rich endemic heritage than the present is. The Monterey pine is unlikely to germinate here anyway, while the council's rapid shift to native use in roadside planting, and a slower but definite trend toward indigenous plants in suburban gardens, is creating positive feedback loops that will give natives the edge here in future. The gully network should extend multiple connectivities, establishing corridors that will allow native birds, reptiles, and insects to move in search of food sources through the heart of the city. They will in turn redistribute the genetic variety of the native seeds they will transport. But the number of such species, and the size of their populations, will be severely limited by continuing mammal predation for the foreseeable future. We have seen, and will see again, that New Zealanders are supportive of pest eradication. But traps for cats and poison grids, even for rodents, would undoubtedly be a bridge too far right here, in the middle of suburbia.

My final restoration stop in Hamilton is Jubilee Park, a 5.2-hectare preserve of fine kahikatea, also known as white or swamp pine. There are also occasional pukatea, another tree that likes the semi-swamp conditions that were typical of the Hamilton basin before drainage and clearance. Much of the restoration here has been done by Weedbusters, a remarkable synergy of public and private initiatives, originally developed in Australia in the 1990s, and supported in New Zealand by organizations ranging from the DOC to a national farmers' federation. Their main job was to clear the understory of pest plants like wandering Jew that had invaded the area. These pine woods are now an oasis of stately calm in the city, buffered by a newly created wetland which would once have stretched all the way—across where the city showgrounds now stand—to the river. Again, Clarkson's pragmatism is evident. Like many urban restorations, he points out, Jubilee Park will be in a permanent state of arrested development. In the wild, most of the pines would probably eventually die off as broadleaved trees colonized the swamp through natural succession. But kahikatea are so few in the city that these will be maintained in perpetuity. "All the processes that created this forest as part of a greater ecosystem stretching from the river to the hill crests are gone now," he says. "We *must* manage it if it is to survive within the city."[13] Sometimes, restorationists have to accept that not all processes can be restored, if biodiversity is the priority. It seems that we must often hold nature back, if we are to continue enjoying some of its riches, in our anthropocene age.

Despite such limitations, the kinds of urban restoration that Hamilton City Council is developing, along with its educational institutions, community groups and individual citizens, may in the long run be as important for the future biodiversity of New Zealand as the much more dramatic restoration experiments taking place on the mainland islands. This is an argument to which we will return. But the beauty of a small(ish) country is that you can see both in the same day.

Making a Mountain into an Island

The volcanic ridges of Mount Maungatautari rise in richly wooded grandeur from the bald pastureland about an hour's drive southeast of Hamilton. From a distance, the forest's utter isolation evokes as much despair as its survival inspires hope. Some tongues of pasture lick greedily up its slopes, until even a New Zealand farmer must have realized that there was no profit in clearing any more trees. But at least one local farmer, David Wallace, saw the mountain not as an enemy to be defeated, much less as a waste of space. Like the Maori *iwi* who previously occupied the region and still hold ancestral rights on Maungatautari, he saw it as a *taonga*, or treasure, to be cherished. Wallace knew, however, that many of the ecological crown jewels of the mountain had disappeared, despite its superficially impressive emerald cloak. New Zealand's iconic bird, the kiwi, no longer probed the forest floor at night for insects nor sent its piercing call through the shadows. Not only were its chicks exterminated by stoats—a single pair will eradicate them over a hundred-hectare area—but the wetas that form part of its diet had been hoovered up by mice and rats. And a whole spectrum of songbirds had disappeared from the canopy—kokako, stitchbird (*hihi*), saddleback, bellbird, and whitehead, among others. Three species of native trout no longer swam in the steep mountain streams. And the forest itself was being stripped of its last vestiges of rata, tree fuchsia and mistletoe by hordes of possums. While the forest had never been cleared for farming, selective logging, from the 1940s to the 1980s, had taken out some of the best of the big trees.[14] Clarkson says you could have ridden a horse through the remainder, so thoroughly had the understory been erased by grazing animals.

By 2010, however, just 7 years after restoration began here, you would have been hard put to walk off-trail through the forest in some areas, and riding there would have been a nightmare. Dense tangles of the liana aptly known as supplejack flourish among a host of other ground level plants. Climb a tower to the canopy, and you may see masses of flowering rata, as though possums had never reached these trees. Fat kaka parrots reveal hidden iridescence as they cross patches of bright light, and make noisy parades through the lower branches. An ambitious reintroduction program has already returned six bird species to the forest and three fish to its streams. Many more are planned.[15] Maungatautari is a daring—some would say hubristic—example of New Zealand's mainland islands. In the 1990s, Wallace used small fenced enclosures on his own land to raise kiwi in safety from predators. He grew up on Maungatautari and used to gaze at the peak from his farm and joke that "We'll fence that mountain one day." That joke somehow morphed into a "crazy but marvelous dream."[16] A great deal of consultation with local people, combined with frantic fund-raising, turned the dream into the reality of the Maungatautari Ecological Island Trust (MEIT). At 3,400 hectares and with 47 kilometers of predator-proof fencing,[17] this is the largest fenced "island" in the country. It is privately owned by the trust, which represents a broad spectrum of local interests, including Maori and adjoining landowners, and MEIT works with the cooperation—and some funding—from local and regional councils and the DOC. Its financing has been problematic. The first appointed staff did good work, but their high salaries raised eyebrows among the generally poorly paid New Zealand conservation community, and they later were laid off amid some controversy. Like so many projects dependent on donations and public funds, it has been easier to raise big money for the spectacular capital items, like the fencing itself, than for its mundane ongoing running costs. MEIT is now very heavily dependent on volunteer labor at all levels. This indicates its broad roots in surrounding communities, but is hardly a secure base for the future of an increasingly valuable refuge of biodiversity.

My own first impression of Maungatautari is not entirely auspicious, though it also gives me my first glimpse of one of New Zealand's most precious avian treasures. As I approach the high and fine-mesh steel fence, two takahe are making use of a feeder strategically placed just inside it. The feeder is right beside a gate that admits the public to the reserve's southern enclosure. Yes, it is exciting to see, close-up, a bird once thought extinct and of which only 230 individuals remain in the wild. But that very phrase immediately raises an uneasy question—are the pair I am looking at *really* in the wild, or are they, as fence and feeder suggest, in some kind of glorified zoo? I pass through the gate—a double gate, to make it more difficult for mammal predators to sneak in with visitors. The zoo impression begins to fade as a trail opens up between natural galleries of tree ferns. They include the lovely silver fern, whose gleamingly reflective underside made it an ideal moonlight trail marker for Maori hunters, and an attractive logo today for the national rugby team, the formidable All Blacks. But that unease does not go away altogether, and will arise repeatedly in varying forms in New Zealand restoration contexts. This is not an aviary, there is no cage between us and the open sky, and the tomtits flitting elusively out of vision are able to move beyond the fence at will. But the flightless takahe, and the much more numerous but also artificially reintroduced kiwi, obviously cannot.

Moving up the trail, Bruce Clarkson indicates many plants that have recovered dramatically with the eradication of possums and grazing animals from the enclosure, including tree fuchsia and begonia. An information point highlights a fine example of rimu—a tree soaring to perhaps thirty meters. Fine as it looks to the untutored eye, Clarkson points out that the twisted ribbing on its trunk shows that it only survived here because it was not worth extracting for its timber. At this point we run into Phil Brown, a photographer who has documented the whole project, and has become a fund of bizarre information in the process. He has found, for example, as he proudly tells us, that a takahe produces twenty-five meters of excrement in a day, if you laid all its pellets end to end.

He is accompanied by Gemma Green, a young woman whose job is to educate schoolchildren about the restoration process, including the eradication of mammals. She gives me a crash course in the merits of various poisons for killing pests as we walk the trail. Most possums within the reserve, she tells me, were killed by an initial aerial drop of 1080, though brodifacoum is now the poison of choice for dealing with residual rodents. If you know what a brushtail possum looks like, you may be wondering what I was wondering. How on earth do you persuade children that killing these creatures is a Good Thing? They are the size of a small fox, have eyes that tug the heartstrings almost as poignantly as a bushbaby's, and fur as cuddly as a koala bear's. What's not to love about them? Everything, apparently, and most of the children know it before she enters the classroom. "The only children who raise objections come from hunting families," she says. And they are concerned about the impact of poison on prey animals like deer and pigs, not about possums, which have very few admirers left in New Zealand. "The kids already see them as nasty beasts," she says. "I have to explain that the damage they do is not the possum's own fault, it is ours for introducing them where they have no natural predators." She explains that the introduction of poison to Maungatautari has been a major operation, with helicopters showering pellets down from buckets with spinning blades across the central zones. Close to the fence, it was applied by direct drops to avoid affecting livestock (and game) outside the boundaries of the island.

We walk on up the mountain as we talk, passing sugar feeders for stitchbirds (*hihi*), a rare endemic species confined to one offshore island until very recently. Maungatautari is one of two sites on the North Island where they have been reintroduced, in this case after more than a hundred years' absence. They do not oblige with an appearance—not surprising as the fifty-nine individuals released a year earlier have a lot of forest to wander in, but the zoo question puts in another appearance in the back of my head. But then something large and utterly unfamiliar flies through the midstory and perches quite close by. It climbs nimbly, with agile claws, more like a monkey than a bird, among the branches. The forest-filtered light reveals tints of shocking pink and brassy bronze in its mostly dull brown plumage. This parrot is the kaka, one of New Zealand's more common rarities, as it were. Which is to say it is still damn rare by most standards, and only thrives on predator-free islands and in the few remaining isolated mainland native forests. It is easy to approach and photograph here, probably because this one was born in a zoo and still associates humans with food. We approach a set of large cages where more kaka, and some yellow-crowned parakeets (*kakariki*), await release. The free-flying kaka come to visit the newcomers—and to feast on a nearby feeder, bringing that uneasy zoo impression back to the foreground.

Clarkson leads me on up a series of steep wooden stairs to reach a viewing platform close to canopy level. That is where I see the entire and ebullient, if typically untidy, crown of a rata for the first time. It rises above all the other trees and stretches itself, luxuriantly magnificent even though its red clusters of flowers are not in season. The crown alone is big enough to seem like a entire tree in itself, growing out of the top of the forest. "That rata would be gone if the possums were still here," says David Wallace, striding to greet me. He is the founding chairman of MEIT, a local farmer who dedicated his retirement years to this massively ambitious and often turbulent project with all the energy of a teenager. The recovery of the tree is impressive, and its loss would probably have heralded a long absence in the forest. Rata and other important native trees like rimu present a conundrum for restorationists. They generally do not prosper when planted artificially, and need an extreme weather event, like a cyclone, to open up enough canopy space for them to regenerate successfully. So should we simulate cyclones by ripping up sections of restored forests to facilitate their return? Or should we just wait a hundred years (on average) for the next cyclone? Clarkson and Wallace discuss this briefly and decide that that is one question they can safely put on hold.

Finding the Last Mouse on the Mountain

Wallace is confident that all the bigger mammal predators are now almost totally eliminated from the forest. But one remains, and its population sometimes explodes because, ironically enough, of the success of the campaign to eradicate the others. The stoats and ferrets, which once kept its numbers down, and the rats that competed with it for food, are no longer present. As always, when you tug one string in an ecosystem, you eventually find it is connected to everything else. "Mice are the thing that now strikes fear into our hearts, how do we find the last mouse on this mountain? That is the challenge now," he declares. Even when the last mouse is found, it will still be necessary to monitor for them, as birds may carry live mice across the fence, and they have also been known to hitchhike in rucksacks and even lunchboxes.

Wallace steers me toward Ally Tairi, a formidable Maori woman, and tells her, rather peremptorily I thought, to tell me how they are going to wipe out the mice. She explains patiently how they have set grids at hundred-meter intervals throughout the fenced areas. At each intersection there is a tracking tunnel box baited with peanut butter ("we tried Nutella but found that the mice were very picky"). The box has a white card base with an inkpad at the entrance en route to the bait, so that everything that enters leaves clear footprints behind it. "There was a time we had only mouse and rat tracks, now mainly the tracks are weta, centipedes, millipedes and other lovely creepy crawlies, the ones we want. The cards are looking busier and busier as the mice decrease and the insects increase," she says, holding up cards covered in indecipherable (to me) hieroglyphs of varied insect traffic. It is not only the indigenous insects that the mice threaten. They compete with birds for many other resources, and also plunder the eggs of ground-nesting birds. They will even gnaw the legs of flightless or unfledged chicks, causing disease and death. The staff used to set the tracking tunnels at twenty-five-meter intervals, but as the mice became scarcer their range expanded, permitting them to be set a hundred meters apart. This is a mercy for the four hundred fifty volunteers, many from as far away as Hamilton and even Auckland, who give a full day's hard labor once a month to check them.

Wallace is robust in his defence of the massive use of poison to create such a unique nature reserve, a kind of ark for almost-extinct species. As a farmer, he has experience of using other methods and seeing them fail. "Possums are a plague here," he says. "We can go out on our farm at night and shoot a possum a minute in the headlamps, and it hardly makes any difference."

"So what is the biggest conservation tool New Zealand had got? I would have to say 1080. It eradicates the possums, and it knocks back stoats, ferrets and rats."

But are there no negative consequences, no collateral damage? "DOC have done lots of tests and found nothing significant. If someone is opposed, though, no amount of testing will convince them, it is an emotive issue. Brodifacoum is even more effective than 1080. But if it gets into liver of a pig, and if a hunter kills the pig and eats the liver, then there are problems." This is not an issue in the reserve, where hunting is not permitted. In fact, the DOC reports that deaths from poisoning do occur among indigenous species, from brodifacoum in particular. But it assesses the losses as minor compared to those caused by the mammal predators it is exterminating.[18] In any case, very rare native species are usually either extremely scarce or entirely absent from islands, mainland or maritime, which are being cleared prior to reintroductions. The evidence is strong that lethal doses of these poisons do not persist very long in soil or water.

But isn't there a cultural issue about the use of poison for Maori, especially on land they hold to be sacred, like some of the land on this mountain? Tairi ponders this, and responds, "This is a small sacrifice for a greater good. That is what we have to weigh up." Or, as one DOC official put it, "It is not poison which desecrates the *mauri*—the life force—of the landscape; it is the introduced mammal pests." There is less unanimity between Wallace and Tairi on the question of Maori land rights, a major issue in contemporary New Zealand's political and legal systems. (See panel 4).

Wallace accepts that farming has done great damage to the mountain but claims that the vision for protecting it also came from Europeans on the land. "Farmers are conservationists at heart," he insists. He remembers one of them, Rex Garland, telling his sons, "young men full of testosterone," that they should not cut any more bush because "'it's too beautiful.' There was a love of the native bush," he continues, "although our fathers and grandfathers cleared a lot of it." Tairi pulls him up, very quietly but very firmly: "Whenever they cut into the wood of our mountain, the *iwi* objected. I would add, David, from a Maori perspective, that this was all Maori land, the Crown confiscated it.

"Confiscated or purchased?" asks Wallace, his voice carefully neutral.

The question hangs in the air. The local *iwi* currently owns fifty hectares within the fenced area, local farmers a hundred and fifty, and the rest is Crown land, but a settlement claim based on the Waitangi treaty (see panel 4) is pending. How would a successful claim to the mountain by the *iwi* impact the restoration project?

"Whatever happens to the claim, there will be no changes in the ecological management at Maungatautari. Everybody wants this to carry on; there is a lot of support from Maori," says Wallace. Tairi concurs. But less than a year after my visit, a damaging and painful ongoing governance crisis split the MEIT wide open. Complaints from the *iwi* about lack of consultation on activities on their land inside the fence—including the use of quad bikes—was a factor. The trust was restructured and Wallace's resignation followed. He and several other farmers with land adjoining the fence became involved in a bitter row with MEIT about access when breaches occurred, putting the whole project in jeopardy. The key issue appears to have been their rejection of Maori rights as interpreted under ongoing Waitangi Treaty

PANEL 4 • Maori, the *Mauri* and the *Maunga*

The history of Maori–European relationships is, like that of all colonial encounters, contentious, painful, and complex. Neither my research nor the space available equip me to do justice to it here. But Maori play such significant roles in the social and political aspects of New Zealand's ecological restoration processes today that some sketch of this history is necessary (for more information, see http://www.nzhistory.net.nz; King 2003).

The Maori hold a unique place among the peoples colonized by the British. Their culture was expert in warfare; they mastered modern weapons and military strategy with surprising ease, given that their technology was still in the stone age when the first Europeans arrived. The British signally failed to subjugate them as completely as they did many other peoples. Ironically, the Maori martial tradition was also their Achilles' heel. It had divided them into dozens of relatively small *iwi* (tribes) pitched against each other in recurrent and ferocious wars that made them unable to unite against the British.

New Zealand's official "foundational document," the 1840 Treaty of Waitangi, is an unprecedented agreement between the British Crown and an undefeated but ultimately colonized people. It guarantees the Maori equal citizenship and property rights with settlers. However, many *iwi* did not sign the treaty, and many who did believed (and believe), with good reason, that its principles were repeatedly breached by the colonists. "It was a gentleman's agreement," as one Maori put it dryly (in the comments book at an exhibition on the treaty I visited), "but it was obvious that none of the British were gentlemen." The ensuing decades were marked by settler land seizures and some very effective Maori rebellions, which the British eventually subdued only with difficulty. Nevertheless, Maori did retain citizenship rights and social acceptance that Australian and African Aboriginals might have envied. Racial intermarriage was not unusual, and Maori language and culture were treated with some respect. Missionaries did succeed in eradicating their religious beliefs at an official level, though they persist today in truncated ritual and cultural forms. Sadly, by the mid-twentieth century, many Maori had sunk into urban poverty and found themselves alienated both from white society and their own traditional communities. But sufficient remnants of those communities survive to maintain a distinct sense of Maori identity and culture.

From the 1960s, Maori developed a new and militant sense of this identity, which again expressed itself through a different strategy to that of most colonized peoples. (Native Americans, for example, can point to no single treaty through which their demands might be focused.) Maori demanded legal redress based specifically on violations of the Treaty of Waitangi.

Since 1985, in a process that continues today, though the deadline for new claims has expired, they have reestablished their rights to many traditional lands, and either reoccupied them or received financial compensation in lieu. By 2006, the New Zealand Parliament estimated the value of settlements to have reached NZ$750 million (New Zealand Parliament 2006).

The relevance of this process to restoration lies in the fact that most restored or restorable land is partly or wholly recognized as the property of, or at least as of cultural significance to, one or more *iwi*. And since Maori culture regards many of the species selected for restoration as their *taonga*, or treasure, no species can be translocated without the permission and ritual blessing of the *iwi*, which is "giving" the bird, reptile, or insect concerned, and of the *iwi* which is "receiving" it. While negotiating these agreements adds significantly to the time involved in any restoration project, and therefore to its cost, all New Zealand restorationists I spoke to welcomed the practice. They did so not only out of political respect for a formerly dispossessed people, but also because it forms a valuable bond between the local indigenous community and the restoration process. And more than one white scientist spoke of its adding a "spiritual dimension" to restoration and conservation.

Maori concepts are built into the vision statement for the ecological restoration plan for Maungatautari: "To restore Maungatautari to a largely indigenous, diverse, fully functional, (near) self-sustaining forest, permanently free of mammalian pests and to strengthen the *mauri* [life principle] of the forest and protect the *mana* [spiritual prestige] of the *maunga* [mountain]" (McQueen 2004, 6). It is indicative of the status of Maori culture in contemporary New Zealand that words from their language have passed into everyday English usage, and appear without translation, or even italicisation, in newspapers and even scientific documents—the translations and italics here are mine.

negotiations. MEIT has happily survived and even prospered, so far, but, as of September 2012, four farmers remain in dispute with the trust.[19]

Tairi's assertion that the local Maori were always opposed to clearance is echoed by Lance Tauroa, who forms half of a two-man team dedicated to fixing breaches in the fence. He is also an MEIT trustee in the Maori interest. "What we are doing here is part of our culture, part of our life. At the *marae* [traditional Maori local assembly] you start [a speech] by introducing yourself, and you say 'My maunga is. . . .' We acknowledge our mountain, our earth, our river, everything that is part of us. And our ancestors that passed before us. We are reconnecting with the landscape."

But what *would* this landscape have looked like in the centuries between Maori arrival and European settlement? "The whole north island would have been covered in bush. The Maori were, if you like, ecologists. We had different marae all over the place, but they wouldn't rape the area, they would take a certain amount and move on. A lot of people have difficulty envisaging the bush all over. For us it is humbling what the MEIT has done, our bush was nearly gone."

We will look in greater detail at the questions raised by claims for Maori environmental stewardship when we consider the broad issues related to traditional (or indigenous) ecological knowledge in the next chapter. It must suffice here to say that the scientific evidence from the landscape is that massive and permanent bush clearances were quite widespread before European settlers ever arrived. Introducing the first mammal predators, rats and dogs, to the islands, Maori colonization resulted in a number of extinctions. Key components of the native ecosystems disappeared in this period, including the giant moa, and the pouakai, an eagle with a three-meter wingspan that was the system's top predator. However, the total isolation of the forest that remains on Maungatautari through intensive clearing, and the chronic bleeding of its native biodiversity, was certainly primarily the work of white farmers.

Whatever the precise environmental history of the mountain, Tauroa is most enthusiastic about working for MEIT. Either he or his colleague are on call 24/7 if the electronic sensor that protects the entire perimeter indicates a breach in the fence. If his pager beeps at 2:00 a.m., he must hightail it immediately to the damaged site and fix it, making the last bit of the journey on quad-bike, on trails specially created for this purpose. Some might find such calls a nightmare, but he describes his role, with unforced enthusiasm, as a "dream job."

Being an ecological island subjects the area to exceptionally intense ecological scrutiny, and Maungatautari has yielded up some serendipitous secrets over the last few years. Volunteers doing an inventory found the first record on the mountain of New Zealand's silver beech, a stand of no less than one hundred trees. Some of them are centuries old, and the species' poor dispersal rate indicates that the stand itself may be a relict of the last ice age, when beech were common on the

Waikato lowlands. The silver beech is an elegant tree with a feathery, almost cedar-like structure. It is host to a wealth of epiphytes. The discovery also adds a new range of insects and fungi, beech specialists, to the mountain's species list. Meanwhile, at the other end of the size scale, the mountain has also proved to be home to several colonies of Hochstetter's frog, an endangered amphibian. And there has been a very recent sighting of a small forest bird, the rifleman. Since this bird does not fly over open spaces, its presence here may have passed unnoticed for decades. If confirmed, it means that one less species will need to be translocated and should flourish once again in the absence of alien predators.

As Clarkson and I exit through the fence, we meet one of the volunteers, an eighty-three-year old man who has just hiked eight kilometers up and down the hill, monitoring tracking stations and traps. He launches into a paean of praise for the enterprise shown by the trust, accompanied by a scatological denunciation of the Department of Conservation for failing to seize such opportunities years ago. In the car, Clarkson cautions me to take this account with a large portion of salt. "DOC staff," he tells me, "get very bad press, but any cost-benefit analysis would show you that they do outstanding work on lousy salaries." One more voice for a chorus I have heard worldwide about hard-pressed national park staff or their equivalents, from Ireland to Costa Rica via Vietnam, while researching this book.

"Doing the *Hakawai*": A Snipe Restored from Celestial Space

As it happened, I had already met a DOC staff member, Colin Miskelly, in Australia the previous year, on a SER field trip. He indeed had an infectious passion for his job, for his country's people and wildlife, and for the relationship between the two. This was reflected in a rather convoluted but fascinating story he told me between stops, as we explored the bizarre flora and fauna of Gondwana. His narrative interwove the restoration of a snipe species to a "muttonbirding" island, and the conversion of a Maori myth about a bird of ill omen into a blessing from the Maori gods. After Bruce Clarkson's invitation, this story was the second lure that confirmed my decision to go to New Zealand. I asked Miskelly to retell it to me in his home in Wellington a few months later. Besides whatever charm it may have in itself, the story seems to me to encapsulate several of the distinctive dilemmas, strengths and weaknesses, of island and mainland island restoration in New Zealand. Miskelly has thought long and hard about these issues, both at practical and theoretical levels.[20]

I think it would not be unfair to say that Miskelly, one of New Zealand's top birders from his youth, is obsessed by snipe. To select these elusive birds for intensive study, anywhere they occur in the world, you would need to be driven by a strong compulsion. In New Zealand you would have to be intrepidly adventurous as well, since they have been exiled by predation to the most remote and inhospitable islands. And when predators reach even these tiny islands, they are further exiled, offshore-offshore as it were, to sheer rock stacks whose rugged crowns provide a last refuge for remnant populations—and whose rocky shores offer no safe landing points for mammals—including humans. There are now three New Zealand species left, the Chatham Island snipe, the Snares Island snipe, and subantarctic snipe. The latter species is itself now divided into three subspecies, distributed across several island groups. They are virtually impossible to distinguish in the field, and only with great difficulty in the hand.[21] There used to be several other full species, including one on each big island. Predation of chicks and eggs by the voracious weka, a large and aggressive native rail, probably meant that most or all snipe species have always been scarce and isolated. But it was the arrival of rats, and then cats, which wiped most species out early after human settlement. The Stewart Island snipe[22] was a temporary exception, hanging on long enough to form one third of the notorious 1964 triple extinctions. Stewart Island lies directly south of South Island, and is New Zealand's third-largest land mass. But the snipe had long been banished to *its* offshore islands, the very last record being a pair on Big South Cape Island. They died after capture in a last-ditch attempt at translocation to a safer habitat. Miskelly was two years old at the time.

Twenty years later, Miskelly's passion for New Zealand's few remaining snipe took him, literally, to the Antipodes Islands and other subantarctic oceanic rocks. He was an unabashed alpha bird lister, and he was also pursuing a doctorate. But he had much bigger ambitions. His long-term dream is to restore snipe to appropriate habitat on islands within much easier reach of other New Zealanders, once they have been cleared of predators. Moreover, his vision is binocular, able

26. **A bird in the hand.** Colin Miskelly about to release a Chatham Island snipe on Pitt Island. This translocation failed, probably because the snipe moved outside a fenced release area and became vulnerable to feral cats and to weka (a locally introduced predatory rail). (Photograph courtesy of Kate McAlpine.)

to focus sharply on culture as well as nature. He was fascinated by the Maori legends attached to many New Zealand birds, some still with us, others extinct, still others, apparently, impossible to identify. Among the latter was the *hakawai*, descendant of an ocean god and a star god. Various Maori spoke of it as "dwelling afar in celestial space, and only descending to earth at night. . . . There is nothing to be seen, but you hear a cry, a dreadful laughter floating down from the heights. 'Hokioi-Hokioi' is the cry, and as it ceases you hear that eerie whistle as a bird swoops down and up again into the blackness and silence of the night sky."[23] One version held that the sound was produced by the bird's choking on the hair of warriors doomed to fall in battle; a more recent and slightly more credible tradition says that it forecast a southerly gale.

Miskelly noticed that the call of the *hakawai* had been heard repeatedly on Big South Cape Island, right up to the year the snipe became extinct. He was not the first to wonder if the "eerie whistle" might be related to a widespread snipe phenomenon: the startling "drumming" or "bleating" sound produced by the rapid vibration of tail feathers as a courtship flight ritual. But he was the first to produce a scientific paper meticulously linking the ornithological evidence for such aerial displays by New Zealand snipe to *hakawai* legends and locations.[24] In some species the display has never been visually witnessed, or even heard, but the presence of characteristic vibration damage to tail feathers is generally accepted as conclusive.

The Maori of Big South Cape Island still remain in very close contact with their environment through the practice of muttonbirding: they have retained the right to harvest the chicks of the very numerous muttonbird (sooty shearwater) for meat. This hunt is an annual event of continuing economic and cultural importance. Miskelly began an intense correspondence with a number of Stewart Island muttonbirders, and especially those of one of Big South Cape Island's own satellite islands, Putauhinu, which was successfully cleared of rats in 1996.

A year later, a remarkable and related thing happened on a much more remote island: after so many subtractions from the New Zealand bird list due to extinctions, there was suddenly an *addition*. A subspecies of the subantarctic snipe, new to science (though not initially identified as such), was found on a rat-free stack off Campbell Island. Rats were then eradicated from Campbell Island itself in 2001. Within two years, snipe were recolonizing spontaneously from the stack. In 2010, Miskelly and Allan J. Baker demonstrated that these birds belonged to a new subspecies.[25] Meanwhile, a New Zealand snipe recovery group had been set up. Miskelly, among others, pioneered the restoration of another species, the Snares Island snipe, to Putauhinu off Stewart Island in 2005, bringing snipe close to the mainland for the first time in forty years. This recalled his research on the *hakawai* with local muttonbirders.

"This guy I had interviewed in 1985, Rongo Spencer, he could remember hearing the *hakawai* on Big South Cape Island in 1930," he says. When Miskelly's paper on the *hakawai* was published in 1987, most Maori were at first reluctant to accept it, he continues. "The *hakawai* was a spirit bird, it wasn't there to be explained by *Pakeha* [Western] science.[26] Nevertheless, over the next twenty years it was accepted to the extent that the muttonbirders, when they wanted snipe back on their island because they had eradicated the rats, they didn't want any old snipe, they wanted a snipe that would 'do the *hakawai*.'" By which they meant a snipe known to perform, albeit elusively, a noisy aerial display. Immediately, there was a problem. The bird selected by the recovery group for the transfer was the Snares Island snipe, which was not thought to "do the *hakawai*." Indeed, Miskelly thought it was evolving toward flightlessness. They had selected the species on good biological grounds: it was both geographically closest and physically most similar to the Stewart Island snipe it was to replace on Putauhinu.

"But the island *kaumatua* [elders], Janny Davis and Rongo Spencer, wanted the *hakawai* back. They independently approached the Chatham Island Maori and asked for their snipe [which *is* known to 'do the *hakawai*']. They got the standard Chatham Island reply: 'No way, it's ours, you can't have it.' So they said to DOC, 'OK, we'll take the Snares Island one.' 'Good, good,' we said, 'agreement is what we like to do.'"

And then, when Miskelly and his colleagues captured thirty snipe on Snares Island, they found something that had not been observed before. Even telling me the story for the second time to me, he still gets so excited that, when explaining what happened next, he goes into a kind of stream-of-conscious mode.

"Blow me down, we found damaged tail feathers on two of the Snares Island snipe, so I bring this news to Rongo, who had heard the *hakawai* sixty-five years earlier, and explain the significance of the broken feathers we have found, and there are tears literally running down his cheeks, this is not just bringing the *hakawai* back to the island, he said, it was their gods choosing when to show us scientists that this bird does the *hakawai*, it was a good omen.[27] So, you know, another layer, we were absolutely thrilled to bring the *hakawai* back but they turned it on its head and said that this was the right time for us to learn that this bird was the *hakawai*."

This story raises two key issues, one sociological, the other ecological: the impact on restoration of efforts to integrate Maori cultural values into restoration projects, and the use of homologue or surrogate species to replace extinct ones in a restoration context. Underlying this is the question, becoming very familiar to us in these pages, of the extent to which restoration in New Zealand, with its very distinctive natural and cultural history, can legitimately attempt to recreate formerly existing ecosystems, and the degree to which it should—or has to—create new systems, or at least systems with new elements, with unforeseen and perhaps unforeseeable outcomes.

On the sociological level, it is clear from Miskelly's story that at least some *Pakeha* DOC employees do take their responsibility to accommodate Maori cultural perspectives in their projects very seriously. I found this repeatedly confirmed in other encounters with agency staff. Miskelly is willing to learn cultural lessons from this interaction that might not square with scientific rationality. Without abandoning that rationality, he embraces an open-minded approach familiar from anthropology and historically aware common sense. He respects the view that the *hakawai* is a spirit bird without necessarily "believing" it. He is genuinely delighted when his scientific observation—those tell-tale broken tail feathers—can build a deeper link with the culture that holds that spiritual view, even if the culture's reading of that link is radically different from his own. On a pragmatic level, the respectful nurturing of this cultural connection to a reintroduced bird will ensure that it is cherished by the indigenous community, and thus greatly strengthen its chances of survival.

How Long Is Your Vision?

But one has to wonder whether he and his colleagues do not find the control exercised by Maori *iwi* over threatened species problematic from an environmental perspective. His account of the Chatham Islanders' response to the muttonbirders' request for some of their snipe shows that they can and do exercise a veto on restoration projects, and this is the case regardless of the ecological appropriateness or urgency of the request.

"Your attitude to this depends on the length of your vision," he responds, adding with a dry smile: "I am reluctant to use the word 'myopic' about colleagues. But some people in restoration do find it intensely frustrating that things that make perfect ecological sense can't be done for socio-political reasons." There is one example that affects him very directly. He has been deeply involved in the restoration of Mana Island, near Wellington, for which he wrote the restoration plan.[28] "A classic question in relation to Mana Island is snipe translocation. The North Island snipe is extinct. As a snipe enthusiast

I would love to bring snipe to Mana Island. The one that would make the most sense biologically is the Chatham island one. But I know that if we went directly to the Chatham *iwi* and asked for them they would say "No," just as they did to the muttonbirders. But part of the reason for this attitude is that on the Chathams, the people there can't see their snipe, they are limited to two closed-access reserves. We tried to move some to a predator-fenced area so that [local] people could go and see them, but unfortunately that did not work. A common perception on the Chathams [about attitudes on] the New Zealand mainland is that everything is 'take, take, take, things only go one way,' so they say we are ripping off their culture. It is very tied into identity." He argues that before they can be expected to change their position there must be a quid pro quo when they are asked to donate birds for translocations. He suggests asking Ducks Unlimited to sponsor the breeding of wildfowl for harvest on the island, reintroducing locally extinct species, a nice instance of lateral thinking by a conservationist.

"Some people are very frustrated by the Chatham Island *iwi* block on snipe to Mana Island," he adds. "But if you look, say, two hundred years hence, and if there are snipe on Mana then, it does not matter much whether they have been there for twenty or eighty or a hundred and eighty years. I am confident that eventually everything will line up, maybe I won't see it, but in an evolutionary time frame there is very little difference. These people want to get everything done quickly. Why rush and do it all now? It takes time; doing it properly is more important than doing it quickly and stuffing it up."

How to "do restoration properly" is the subject of much lively debate in New Zealand. In this chapter I have stressed species restoration, especially linked to islands or "mainland islands," because this is such a distinctive feature of ecological restoration in New Zealand. But there is also much lively debate about the context of such restorations, about which landscapes should be prioritized for restoration, and about the most appropriate targets for ecosystem restoration. Because New Zealand was colonized by humans so recently, and because some landscapes have—so far—remained relatively "pristine" in comparison with those elsewhere, there has been a natural tendency to prioritize the preservation of such areas and to view them as historical reference systems for restoration elsewhere. Is New Zealand, then, the one place in the world where we can talk with some accuracy about restoring the past, because we really do know what it looked like? It is a tempting thesis, but many restorationists, including Colin Miskelly, believe that even here we are always compelled to restore the future: "We have lost so much that we fully acknowledge that we cannot restore what's gone. The best we can do is bring back ecosystems that are of predominantly indigenous character. We are not going to have landscapes with huia or moa or giant eagles once again."[29]

But would it not be possible to restore something approaching presettlement *vegetation* communities in some areas, since far fewer plant than animal species—only five—have been lost to extinction?

"Yes, and those plants were probably always rare, so you can restore the basic vegetational character that was there. But still there are so many pervasive invasives, like the blackbird or chaffinch, that are ubiquitous and relatively benign . . . nobody talks about bringing back the whole system, you are setting yourself up for failure if you try to do that." Then he pauses, and comes up with an exception, which nonetheless proves his rule: "The Campbell Island rat eradication is the biggest in the world to date, and now that island will restore itself. We found one parakeet bone there, so we will restore a parakeet species, maybe a different one but something similar to what was there. Everything else is already on the offshore stacks. We are thinking of putting in sound systems to attract the petrels back, but we won't have to translocate them. These islands had limited species, there were never any moa or huia there, so it will be restoration to very close to the original pre-human condition, you can do that there. But [such] islands are only a tiny subset of the New Zealand condition as a whole."

Miskelly's willingness to use a surrogate species, similar to but not identical to the species one would ideally restore—if it still existed and were available—is indicative of flexibility within New Zealand restoration principles. It is hardly controversial where we are dealing with very similar species or indeed subspecies of snipe or a parakeet where the original species was unidentifiable. It has become accepted practice with the takahe, where the North Island species has long been extinct, and its South Island analog, a very closely allied species, is introduced to restoration projects on suitable North Island sites. Miskelly probably went a little further when he proposed introducing the rock wren to Mana Island to replace the extinct bush wren.

There had been speculation ("rubbish, we know that now") that they were the same species, the former adapted to high altitudes, or simply driven there by predation. But then he found out that translocations of rock wren, even to alpine zones in Fiordland very similar to its current home range, had failed. It was hardly going to do well on a low-lying island. "As it is quite a rare bird anyway, and you could be mucking up the source population, that is an example where subsequent learning was the death of clever idea."

We get into considerably more difficult territory still with proposals to introduce surrogates from abroad. For example, it has been suggested that any of several species of raven, all alien to New Zealand, might fill the ecological niche vacated by the New Zealand raven, which became extinct between the arrival of the Maori and white settlers. The jury is still very much out on this question.

Where and What to Restore: Trophy Birds or Whole Systems?

Moving from species to ecosystems, what areas should be prioritized for restoration, and what targets should such projects have? The selection of sites for mainland islands, both within the DOC program and in the burgeoning private initiatives in this field, has come under lively and revealing criticism, some of it commissioned by the DOC itself.[30] Six mainland islands were selected by the DOC in 1995–1997, but a review the department published as early as 2000 found that "the precise reasons . . . for the initiation of these projects was unclear."[31] Despite the apparent intention to move toward broad ecosystem restoration targets with this initiative, Alan J. Saunders found that "the greatest weighting [for selection] was given to the potential to recover threatened species."[32] It is evidently hard for New Zealand to escape this species-driven priority, though the mainland island concept is obviously a much broader initiative.

One factor that underlies this failure to find a broader focus in developing and defining selection criteria is the relative immaturity of restoration ecology as a science, in New Zealand as elsewhere. New Zealand's recent experience shows that the ecological restoration movement has all the vigour of youth; it is rich in innovative capacity, but that it has not yet bedded down clear principles that are familiar to or accepted by the whole conservation community, much less absorbed by the general public. Saunders makes this quite explicit, commenting on "*the lack of any policy on the overall goals of ecological restoration, or guidelines on how restoration projects differ from other conservation activities*" within the DOC.[33]

There is no shortage of productive thought on these questions by Clarkson, Miskelly, and many others, but basic concepts are still being shaped, here as elsewhere. Until those shapes are firmer, they will not find full traction among policy makers. There is a natural tension between academics and practitioners in restoration, but Bruce Clarkson's record outside academia, spanning restoration on his own land, through his urban initiatives in Hamilton, to advisory roles on many DOC and other schemes, puts him in good position to have an overview. He offers a telling summary on the mainland island site selection issue. "There was an exercise within DOC where people made suggestions about where to put the 'islands.' It was not driven by science, it was driven by people's drive and passion,"[34] he says. And, almost inevitably, as we saw with Working for Water in South Africa, where public money is being spent, regional and bureaucratic ranking systems will trump ecologically motivated preferences. Clarkson continues: "'One in our area, one in yours,' that kind of thing went on. We have the scientific tools to do a very thorough analysis to maximize biodiversity protection with these selections, but it has not been done. It is obvious that somewhere like Te Urewera should have been chosen, but there are others you might wonder about."[35]

The proposal that mainland islands were ideal "learning sites for conservation management"[36] was built into the DOC projects, but the potential for research has not always been exploited effectively. Says Greg Moorcroft in Te Urewera, "Despite the research intention behind the 'islands,' science is frustrated by failure to standardize." For example, he says that trapping methods for the same species are different at different DOC sites.[37] "The problem is that people want to see the outcome, the iconic bird in the forest," says Clarkson. "They forget [the process of restoration], what you need in terms of learning for the future that could be applied elsewhere." Then he relents a little, keenly aware that such failings in DOC are often "a question of resources," and continues: "learning . . . does happen. If there is a new method in Te Urewera, in no time at all everyone is talking about it, the best practice is being picked up, but probably not in as formal a way as it should be."

Selection of private sites has been even more contentious, and there is a feeling, especially among DOC staff, that too many are being set up too quickly—well over forty have been established over the last decade—and that some may collapse, with negative ecological outcomes that could also damage public perception of an inspirational idea. Clarkson takes a more indulgent line on this issue: "Ideas like this don't necessarily come about in a rational way. Was Maungatautari the best place to spend that amount of money on a predator-proof fence? Not necessarily, but you had a willing group of people who put up their hands and said 'we are going to do it.' It would be ridiculous to pour cold water on that. So when a Maungatautari comes up, you encourage them to monitor and do the right things, but you don't say this is a total waste of time. Private schemes are no more at risk than government ones, all are at risk of collapse." But he acknowledges that private "islands" have special problems, reflected since we spoke in the crisis at Maungatautari: "They will only survive if there is a succession plan in place [for leadership], people must hand them on when they get exhausted," he adds.

These views tie in with Clarkson's insistence that one of the vital roles of restoration projects is the building of citizen awareness and practical activism. "We need to teach stewardship, and that is only possible if we engage people with the environment in their own back yards. That is the motivation for my work in Hamilton. If we don't do this locally and nationally, our heritage will be gone forever. The DOC has criticized the Hamilton projects because they are not making a difference to survival of this or that bird, but that is not the reason I am doing it, I am doing it so that people will learn, locally and nationally, that if we don't act our heritage will go. It [the choice of sites] can never be just a scientific direct cost-benefit analysis. The most important thing is to influence people to be stewards of their *own* environment."

Perhaps the most radical and fertile critique of New Zealand's restoration practice was made ten years ago by John Craig and a team at the University of Auckland in a paper entitled "Conservation Issues in New Zealand."[38] This essay brims with provocative ideas, which remain very relevant. Its core argument is against the complacency that might be engendered by the country's relatively high levels of conservation. True, 30 percent of the land mass is protected, and there is no unsustainable harvesting of native species. There is a strong public constituency for biodiversity, and the DOC is the only state agency that administers conservation nationally and advises on all related issues. It is a rude shock, then, to find that "government funding allows less than 5% of the protected lands to be managed sustainably,"[39] and biodiversity continues to suffer a chronic and apparently irreversible pattern of decline, both within reserves and on private land.

A Robust Critique of "Preservationist Ideals"

Craig and his colleagues are especially critical of "preservationist ideals," which they argue lead to "a management dichotomy based on extremes of market forces versus preservation." Seventy percent of the land is "managed unsustainably" by private owners, while "30% is locked up in reserves." They argue that these ideals not only fail to address the crucial issue of restoring biodiversity on private land, they also enshrine outdated ecological concepts of eternally stable ecosystems. And, like Daniel Janzen,[40] they note that the land that has been developed for direct economic exploitation usually once held the highest biodiversity, and the areas saved—at the last minute—for preservation are often the poorest in ecological terms. So a new focus on restoration in agricultural, industrial and urban areas is key to conserving native biodiversity and arresting its overall decline. Since this critique points up issues far beyond the shores of New Zealand, I think it is worth quoting some of their points verbatim:

- "Under preservation ideals, native biodiversity cannot be used [by humans] and hence has no recognized economic value."[41]
- "Since it cannot be used, native biodiversity on private land . . . receives no economic recognition for the ecosystem services that it provides to society. In the absence of realizable value, landowners carry all the costs of pest control, and it is economically rational to replace natives with exotic species that have economic value."[42]
- "Most conservation land is either in the super-humid regions or

the uplands and montane areas that are not useful for production. In contrast protected areas in the fertile lowlands tend to be small, fragmented, isolated, extensively modified, and generally poorly managed."[43]

- "Reintroductions have traditionally been conducted as one-off, nonreplicated events (trials) to locations that mirror source habitats. Reintroductions designed as well-planned experimental comparisons will more rapidly advance knowledge."[44]

The authors argue that the requirement of the Reserves Act (1977) and the Conservation Act (1987) that preserved land should represent the "original" character of the country is a false target, which denies

> the dynamic nature of ecosystems as confirmed from historic ecology. Few areas of New Zealand have had a stable forest composition for more than a few tree generations. Change, rather than stability, in composition appears to be the rule, at all scales. This is not to say that there are no relict areas of formerly more widespread vegetation types, nor that preservation of forests that have remained relatively unchanged since before European colonization is not a laudable goal in some areas. However, *it is not possible to maintain representative areas unchanged in the long term, nor is it possible to define a primeval restoration goal, except in a very general sense.* Moreover, the dynamics of historical ecosystem change also weakens the dichotomy between what is regarded as original (and therefore desirable) versus modified. It provides the basis for a new paradigm in conservation that would accept greater merging of the indigenous and exotic biota and provide a class of conservation land in which management goals would recognize the need to integrate the protected and productive components of the landscape.[45]

While the authors do not use the phrase, the thinking here comes close to that of Richard Hobbs and his colleagues on "novel ecosystems" a few years later. And they add: "Such a change would also be more accepting of people as part of the landscape and would more closely align with Maori values."[46] The essay goes on to make proposals that are equally challenging to traditional conservation taboos, such as the reintroduction of the hunting and harvesting—at sustainable levels—of native animals and plants, on preserved as well as private lands.

These ideas continue to arouse robust debate. But Craig and company were not indulging in an exercise in iconoclastic grandstanding for the hell of it. This is a considered piece that recognizes the significant achievements of conservation and restoration in New Zealand to date—especially in the field of species reintroduction, and, with reservations, of the mainland islands.[47] But it points to the much broader context in which restoration must operate if it is to make the impact required to halt the decline in biodiversity. The views they have expressed have been endorsed by other influential restoration writers.[48]

These are the big issues of restoration, and we will return to them in our final chapter. But it is in the ecological concepts related to islands that New Zealand has made its most distinctive contributions to restoration so far, and is recognized by colleagues in Hawaii, California, and Mexico as a world leader.

As we have seen, the development of island-based restoration here was born out of desperation, as more and more native species teetered on—or fell over—the brink of extinction. The idea that mammal-free islands are "arks" where endangered species could be saved from the flood of alien predators dates back to the 1890s, when Richard Henry moved two species of kiwi and the kakapo to offshore islands.[49] But even remote islands were far from secure, and many more fell victim to introduced or escaped mammals in the twentieth century. The doyen of New Zealand nature writers, Herbert Guthrie-Smith, knew in the 1930s that the security of the Stewart Island snipe and other species on Big South Cape Island was more apparent than real. "Always hangs overhead the sword of Damocles," he noted. "Should rats obtain a footing, farewell to Snipe, Bush Wren and Saddleback, none of which species are able to adapt themselves to novel conditions. As on the mainland these . . . interesting breeds would disappear."[50] Sadly, the sword fell and rats indeed arrived on the island in the 1960s. These species, along with the Stewart Island robin and fernberd, were lost to the island, and the snipe and bush wren, along with the greater short-tailed bat, were lost to the world, in that troika of extinction recorded in 1964.[51]

But it was also in the 1960s that the New Zealand Wildlife Service, the precursor of the DOC, found much more reliable techniques for capturing and translocating birds safely—the invention of mist nets was a big help—and for eradicating rats from small islands. The first complete rat eradication was achieved twenty years later. The remarkable successes in restoring species on island sanctuaries over this period sparked the concept of mainland islands in the 1990s. Some used the rapidly advancing technology of predator-proof fences, others used cordons of traps, poison bait stations and tracking tunnels, to mimic the barrier effect of the sea.[52] Overall, relocations to "island" sites, maritime or mainland, had assisted the restoration of seventy-six animals, ranging through forty-three birds, seventeen lizards, ten insects, a snail and a slug, by 2009.[53] Curiously, success rates with threatened plants on these sanctuaries have been much lower than for animals.[54] A rich spectrum of knowledge has been garnered from this experience, and many fertile questions, often still unanswered, have been provoked. For those who would like to explore these issues in more detail, a selection of notes related to my visits to, or conversations about, a number of these sites is available online at http://www.press.uchicago.edu/sites/woodworth/.

A Warning Light and a Beacon

New Zealand is both a warning light and a beacon for conservation. Its experience shows us just how rapidly and chronically biodiversity can decline. What is true of one archipelago or small continent over several centuries may well become true of the whole planet over the next fifty years, as human populations soar and pressure on natural capital and ecosystem services becomes ever more extreme. But the New Zealand experience also demonstrates that bold and committed initiatives, informed by science and constantly subject to reevaluation and adaptive management, can bring species back from the very edge of extinction in the short term. In the long term, only the restoration of ecosystems, incorporating novel and alien elements where absolutely necessary, will bring back the promise of restoring biodiversity and its evolutionary potential.

In the meantime, the price is often grim for those of us who love nature and wild places, and wish to do no harm in the world: we must pay with possums hanging broken-necked from traps, with national parks protected by poisons, and with critically endangered birds dependent on feeding stations; with intrusive interventions—loudspeakers on windy cliff tops, plastic burrows beneath them, hand-fed chicks—to lure back seabird colonies. We will always be conscious, more conscious than even the prophetic Aldo Leopold could have imagined fifty years ago, that we are living in a world of wounds.

But the rewards remain priceless. As the rising sun ignites a blaze of rata flowers in Maungatautari, as the dawn chorus resonates anew through the canopy at Te Urewera, or as a snipe performs the *hakawai* in the night sky over Putauhinu Island, we learn that we are also capable of healing at least some of them.

12 *The Mayan Men (and Women) Who Can (Re)Make the Rain Forest*

What is idyllic and simple is almost never true. * JULIO CARO BAROJA, "The Basques"

The Maya have learned to read the wind . . . the real book of the Maya is in their minds. * LÁZARO HILARIO TUZ CHI

I can't write, I can't use a computer, but in my head there is a school. * DON MANUEL CASTELLANOS, *leading* hach winik *("true man") of Lacanhá Chansayab*

Lacandon Maya culture has been successful, or at least it has known how to survive, not because it exploits the rain forest but because it manages the *acahual* [the four diverse "fallow" stages after cultivation]. They manage the forest by living in it. They have domesticated the environment. But here we are thinking like westerners, separating things out in a way that would not make sense to them. * SAMUEL LEVY TACHER, *ethnobotanist and leader of an ecological restoration project in the Lacandon jungle*

We are not substituting one system for another, but using the best of both to make something new. * SAMUEL LEVY TACHER

* *

"*Vivo! Vivo! Vivo! Vivo!* [A pause] *Ah, Muerto! Vivo! Vivo! Vivo! Vivo! Vivo!*" Hidden from view in a massive tangle of bracken, Antonio Gómez Sánchez (Toño) is calling out the survival rate of tree seedlings in an abandoned pasture. Though a sea of ferns still dominates the entire plot, the seedlings are doing very well: nine out of ten in this row are still alive. The midday heat is punishing—fit only for Kipling's mad dogs and Englishmen, I find myself thinking. I am obviously wrong. Toño's chant of life and death continues until two hours past noon, long after I have had to withdraw to a nearby tree for shade.

Pancho, a native of the nearby village of Nueva Palestina, follows Toño with a notebook. He painstakingly makes four vertical marks and a stroke-through for every five tiny trees still living. Samuel Levy Tacher, somehow still immaculate in white hiking gear, an elegant Tilley hat, and a jaunty red bandana, supervises the operation. He is happily surprised with the outcome of today's monitoring, which shows a final tally of more than 70 percent survival. The trees had been planted months further into the rainy season than they should have been due to a delay in funding, and so they are now prematurely exposed to the relentless January heat at the start of the dry season. The plot we are surveying is one of a hundred and fifty (and counting) restoration experiments around Nueva Palestina, part of the territory of the Tzeltal Maya people in southeastern Chiapas, Mexico, close to the Guatemalan border. Levy Tacher has been working here since 2005. There is no rain forest left where we are standing, but the village is, at least in theory, part of the Lacandon jungle, one of the most biodiverse areas in North America. It is located right on the edge of the Montes Azules Natural Park, a UNESCO Man and the Biosphere reserve.

But the land around the local Tzeltal villages has been extensively cleared of rain forest for pasturage and is badly degraded by the introduction of African grasses and subsequent overgrazing. The degradation takes several forms. In this case we are looking at a very successful invasion, post-grazing, of bracken that completely covers two hectares of hilly ground.[1] As well as shading out all competitors, the two-meter-tall ferns have generated a thick mat of rhizomes beneath the soil, presenting an almost impenetrable barrier to any seeds from nearby forest remnants that might penetrate the first obstacle of the ferns' tightly meshed canopy. The farmer who owns the land that Toño and Pancho are monitoring has twice attempted to burn the ferns out, only to find them surging back, each time stronger and more dominant than before. In this scenario, the seedlings, all still less than ten centimeters tall, look like a very poor bet to beat the ferns. However, Levy Tacher is basing his experiments on the traditional agricultural practices of a neighbouring people, and he is quietly confident of success. Not too far from Nueva Palestina, there is another world, that of the Lacandon Maya, a living laboratory of local knowledge that Levy Tacher believes offers multiple keys to the restoration of the rain forest and to sustainable development for its human inhabitants.

We have driven to the Selva Lacandona from San Cristóbal de las Casas, one of the oldest and most beautiful Spanish colonial cities in the Americas. This is where Levy Tacher has his office with Ecosur, a study center dedicated to sustainable development in the Mexican south. En route we have traversed mountains well over two thousand meters high whose upper slopes are capped with temperate Montezuma pine forest. This is the land of the Ch'ol people, one of several linguistic groups once encompassed within Mayan civilization. Many of the women, though generally not the men, still wear colourful traditional dress, even in quite large towns. Smaller towns often bear emblems of a conflict the world has half-forgotten: their entry roads are flanked by crude but sturdy wooden billboards which proclaim: "You are now in rebel Zapatista territory. Here the people rule, and the government obeys." The 1994 uprising by the guerrillas of the Ejército de Liberación Nacional Zapatista (EZLN) still reverberates here. Their charismatic ideological leader is a university professor who has reinvented himself, complete with balaclava mask and incongruous pipe, as Subcomandante Marcos. The initial military confrontation was bloody but brief, as the Mexican army easily drove the insurgents back from the cities of Chiapas into the countryside. Even today, however, the Zapatistas remain a shadowy yet significant presence, organizing indigenous rural communities and playing the role of Internet poster boys (and girls) for the antiglobalization movement. A de facto ceasefire with the government has proved remarkably stable. Even the EZLN's critics recognize that the status of indigenous peoples, and especially of indigenous women, has been advanced by their activities. Prior to 1994, a Ch'ol or a Tzeltal had to step off the pavement to let a white person pass by in San Cristóbal de las Casas. Not today. Indeed, in one sense we have to thank the Zapatistas for the very road on which we are traveling, because much of it was unpaved before their rebellion. Now it offers (relatively) rapid transport to the Mexican army, which maintains several heavily armed but relaxed checkpoints along our route.

Dropping down from the mountains toward the ancient Mayan

capital of Palenque, there is a steady shift from the open pine woods on the heights toward the dense vegetation of tropical rain forest, rich in bromeliads, epiphytes, and vines. The forest is interrupted by occasional cornfields and an increasing number of bigger sites dedicated to monocultural African palm production, chasing the new grail of biofuel dollars. Swinging southeast away from Palenque, the vast mass of the Selva Lacandona is now on either side of the road, but almost always separated from us by irregular patches of pasture or cultivation up to five kilometers deep, or by clusters of often flimsy homes of plywood and corrugated iron. Most of the inhabitants are Tzeltales of Mayan origin, but there are now many immigrants, from Guatemala or from other parts of Mexico hoping to scrape a living from forest clearance and cattle farming who have been attracted by the accessibility of the new road.

The scenery is similar all the rest of the way to Nueva Palestina, but twenty kilometers after that it changes abruptly. The jungle now comes right down to the roadside. Where human dwellings are visible at all, they are often half-hidden by trees and undergrowth. This is Lacanhá Chansayab, home to the largest remnant of the Lacandon people, another Mayan group. There are only about seven hundred of them here, with a handful in two other centers, and this is the last place where they constitute the majority community. Their relationship to the jungle is radically different from that of their Tzeltal neighbors. It is from their wise men and women, or *hach winik*, that Levy Tacher has been studying the traditional Lacandon cultivation techniques. Their methods, he believes, could make possible the restoration of much of the forest edge, all the way back up the road toward Palenque. *Hach winik* translates literally as "true man," a hint that, for the Lacandones, human potential is only really fulfilled when it is engaged seamlessly with the forces of nature.

Paradise Lost?

"Lacandon" is a name to conjure with, echoing with resonances of noble primitives, preserving an ancient culture in remote jungle fastnesses, uncontaminated by industrial civilization. They were immortalized in the exquisite black-and-white photographs of Gertrude Blom, taken mainly in the 1940s and 1950s. Eyes dark as the forest in which they dwell, liquid as the misty waters on which they canoe, huge as the moon that the Lacandon claim as mother, these are seductive images of Eden before the Fall.

But the history here is, inevitably, a little more complex than Paradise lost. This jungle was Mayan territory when the Spanish *conquistadores* arrived. The Mayans were a multicultural, multilingual civilization, and the main inhabitants at that time were the Ch'oles. It was they who left behind the region's monuments of classic Mayan architecture and art, like Palenque and Bonampak, which had been abandoned centuries before white colonization. The Spanish called these Ch'oles "Lacandones" and moved them to the forest edge, where they could be more easily controlled. Yucatec-speaking Mayans then moved into the jungle from the south and inherited the Lacandon name, until they too were relocated by the Spaniards. Finally, in the eighteenth and nineteenth centuries, more Yucatec speakers moved in from Guatemala, and these are the ancestors of today's Lacandones. And they indeed survived in relative isolation, deep in the jungle, maintaining their cultural and agricultural traditions fairly intact. Their only contacts with the outside world were loggers and, rather bizarrely, agents from the chewing gum industry—the tree that produces its traditional raw material, chicle, is common in the jungle. In the 1940s, however, the Lacandones came under pressure from descendants of the original Ch'ol inhabitants. Along with Tzeltales, the Ch'oles moved back into the jungle by the tens of thousands, attracted by agrarian laws that encouraged clearing forest for pasture. The Lacandones sought more remote living spaces but were eventually resettled in three areas, including Lacanhá, by the Mexican government.[2] They enjoy significant government subsidies and, unlike the Ch'oles and Tzeltales, have shown little interest in the Zapatista movement.[3] But their numbers, already counted only in hundreds, are dwindling, and their traditional knowledge is in danger of being lost within the current generation.

Levy Tacher has been working closely with the Lacandones for eighteen years, and especially with Don Manuel Castellanos Chankin, a remarkable *hach winik* originally from Naja, one of the other two Lacandon settlements. Don Manuel has become an active participant in the restoration process in Nueva Palestina, despite a legacy of bad feeling over land rights between Lacandones and Tzeltales. Our first encounter with him, however, reveals just how precarious even the best-cultivated relationships between scientists and indigenous people often are.

We arrive at his household, a group of barn-sized wooden structures grouped loosely around a fish pond in a jungle clearing, just as darkness falls. Several noisy macaws are still scrambling through nearby trees, but my excitement abates a little when I realize that they are tethered on long leashes. We stand some distance from the building, as local good manners dictate. Levy Tacher makes a resonant hooting noise to formally announce our arrival—though the noise of our Jeep must already have alerted Don Manuel and his family. We wait in silence for several minutes, and then I catch a glimpse of a small figure dressed in white, scurrying in the dusk from one building to another. It is Don Manuel, in the long, white V-necked tunic that is traditional to Lacandon men. He emerges again a few moments later, approaches, and greets us rather cooly, though smiling. He moves as if to shake hands, but his own fingers remain rigid, so that there no mutual grasp, a curiously unsettling gesture that I will find is common to all Lacandones I meet. He cannot be much more than five feet tall, and is light-framed to match, giving an impression—very deceptive, as I later discover—of boyish frailty. This impression is somehow accentuated by his long pale reddish hair and faint mustache and beard. His eyes are crossed so severely that I take him to be partially blind at first, though once again I am way off the mark.

We are invited in to a spacious, high-roofed wooden building to sit down and talk. We take our seats at long tables on long benches covered with oil cloth. Don Manuel vanishes again while family members peer in from the next room. He reappears and takes a seat at the head of one of the tables, now wearing a heavy woollen cardigan over his tunic. He constantly examines and picks at the sleeves, as if looking for flaws, and makes little eye contact with us. Levy Tacher formally introduces me, describing me as a journalist and explaining that I would like to visit his cornfields and learn about his techniques for cultivating and restoring the forest. Don Manuel responds abruptly, repeatedly stating that journalists have come here many times and often seem like good people. But then they write bad things about the Lacandones. His attitude surprises me, because I have seen Don Manuel on video, enthusiastically collaborating with the media in discussing his eco-agricultural methods, and I had understood from Levy Tacher that he was very comfortable in this role. Perhaps something else has happened more recently to change his views; his words tonight suggest a general diatribe against outside visitors, without exception.

A Wasted Journey?

But I find it very hard to grasp the detail of what he is saying, as his already idiosyncratic Spanish is further blurred by his habit of speaking with a hand in front of his mouth, and at times he seems to be rambling incoherently. With a sinking heart, I begin to doubt Levy Tacher's judgment that this man is a font of botanical knowledge, or at least that he is still willing to share it. Perhaps this long journey has been wasted. An uneasy situation is made more awkward by the sudden arrival of two young women. I first take them to be Don Manuel's daughters, but they turn out to be Mexican backpackers wearing traditional ponchos. They are ushered in by a local ecotourism agent. He knows Samuel well and, understandably, thought his arrival might be a good moment for his guests to gain a privileged insight into traditional life. Our strange standoff perhaps gives them more insight than they had bargained for. They look awestruck, baffled, and uncomfortable, by turns.

Levy Tacher does his very best on my behalf, putting his own reputation with Don Manuel on the line in defense of the quality of my work. His manner is usually laid back and gentle, and always exquisitely courteous, but by now he is fairly quivering with tension. This is not what he expected either, then. Toño, who has spent many nights in Don Manuel's household, also makes an eloquent intervention in my favor. I fear that my own hesitant attempts to describe what I am doing are abject failures in communication. Coming from a profession that includes such luminaries as Rush Limbaugh and Ann Coulter, there is no way you can prove your good faith as a journalist to a stranger before your work about them is published. Finally, Samuel rises and announces that he and I will stay elsewhere, though our plan had been to stay here; indeed, the rent had already been paid. This tactical withdrawal seems to surprise Don Manuel, and his attitude softens slightly. It is agreed that we will at least meet again. Miguel, the ecotourism agent, takes us to some cabins in an "eco-lodge" that has not yet opened and is not yet finished. It is pretty basic, with half-hung doors open to the creatures of the night, but at least there is hot water. Better again, the large windows reveal stars blazing with limpid intensity through the jungle canopy. The multiple cascades on the nearby Lacanhá river make night music so soothing that I find it easy enough to believe Levy Tacher's assurances: tonight's debacle will all be sorted out in the light of day.

Not much of that light has filtered through the thick mists that lie over the early morning landscape when we are on the road again. We have breakfast—scrambled eggs with the—mercifully optional—condiment of an incendiary chili jam. We are in an informal café, in the house of another Lacandon, Enrique Paniagua. Here we are made most welcome by his entire family, and in a quiet moment Paniagua agrees willingly to bring me to see his cornfield on another day. With that agreement for insurance against a prolonged impasse in relations with Don Manuel, we head to Nueva Palestina, where Samuel and Toño will teach Pancho how to monitor the restoration plots in their absence.

Traditional Ecological Knowledge: Reality or Romantic Fantasy?

I had come to the jungle with some skepticism about the current of ideas in ecological restoration that swirls around the concept of traditional ecological knowledge (TEK). The notion that peoples whom the white imperial powers regarded as "primitives" are in fact the guardians of ancient and sophisticated environmental wisdom is superficially attractive. But it has been romanticized so often that I think we have to approach it with caution, or at the very least with the same scientific rationality with which we approach other socioecological issues. It may be comforting to believe that the colonized are always more virtuous than the colonizer, that the dispossessed have an innate superiority to the plunderers, that in some way the often genocidal agression of the West's empires against subject peoples is also the sole cause of humanity's rape of the planet—in short, that only industrialized socities degrade the environment, and that preindustrial peoples are universally the protectors of Mother Earth.

Unfortunately, even a cursory glance at the broad picture of human history reveals that *Homo sapiens* has been well capable of fouling her and his own nest millennia before the rise of Euro-American industrial power. No universal certificate of ecological wisdom attachs to hunter-gatherers or early agriculturalists. The more we learn about previous civilizations, the more it is evident that they, too, were often irresponsibly spendthrift with their natural capital accounts and showed scant concern for the carrying capacities of the ecosystems that supplied them with vital services. Think of the Native Americans stampeding hundreds or thousands of buffalo across a cliff edge to harvest a few dozen carcasses; think of the litany of extinctions recorded as the Polynesians advanced across the Pacific; think of the massive erosion probably caused by Neolithic overgrazing on Ireland's Burren landscape.[4] The list goes on and on and is not mitigated by the fact that the populations of earlier societies rarely reached the kind of critical mass that make such human impacts truly catastrophic on the environment. As Daniel Janzen says, this is what human beings most characteristically do: we change our environments in what we see as our own best interests, usually paying little heed to the consequences on other species or ecosystems, until we notice that those consequences are becoming negative for our own survival.[5] By which time it has often become too late to redress much of the damage.

Wishful Thinking, Willful Ignorance

Romantic proponents of TEK claim to hold preindustrial peoples in the highest regard. Yet their failure to appreciate the impacts that these peoples actually made on "pristine" landscapes—which often turn out, on closer examination, to have been culturally and ecologically molded by humans for centuries—suggests little real respect. The underlying sentiment is wishful thinking, tantamount to willful ignorance. Perhaps I have been unfortunate in my choice of TEK sessions at SER conferences, but I found most of them to be at best infuriatingly vague and wooly. At worst, they manifested symptoms of a blind flight from science to a gooey mysticism. A particularly clear-cut example had arisen at the SER conference in Perth in 2009. At one TEK session, a young Maori, Aareka Hopkins, spoke about his restoration work in New Zealand. A highly articulate young man, he painted clear pictures of his projects, and described movingly how he conducted them in harmony with Maori cultural practices, involving the local *iwi*, or tribe, in traditional ceremonies at crucial moments in the restoration process. This might, for example, involve the ritual release of a reintroduced bird species.[6]

Working within the cultural envelope of people local to any restoration project, from whatever background, is obviously a good idea, reflecting respect, sensitivity and democratic accountability. Hopkins's work seemed admirable in this regard. And so did the quality of his restoration work. But it struck me that there was no traditional Maori *ecological knowledge* involved in the latter—the projects he described were all based entirely on Western scientific principles. I raised this issue with him at the session, and he agreed that I was absolutely right, but then offered a rationale that raised a lot more questions than it answered. He explained that his people's ecological knowledge never extended to restoration, because "everything was pristine in our ancient culture. We never had to restore, because we looked after it well." This was a truly extraordinary statement for a New Zealand ecologist to make, but no one challenged it. It is abundantly clear that the impact of the Maori introduction of dogs and rats was disastrous for many New Zealand species and that unsustainable hunting and forest clearing drove others to scarcity or extinction.[7] Of course, the subsequent European settlement, which introduced a far wider range of mammalian predators and cleared native forest at a far faster rate, had an even bigger environmental impact. But damage done by white settlers does not erase the damage done by their Maori predecessors from the record. To pretend that it does is a poor service to history, and to science. Which is only to say that Maori and Europeans belong to the same ferociously destructive species, no more and no less.[8]

Ironically, to erase this record is to fail to appreciate the real value of the ecological knowledge that the Maori had indeed acquired, though it is lodged in a word English has borrowed from their language. As an island people, they had learned the hard way, through a string of early extinctions, that they were in bad environmental trouble. They gradually grasped, to use today's ecological jargon, that they were exceeding the carrying capacity of their environment and losing species they valued as a result. So when the New Zealand pigeon (*kereru*), for example, became noticeably scarce, the local elders would declare it to be *tapu*: that is, it could not be killed for a specific period and/or within a specific geographical range. This was a very early version of the European and North American legislation for sustainable hunting based on seasonal or species-based bans. Hence, though we use it in a different sense today, the relatively new English word "taboo."

There were many well-informed restoration scientists and practitioners at Hopkins's session in Perth. But no one present, myself included, questioned his assertion. It seemed that was another kind of taboo in operation here: that one must not challenge statements about the good ecological stewardship of traditional or indigenous peoples. I felt bad about my own silence for weeks afterward. It would have shown much more respect to Hopkins, and to his people, to have taken issue with him, as I would have done had the speaker come from a European background.

But I tried not to let my frustration at this kind of experience polarize me into thinking that there is *no* traditional ecological knowledge. I was also aware that there is strong evidence that some—not all—preindustrial peoples did have an acute sense of the mutual interdependence of humanity and the rest of the natural world. Sustainability is not a new idea, though it has rarely been dominant in our societies.[9] On a literary level, the prophetic language used by some Native American leaders to denounce the environmental disasters they foresaw, as hordes of white settlers swarmed westward, is resonant in its poetry and in its prescience. But the dominance of True Believers at these TEK sessions almost always left me uncomfortable and disappointed.

A single session at the second International Symposium on Ecological Restoration, in Santa Clara, Cuba, in 2007, changed that disappointment into intense curiosity. Samuel Levy Tacher presented a DVD in which Don Manuel Castellanos made the remarkable claim, "I can make the forest." And according to Levy Tacher, that is what the Lacandones do, restoring their cornfields to rain forest in a meticulously planned five-stage process based on a knowledge of local botany and plant ecological functions that any scientist would envy. And each phase of this management of the forest yields distinct and diverse harvests, ranging from food to medicine, tobacco to beads, construction materials to clothes.[10] He and his colleagues from Ecosur are engaged in studies that they describe as "bridge-building" and "translation" between this traditional ecological knowledge and conventional science.[11] Driving down from San Cristóbal, Levy Tacher tells me how he sees the study of TEK having practical applications in the field of contemporary ecological restoration. He is an ethnobotanist by training. He describes his discipline as "the daughter of anthropology, not biology"; it always takes into account the cultural context of the indigenous botanical practices it studies. Mexico's remarkable double endowment in biodiverse ecosystems and diverse indigenous human cultures makes it an especially promising focus for such studies. He was trained by a founder of Mexican ethnobotany, Efraín Hernández Xolocotzi, but he believes it is time for the discipline to move beyond his *maestro*'s exclusive focus on description to a new level, that of experimentation:

"It is not just a matter of studying 'what the Indians know'; we need to find out how we can use that knowledge to protect the things they consider important, to see what use TEK may be in the present and in the future. Our responsibility is to find the factors in their knowledge that may enable us to develop our science of conservation and restoration. This is very difficult, because we see the world as made up of separate things, while they see it as a whole."[12] But one of the most important points he makes is that those of us coming from a scientific background to TEK also have to separate two strands in our approach: "In order to draw on this reservoir of information ecologists must learn to respect its cultural significance while evaluating its predictive power."[13] In other words, ecologicists should not confuse an entirely appropriate sensitivity toward the belief systems of indigenous groups with an uncritical approach to the scientific veracity of their ecological practices.

A scientific experiment, Levy Tacher says, is an artificial way of finding an answer to a problem, and in the Lacandon view it is a clumsy device: there are always more factors in nature than we can include in an experiment, so that that the experiment may actually hide the interaction of all these factors. Lacandon agriculture is in fact a cumulation of the practical experiments of many generations, and they can find it absurd that we need to replicate experiments to prove something they already know to be true. "Our work is not folklore or cultural do-gooding," he continues. "We have to teach them to separate out their united reality so that they can guide us to design experiments. We are not substituting one system for another here, we are using the best parts of both to make something new."[14]

A Seamless Web of Cultivation and Restoration

The challenge he has set himself in Lacanhá and Nueva Palestina is formidable. He is attempting to learn everything, cultural and technical, about Lacandon agroforestry. Since they exploit the forest without destroying it, you might say that there is nothing for them to restore. Or, rather, that they are engaged in a form of ecological management and manipulation of the forest that constitutes a seamless web of cultivation and restoration, in which they see no need to unstitch these two concepts.

However, there is clearly a need for restoration of the land farmed by their Tzeltal neighbors. Their practices, stimulated by the government policies that have subsidized forest clearances for cattle pasturage since the 1940s, are based on classic slash-and-burn techniques. And when the native grasses that replaced the forest proved inadequate for raising cattle, they planted nutritious and fast-growing alien species like African star grass, only to find that they rapidly exhausted the soil. Levy Tacher began to wonder whether it might be possible to transfer the Lacandon techniques to these degraded pastures, and at least partially restore the rain forest on Tzeltal land. After all, the gulf between their agricultural practices might not be unbridgeable. The Lacandones, too, slash and burn the forest. But they do so on restricted spaces *inside* the jungle, rather than turning it into open land. As two veteran observers of the Maya succinctly put it, "The Lacandones farm in the forest, they do not replace the forest in order to farm."[15] And they cultivate their clearances for corn until they see from a shrinkage in the cobs that the land is getting tired. Then they not only move on to a new clearance but, crucially, they accelerate the fallowing process on the exhausted cornfield through replanting a sequence of forest species. Finally, depending on circumstances, they either allow it to return to something approaching approaching the condition of the surrounding old-growth forest,[16] or they clear it again to start a new cycle of corn cultivation.

PANEL 5 • Farming the Forest

The Lacandones traditionally describe a five-stage farming cycle within the forest in their own language, but it is often simplified into two broad categories when they discuss it in Spanish: the *milpa* (cultivation phase) and the *acahual* (fallow phase).

But this translation can mislead, because *acahual* (and "fallow") implies an abandonment to natural succession, in which the land "rests" and restores itself without little or no management. But there is usually some degree of cultivation within each of the four stages subsumed by the Lacandones under *acahual*, and always a significant degree of harvesting. However, the fifth stage approximates more and more to the condition of the surrounding old-growth forest with every passing year.

And while *milpa* is usually translated as "cornfield," and corn (*maize*) is indeed usually the dominant crop, the term gives little indication of the exceptionally varied cultivation that a single Lacandon *milpa* contains. Additional crops range from squash to coriander, from papaya to sweet potato, from tobacco to bananas. Up to fifty species of vegetable and fruit may be grown in a single *milpa*, and rarely less than twenty:

"The ground between hills of corn is covered with squash and the vines and leaves of sweet potatoes, yams, and jicamas. Small groupings of garlic, chiles, tomatoes, and sugar cane are dispersed throughout other areas. Standing above this surface growth are maturing trees of papayas, bananas, plantains, and varieties of wild plant species. Finally, root crops lie at varying depths below the surface: taro and sweet potatoes a few inches beneath the soil, manioc below them, and yam tubers below the manioc. In this way crops utilize available space, water, and soil nutrients in a highly efficient manner" (Nations and Nigh 1980, 11).

Sowing and planting seasons for some species are determined not by a rigid calendar, but by the flowering of particular plants in the jungle that indicate propitious climatic conditions, which may differ, marginally or significantly, in any given year (Nations and Nigh 1980, 9). Planting patterns are very sophisticated. Several clusters of the same species are widely dispersed in each *milpa* to minimize the impact of disease or predation; but there are enough individuals in each cluster to ensure that genetic variation is facilitated.

Some plantings are deliberately manipulated to make them attractive to mammal predators, partly so that they will ignore other plantings, and partly as lures so that they can easily be hunted in this relatively open space. "The *milpa* produces meat as well as fruit and vegetables," Don Manuel told me.

Though fewer species are harvested from the various *acahual* stages than from the *milpa*, the variety is still remarkable, with the Lacandones finding uses for as many as 60 percent of the trees. Nations and Nigh suggested that "orchard garden" conveys the nature of these phases better than "fallow" (Nations and Nigh 1980, 15).

The *acahual* is also home to many wild animals and birds, attracted by the rich crop of flowers and green shoots the growing trees produce. Many of them will also regularly end up in the Lacandon cooking pot, to the extent that some observers have described them as "semiwild" creatures (Nations and Nigh 1980, 17).

Even among the few practitioners of this form of agriculture remaining today, there is considerable individual variation. One researcher, Francisco Roman Danobeytia, has described how his original conception of "traditional" ecological practices has been revolutionized by encounters with the Lacandones, whom he sees as "constantly innovative and experimental" in their approach. Like all good farmers, what we would call "adaptive management" is an essential part of their relationship to their land. Despite their reverence for ancestral custom, he says, "they are the vanguard, not us" (Román Dañobeytia 2011).

In any case, a single *hach winik* will vary her or his own cycle according to need, and local soil and conditions. Their skill lies in their ability to "direct and accelerate succession" in each of the fallow phases to best respond to these needs and conditions (Douterlugne et al. 2010, 322).

Sometimes there is a need to bring an *acahual* back into *milpa* production relatively quickly. In this case it will be sown with a tree like balsa, known by the Lacandones to rapidly enrich the soil with its leaf litter. Moreover, balsa shades out weedy plants, and is easily cut down, so that clearing the site for renewed and relatively intensive cultivation is easy. Making entirely new *milpas* obviously involves felling and burning old-growth forest, a laborious task that is avoided as far as possible, in favor of old *acahuales* being brought back into production as *milpas* in rotation.

Sometimes *milpas* and *acahuales* are scattered in a necklace around a homestead, each enclosed in old-growth forest. But the cycle of five stages may also move back and forth laterally across a single area, with the *milpa* advancing in one direction every several years, leaving bands of *acahual* in successive stages behind it, before eventually reversing direction, or starting again at the oldest *acahual*. The *milpas* (and on occasion the *acahuales*) are weeded with high selectivity. Given the range of species involved, the Lacandones botanical and taxonomic expertise always amazes visiting botanists.

While the Lacandon system is based on the intensive manipulation of

PANEL 5 *(continued)*

forest plants, it is deeply dependent on the close presence of old-growth forest in order to function. The nearby trees are the vital seed source for the regeneration of the *acahuales*, and also provide services to the animals that feed there that an orchard garden cannot supply. For this reason the plot sizes are always small, with a *milpa* rarely larger than a single hectare.

Social change can impact the *milpas* in unexpected ways. Since industrially produced cigarettes have become easily available in the region, the Lacandones are growing much less tobacco. And because tobacco requires especially intensive weeding to flourish, this shift in cultivation has also contributed to an increase in weed invasion. More recently, the expansion of ecotourism is assumed to be a factor in the increase in cultivation of *Canna indica* (Indian shot), source of the berries from which the Lacandon make decorative beads. Since their own population is declining, the tourist market must be driving the rise in *C. indica* production (Diemont and Martin 2009, 262).

These recent changes add to the urgency already created by chronic decline in population in making it vital that the unique repertory of knowledge held by the few remaining *hach winik* is understood and recorded as fully as possible in this generation.

Levy Tacher knew from the outset that his proposed knowledge transfer would be a delicate task. There was little love lost between the neighbor peoples. The Lacandones, already reduced to a tiny population, resented what they regarded as a massive Tzeltal and Ch'ol incursion into their traditional territories, and saw their destruction of the forest as barbarous, even sacriligious. The Tzeltales resented the iconic status the Lacandones had aquired as a pristine jungle people in the eyes of anthropologists and cultural travelers, which had brought them subsidies from the Mexican government and an increasing flow of ecotourists.

It was ecotourism, however, that gave Levy Tacher an opening. Rain forest was a big attraction, so some budding entrepreneurs on the Tzeltal side realized they needed to bring the forest back if they were to get a share in this market. They accepted Levy Tacher's assistance in doing so. In time, a number of farmers also recognized that they needed assistance in bringing back fertility to their degraded fields, and signed up to experimental restoration programmes. Levy

Tacher's long-term vision also draws on traditional ecological practices still maintained, or only recently abandoned, by the Tzeltales (and Ch'oles) themselves, which could be crucial in developing connectivity between restored areas, and indeed between restored areas and old-growth forest. Ultimately, they could even be of considerable significance in actualizing the macro-vision of the Mesoamerican Biological Corridor, an ambitious but deeply troubled plan to reconnect ecosystems from Mexico to Panama.[17]

All the local Mayan peoples had the custom of leaving a belt of forest about two kilometers deep around the village before they would commence cultivation or pasturage. This was recognized in Spanish law as the *Fundo Legal*, an area of commonage that could be lightly exploited for firewood and hunting, but otherwise remained intact. The villagers saw one of its additional benefits as the regulation of temperature, as the proximity of dense vegetation reduced the suffocating heat of the Yucatan peninsula by several very welcome degrees. Secondly, there was a strong tradition of leaving a strip of forest, about twenty meters wide, on either side of all paths, around fields, and on the banks of rivers, ponds, and canals. Again, this was valued for shade and shelter, and there was also an understanding, articulated to me in detail by Don Manuel, that these *tolches*, or "tree rows," prevented erosion and flooding along watercourses. And there is a third enlightened practice, found especially in Tulija, north of Palenque, where the Ch'oles still retain a scattered tree cover, savanna style, in their pasturage. This halfway house between intact forest and total clearance maintains some connectivity among forest remnants and retains many niches for jungle species, while the root systems and shade spread by the remaining trees offer some defence against soil degradation and erosion.[18]

Despite these elements that point to common traditions, Levy Tacher needs skills more associated with high diplomacy than with ethobotany in his efforts to effectively bridge TEK and science, and also to mediate between two closely related but mutually suspicious indigenous cultures. Gradually, however, Don Manuel has accepted a new role as tutor of the Tzeltales, and his guidance has been accepted by a number of his neighbors. It will be easier, of course, when some of the Tzeltales become proficient in Lacandon techniques, and one of Levy Tacher's most promising strategies has been to establish a high-visibility base in the local *preparatoria* (high school), Pancho's alma mater.

This had been one of our first stops on arrival in the area. The light was already fading, the school long closed for the day, but its grounds stood out from the surrounding cattle pasture because of the presence of many trees. Levy Tacher led me into the plantation as a pair of bat falcons flickered out to hunt through the dusk. He could not wait to show me the star performer of his restoration projects, known as *chujúm* to the Lacandones. *Ochroma pyramidale* is familiar across the world as balsa, the ultra lightweight wood invaluable for a range of products from model airplanes to water sports equipment. Indeed, the word *balsa* means "raft" in Spanish, because the colonists found that the Maya and other peoples used the wood for short-haul river transport.

An Ideal Pioneer

But Don Manuel has shown Levy Tacher and his colleagues that *chujúm* has other remarkable properties, all related to renewing the soil in an *acahual*. Levy Tacher and his colleagues have made a detailed study of its functions, and have confirmed that it punches well above its weight in its native forest ecosystem.[19] Balsa is an ideal "pioneer" tree for restoring soil, and Levy Tacher stresses that in his view soil is, literally, the foundation of all effective restoration. "Ecologists think restoration is a matter of planting trees," he told me that evening, "but you must first work out how to give back to the soil what it has lost."[20] If you can change the composition of the soil, he says, the plant community it supports will change spontaneously, as long as your site is close to a seed source.

Balsa grows very fast, up to four meters in its first eight months, and can reach its maximum height of twenty-five meters within two years. Those frail-looking seedlings Toño and Pancho were monitoring in the bracken patch are balsa, and it is in these situations, where an aggressive invasive has turned degraded land into a monoculture, that balsa's role in bringing back biodiversity is so dramatic. If previous experiments—and generations of Lacandon experience—are reliable, in two years' time the bracken on that site will have almost totally vanished, and an increasingly diverse open understory will be developing under a thick deep-green balsa canopy. The same leaves that have shaded out the light-dependent ferns then perform two more vital functions, as they fall in copious quantities throughout the year. At first, their large drying fan shapes merge to make a natural filter. They prevent the lightweight seeds of weedy pioneer grasses reaching the earth, and the seeds wither on these parched surfaces. But they allow the heavier seeds of large trees to penetrate their layers. And when they do, they encounter a rapidly forming leaf mould litter busy building a soil in which they can easily flourish. Balsa is just one of many plants which the Lacandones use to "direct and accelerate succession in order to enhance production and allow depleted soils to recover," in scientific terminology,[21] or to "make the forest" in the words of Don Manuel. But to transfer the Lacandon technology effectively, Levy Tacher felt it would be best to simplify it, at least initially, and just use one species if possible. He recognizes that this approach begs a big question.

"Can you recreate forest biodiversity on degraded land with just one plant?" He gives his own answer: "Yes, you can, but not just anywhere, and not in any old way. You must use the Lacandon methods of clearing, planting and weeding, and you must be in, or close to, a megadiverse forest so that natural seeding will occur."

To decide on which plant was most effective for this role, Levy Tacher and his colleagues conducted an elaborate multivariate analysis of forty late successional species found in the jungle and recommended by Don Manuel. They evaluated them for factors like rate of growth, production of leaf litter, profitability as a product, resilience, attractivity to fauna, and so on.[22] When they had finished, he asked Don Manuel to give them his own top-forty list, and the results were extraordinarily similar, with balsa topping both charts. Don Manuel found it amusing that the scientists had had to create so many complex graphs to establish what was, to him, common knowledge. Once again, Lacandon lore had worked along a parallel but separate track to scientific method, and reached the same destination.[23]

While waiting for a new personal convergence with Don Manuel, after the mysterious rupture of our initial visit, we continued to look at other projects. I got to visit one Lacandon *milpa*—not Enrique's, as it happened—and while the general impression of farming within the rain forest was certainly striking, neither the variety of crops nor the quality of weeding met the standards I had been led to expect. The owner confessed that his new job as a taxi driver, and a sideline as an eco-guide, allowed him little time to give it the attention it needed, and that his children had little interest in taking on its undoubtedly heavy responsibilities. We had lunch that day in a Lacandon restaurant. The young man who owned it wore the brown Lacandon tunic, and his hair in the very long and straight traditional style. But he said he had no desire to maintain a *milpa,* and hated visiting them. "Too many ants there for me," he told us. But that night we met one of his contemporaries, Adolfo Chankin, a much younger stepbrother of Don Manuel, and heard a different story. From his T-shirt and jeans he might have been from any Mexican urban community, but he expressed great interesting in preserving the ancestral agricultural system. He was annoyed that a jaguar had recently taken one of his pigs, but he was happy about the presence of the *tigre,* as it suggested that the forest was still healthy.

Behind the scenes, Levy Tacher had been repairing his relationship with Don Manuel, and the following morning I found myself graciously invited to visit his late father's *milpa,* currently maintained by his stepmother and an aunt. While women used not be regarded as potential *hach winik,* the shrinking numbers of men have thrown them into the role, and now no one thinks it exceptional to find them busily weeding in the jungle, as Doña Rosa and Doña Maria were doing that morning.

Chanting a Horticultural Cornucopia

We walk to Don Manuel's home past houses with Sky TV dishes, and a well-equipped school. An electronic communications tower bristles above a canopy of tropical trees, cecropia, cedar, and ceiba. Don Manuel's wife, Carmita, brings me coffee while I wait for him to arrive, and she quickly dispells any romantic notions I have left about how the Lacandones actually live.

"My sister has a *milpa* because it supports her children," she tells me, "but her children won't go there, they will go to school. Before there was nothing here but meat from the forest, and fish, and the *milpa.* My father as president of our community brought in flour, flour was wonderful, and food in cans. Before that there was nothing, no roads." A little taken aback by the notion that a rain forest without roads or canned food was "nothing," I stumble into a bland comment about the benefits of direct contact with nature. "Do *you* have your own *milpa*?" she shoots back at me, black eyes glinting with knowing humor. "I don't think so. I don't think you do. I don't think you could stand the heat."

Put gently but firmly back in my place, I follow Levy Tacher and Doña Carmita's *hach winik* husband into the rain forest. I am further humbled by trying to follow this frail-looking man as he dances—there is no other word for it—barefoot along paths I cannot see. He glides on, across a dozen streams on slippery stones or single-trunk bridges, until we reach his father's *milpa*. Here indeed is a true cornucopia. Don Manuel gives all the credit to his favored tree—and to the sweat of his family's brows.

"*Chujúm* gives richness to the earth for all these crops, and it never fails," says Don Manuel. And he chants their names in a sing-song voice, pointing each one out like old friends he loves, a litany of horticultural bounty: "*Tomato, chile, comote* [sweet potato], *chayote* [squash], *jicama* [a root vegetable like a turnip], *papaya, cilantro, piña, plátano* [plantain], *frijol, banana, maiz.*"

Balsa is prominent on this *milpa*'s forest edge, and where it has been cleared to remake the *milpa* from an *acahual*, its trunks are usually allowed to decompose directly into the soil and provide homes for myriad insects and small reptiles in the meantime. Sometimes they had been burned and the ash spread to enrich the soil with nitrates. One edge of the *milpa* had been fallowed, and dead trunks are left upright as perches for birds. They defecate a steady supply of forest seeds, and occasionally present an easy target for a rifle (even Don Manuel no longer hunts regularly with a bow and arrow, though he knows how to). Nothing seems to be wasted here. He stresses the amount of work required to maintain the *milpa*, "day after day, month after month." There is no good restoration anywhere, it seems, without a tribute of human sweat. The two women, who must both be well into their sixties, bear out his words. They barely pause for shy introductions, moving through the crops constantly, pulling up weeds, which are laid out in neat piles for their roots to wither, almost visibly, in the morning sun.

We move back into the forest, and every so often stop to study an *acahual*. I have to confess that I could rarely have detected the merging boundaries between mature *acahual* and old-growth forest on my own. But they are clearly visible to our guide, who comments ceaselessly on the particular fruits, nuts or timber that had been or could be extracted from each one. His claim to "make the forest" seems less like hyperbole with every *acahual* we visit. Finally we reach a pair of plots where he and Levy Tacher had conducted experiments on the impact of *chujúm* on bracken, which had found a foothold in this part of the jungle after a fire. And here the differences are impossible to miss. The block where *chujúm* had not been planted is a single impenetrable thicket, thirty years old, where even Don Manuel's redoubtable machete can hardly penetrate. But he uses it deftly to hack away a few big fronds and then to dig down into the underlying soil. The handfuls he brings up are always more rootstock than humus, where the aggressive rhizomes have left little space for competitors. Yet there is so little trace of bracken in the nearby plot under balsa that it is very hard to believe that five years earlier the vegetation had been almost identical to the control plot. Hard, that is, until a few deep jabs of the machete reveal rich dark earth, in which a number of other tree species have now set seed. And if you look carefully enough, you can still find the decomposing skeletons of the rhizomes, already almost absorbed in this freshly fertile soil. *Quod erat demonstrandum*, indeed.

Can You Eat Concrete?

It is midday, and time for a rest. Levy Tacher and Don Manuel stretch out side by side, an unlikely couple in everything except their love for the jungle, and united again, after whatever storm had divided them briefly, by their respect for each other's ways of knowing it.

I ask Don Manuel why he thought the Tzeltales and Ch'oles had lost the Lacandon art of sustainable agroforestry. He cannot find an answer in the past, but is adamant that in the present their use of herbicides was disastrous: "I never use poison, it kills everything below the surface, then it burns the earth, and we kill the earth after several uses. They told us it was good, I used it to see how it worked. All the insects and bugs under the earth die. It frightens me that the Tzeltales use it, all the roots rot, it is like cutting the body and drain-

27. **The *hach winik* and the ethnobotanist.** Don Manuel Castellanos and Samuel Levy Tacher conduct a leisurely master class on restoration in the Lacandon rain forest in Chiapas, Mexico. (Photograph by Paddy Woodworth.)

ing it of blood. What will I say to God if I have destroyed everything with chemicals?"

What does he reckon about the prospects for Levy Tacher's restoration projects in Nueva Palestina and beyond? "In Nueva Palestina, there is nothing but open and degraded fields, one after another after another. The solution will be complicated. . . . But I know we can . . ." And his thin voice rises and firms up, with suddenly emphatic authority: "We can do it there, like in the native forest. But if they don't take care of their fields, month after month, it won't work. It will take a huge effort. But it can be done."

I had learned from Levy Tacher that Don Manuel is the pastor of a local Protestant church, yet I notice that he makes frequent references to Mayan deities and spiritual beliefs as we move through the jungle, speaking of them as naturally as the plants and animals he comments on, often in the same sentence. I attempt to tease out how his evangelical faith can coexist so comfortably with the old beliefs, but my question is either inappropriate or too clumsily put to be understood. But it prompts a fervent response about the relationship between humanity and the natural world.

"I struggled a lot for the forest, for the river. Why? Because I wanted to obey the law. God said the land is yours, but use it well, care for it, do not use it badly. The old people said the forest was life, was breath, forest-life-man, but not just man, the animals, the toucans, the fish. . . ." And after a pause he does address my question more directly: "Yes, I believe in these [Mayan] spirits. That is why I care for the jungle. In the Bible, God gave us the world to care for it, not to dig in the forest, not to make a quick profit, but to care for the trees and the animals that give us our living." Again, that timbre of authority puts backbone into the fragile voice: "For this reason, *I don't want them to clear it.*" He pauses and smiles sadly. "Can you eat concrete?"

But what hope does he have, with the Lacandon population threatened by extinction and rapidly modernizing at the same time, for that relationship with the jungle to continue? "I tell my son to plant a *milpa*, but I don't know if it is going to work. I tell him if he does not take care of it, he will be crying. I don't know how to write, I can't use a computer—but in my head there is a school."

Can Don Manuel's ideas put down durable roots among his Tzeltal and Ch'ol neighbors? Can he even pass on his school of knowledge to another generation of the Lacandones? Both questions must remain open. But there is no doubt that one of his most devoted students and collaborators is sitting beside him, a scientist determined "to convert this rich source of traditional ecological knowledge into scientifically validated tools for ecological restoration."[24] Between them, they have created a remarkable opportunity for the use of traditional ecological knowledge in a scientific framework—or should that be vice versa?

13 *Making the Black Deserts Bloom: Bog Restoration on the Brink of Extinction*

We have no prairies
To slice a big sun at evening—
Everywhere the eye concedes to
Encroaching horizon,

Is wooed into the cyclops' eye
Of a tarn. Our unfenced country
Is bog that keeps crusting
Between the sights of the sun.
SEAMUS HEANEY, "Bogland"

I can't restore Atlantic blanket bog here. I certainly can't guarantee that the same climate that made it happen will come back, the whole process is so long. What I can do is very active rehabilitation through management, getting back peat-forming conditions, so that this is like a juvenile bog, a teenager, something that needs to find its own identity. * CATHERINE FARRELL, *Bord na Móna ecologist, on restoration at Bellacorick bog*

Restoration is only successful when the whole [raised] bog is treated as a unit. If this is not done, one is at best simply slowing down the rate of loss of active bog. * JIM RYAN, "Raised Bog Restoration in Ireland"

* *

An Irish writer—any writer—takes issue with Seamus Heaney in some trepidation. "Bogland," the first of his several fertile meditations on this emblematic Irish landscape, has opening lines that beautifully encapsulate the intimately hilly nature of much of our peatland terrain. But they don't match the vast landscapes of the Erris Peninsula of northern Mayo. This is said to be the next parish to America, after all, and the emigrants who made it from here to the Midwest would have found the prairies not entirely unfamiliar. For sure, this ain't Iowa or Illinois, but the softly rolling expanses of Atlantic blanket bog spread out here to form the broadest horizons you will find anywhere in Ireland. They can do a fair job of slicing big suns, too—but only when the perennially rainy climate concedes an occasional cloudless evening.

For centuries, this was one of the poorest parts of one of Europe's poorest countries, where emigration offered the best and often the last hope to the best and brightest in many peasant families. Local people often associated their poverty with their environment and saw the ubiquitous bogs that sank almost all attempts at cultivation as wastelands. True, they provided fuel for home fires, but that was not much good if there was no food to cook on them. Cutting turf by hand is an ancient tradition in these parts. It remains a remarkable sight in spring, though a rare one now, to see a skilled worker cut the peat out of a bog bank as if it were butter. In the same elegant movement, the turf-cutter will flip each damp sod onto a neatly symmetrical stack where, with luck, it will dry out over the summer. But it is very hard to produce a profitable surplus from hand-cut turf.[1]

In 1949, a Mayo politician, Michael Kilroy, made a revolutionary proposal that he hoped might stem the hemorrhage of emigration—and that ultimately had a massive impact on the landscape. He called on the government to build "a turf-fired station for the generation of electricity" in the area. The industrial exploitation of turf as a resource was gripping the struggling country's imagination. A national peat board, Bord na Móna, had been set up three years earlier. This agency had already begun to strip the raised bogs of the Irish midlands. Its machines—bizarre monsters Heath Robinson would have been proud to invent, but very efficient at stripping turf—arrived in Mayo a decade later. Kilroy's proposal saw five thousand hectares purchased, to supply turf to a new power station at Bellacorick.[2] Its tower rose like a giant and incongruous brown egg timer, a striking icon of the new era, above the gently undulating blanket bog that surrounded it.

Over the next fifty-five years, the tower remained an eyesore to some and became a well-loved landmark to many others. The peat that fed it was sliced off the massive Oweninny complex of bogs around Bellacorick. That complex includes a second group of sites, making up an additional fifteen hundred hectares some twenty kilometers to the west, around the town of Bangor Erris. Hundreds of local people were employed, but the project was always more a social program than an economic proposition. In 2005, at the height of the Celtic Tiger economic boom, the station was closed down. The tower was demolished in a spectacular implosion for safety reasons. Some of the local spectators wept to see it sink into the earth.

Few shed tears, however, for the Atlantic blanket bog that had been disappearing all around the power station for decades. The exploitation of a bog requires aggressive drainage to make the peat dry enough to cut and stack, whether by hand or machine. This radical shift in hydrology immediately damages the very vegetation community that forms the peat. The subsequent removal of the surface layer, and of as many layers of peat below it as may be conveniently extracted, cuts vast black patches into the landscape. This highly visible evidence of loss of biodiversity—and of a pleasing landscape—did not provoke any great outcry. Like the monocultures of alien conifers that had begun to carpet our uplands in the 1950s, it was generally accepted as the price of progress in the midlands, and, in the case of Oweninny, as the price of the survival of a human community in the west.

A Waste of Space or a Place of Beauty?

But even from a purely environmental point of view, ecologists—with some honorable exceptions—paid little attention to the Irish bogs until the 1980s. By then, most of them had already been irrevocably altered. It was as if Irish science shared the popular view that a bog was indeed a waste of space. That view was not universal. Patrick Kavanagh celebrated the beauty of bog landscapes (though not Mayo's) in the 1960s, as did several other poets and some painters. One of Kavanagh's best-known poems, "The One," sheds a lovely but rather fuzzy light on a complex reality:

> "Green, blue, yellow and red—
> God is down in the swamps and marshes

Sensational as April and almost incredible the flowering of our
catharsis

. . . .

Yet an important occasion as the Muse at her toilet
Prepared to inform the local farmers
That beautiful, beautiful, beautiful God
Was breathing His love by a cut-away bog."[3]

There is an irony here, environmentally speaking at least: "cutaway" bogs are not pristine expressions of these remarkable ecosystems, but bogs that have been stripped of at least some turf.[4] And natural colonization of cutaway peatlands by marsh, meadow, and woodland plants can indeed produce remarkable bursts of floral biodiversity, given the right circumstances. This provides a knotty conundrum for restorationists: these new communities are valuable in themselves, so should they be destroyed in the difficult and uncertain process of restoring increasingly scarce bog communities? Once again, the painful issues of triage and ecological cost-benefit analysis raise their heads.

It is certainly true that the original bogland is a much subtler—and rarer—creature than most of these floral successors. But you need to look at it rather closely to appreciate its beauty, its extraordinary ecological structure, and the remarkable services it provides us with. Which is probably why very few people shouted "Stop!"—no more than they did for prairie in the United States—as so many bogs disappeared beneath our feet in the twentieth century.

Irish peat bogs were created very slowly, from materials at once very soft and very resilient, over long periods. They are built mainly of moss—especially Sphagnum moss species—and water, lots of water, which Sphagnum absorbs with ease. Up to 95 percent of the surface mass of a bog is water. There are, broadly speaking, four types in Ireland: fen, raised bog, mountain blanket bog, and Atlantic (or lowland) blanket bog (see panel 6).

A broad range of peat-forming ecosystems is found worldwide. Parts of Europe and north America have—or had—broadly similar bog types to those in the Irish quartet. But Ireland, slow to industrialize and sparsely populated since the Great Famine of the nineteenth century, kept its bogs longer and more extensively than most other countries, except Finland, Estonia, and Canada, though comparative

PANEL 6 • Peat-Forming Systems Found in Ireland

Atlantic (lowland) blanket bog: Exclusively on western sites near the coast, below 150 meters. Excess of rainfall over evaporation keeps them waterlogged. Many were once extensively forested, but the advent of a cooler, wetter climate 6,000–4,000 years ago led to acidification and killed off the trees. This process was accelerated in many (but not all) cases by Neolithic human clearances. Human intervention has continued, with burning regimes (to promote whatever sparse grazing was available) as a traditional agricultural practice right up to the present. The plant community was probably first dominated by Sphagnum, but black bog rush and purple moor grass are now the characteristic dominant species. Example: Ballycroy National Park, County Mayo (http://www.ballycroynationalpark.ie/Habitats.html#BlanketBog).

Mountain blanket bog: Above 150 meters, mainly in the west but with some good eastern sites. Process of formation broadly similar to Atlantic blanket bog, but altitude, climate, and less intense burning regimes have produced distinct plant communities, usually dominated by deer sedge and bog cotton, and with much more heather. Example: Liffey Head Bog in Wicklow Mountains National Park. (http://www.wicklowmountainsnationalpark.ie/BlanketBog.html).

Fen: Often a precursor of raised bog, and can also occur in hollows in blanket bog. Persisting fen systems are fed by mineral rich ground waters that can support a variety of diverse plant communities, and do not allow Sphagnum to dominate. Peat formation is slow and shallow, rarely more than 2 meters deep, whereas raised and blanket bogs may reach peat depths of up to 12 meters. Example: Pollardstown Fen, County Kildare (http://www.iwai.ie/releases/pollardstown.fen.phtml).

Raised bog: Mainly found in the Irish midlands. Originally lakes, which gradually filled in with vegetation, usually passing through a fen (but not a forest) stage; then overwhelmed with Sphagnum mosses, which finally built a characteristic dome above the surrounding landscape, raised up on the accumulating peat below the surface. Its rich plant community typically also includes some heathers and bog cotton, but it tends to lack the grasses so common on blanket bogs. Example: Clara Bog, County Offaly (http://www.npws.ie/naturereserves/offaly/clarabognaturereserve/).

"Active," "degraded," "intact," and "high" bogs: An "active" bog is defined by its capacity to form new layers of peat over time, though the process may be interrupted by recent fires, or by a period of drought. A

PANEL 6 *(continued)*

"degraded" bog has lost this capacity, usually due to human interventions such as drainage, but it may retain all or most of the typical bog plant communities and may be capable of restoration to active status, for example, by rewetting through blocking drains. Because a raised bog is a remarkably self-contained ecosystem, it is said to be "intact" if it is completely unmodified by human agency, and its "lagg"—usually a fen-type zone intermediate with the surrounding mineral soil ecosystem—is also present. There are probably no intact raised bogs left in Ireland. The term "high bog" is often used rather vaguely, but usually describes parts of a raised bog, generally the dome area, where no cutting (peat extraction) has been done, though drains may have been put in. (For further discussion of bog definitions, see Renou-Wilson et al. 2011, 9–16. Visit also the National Parks and Wildlife Service's website for their 2007 and 2008 conservation status reports, especially the glossary and "short reports" in the active raised bog and blanket bog sections: http://www.npws.ie/publications/euconservation status.)

Active raised bogs in Ireland, supposedly the stronghold of the Atlantic or Oceanic variety on a European scale, are close to extinction. On its website, the National Parks and Wildlife Service estimates that "less than 1% remains of the living, growing bog and this is rapidly being lost"

figures are notoriously difficult to establish. Close to 20 percent of Ireland is peatland, though much of it is very degraded.[5] In 2007, the National Parks and Wildlife Service (NPWS) rated the overall conservation status and future prospects of the bog types designated for priority protection by the EU as "unfavorable/bad."[6]

"A Soggy Brown Cloak to the Hills"

Ireland's island location and exposure to the Atlantic gives our bogs unique qualities, especially the blanket bogs of the west coast. They are not particularly rich in birds—they hold far fewer breeding waders and divers, for example, than Scotland's blanket bogs. But they are important for dragonflies and damselflies, and above all for plants. It can be a little difficult to grasp that these biodiverse communities could be nourished by rain alone. Michael Viney, the doyen of Irish natural history writing, evokes the nutrition system—and the special atmosphere—of the western bogs in this passage: "A Connacht mountainside veiled in rain dramatizes the role of soft but steady moisture in the rise of the blanket bog as a soggy brown cloak to the hills. At a windy shore of oceanic bog, with the taste of salt on one's lips, the flow of mineral nutrients in Atlantic showers is that much easier to accept."[7]

The Dutch were among the first European peoples to drain and "reclaim" their bogs. They had destroyed almost all of them by the 1980s. Then they pumped the last nutrient-poor peat full of nitrogen fertilizers for agriculture. Their famous horticulture industry needed peat biomass to produce blossoms on an industrial scale, and in the end they had to import moss peat from Ireland to provide beds for their blooms. Around this time, Matthijs Schouten, a postgraduate student from Nijmegen University, undertook a study of Irish bogs and was shocked to find that very few of them were intact. Even those few were in imminent danger of degradation. Alarm bells rang in both countries, and a remarkably effective international campaign was launched by new conservation foundations in Ireland and Holland, supported by media-friendly figures like David Bellamy and powerful media like the *National Geographic* magazine. The campaign persuaded Bord na Móna to hand over a valuable raised bog in Clara, County Offaly, in 1986 to the forerunner of the NPWS for conservation.[8]

The Dutch foundation went on to purchase four Irish sites. In a well-judged propaganda gesture, Prince Bernhard presented their property deeds to an Irish government minister at a ceremony in Holland in 1990.[9]

Other pressures from overseas pushed Ireland's approach to bogs in two contradictory directions during this same period. Dublin's entry into the European Economic Community (the European Union) in the early 1970s came very close to wiping out our mountain blanket bogs, previously the best preserved of our bogland systems. As part of a well-intentioned effort to keep small farmers on uneconomic holdings, Brussels offered "headage" payments that subsidized small-holders for increasing their sheep flocks. Since Irish farmers, especially in the west, grazed their animals largely on commonage, the sheep population soared without regard to the carrying capacity of land. In a classic illustration of the "tragedy of the commons," the mountain blanket bogs were stripped of vegetation by increasingly numerous ewes.

Because the bogs belonged to everybody, nobody took responsibility for them.[10] Overgrazing laid bare the soil. Sheep cropped the vegetation right down to the root stock. Their increasingly anxious hooves, scrabbling for any little bit of sustenance, broke up the surface peat into unstable fragments, then to dust. These bogs turned into true wastelands, black and bleak. And things just kept getting worse as rain swept the fragmented peat into river systems, which became so acidified that precious populations of trout and salmon collapsed, along with the populations of tourists who had come to fish for them. A subsidy intended to sustain rural communities ultimately undermined both their ecological and economic fabric.

Meanwhile, another state-sponsored initiative was attempting to clothe the bare black skin left by many industrial cutaways with green trees. Coillte, the state-owned commercial forestry company, was welcomed onto Bord na Móna lands to test solutions to a problem that had baffled the peat board's best minds: what productive use could be made of boglands once all the extractable peat had been removed?[11] For a while it looked as if plantations, both native (mainly willows) and alien (conifers, mainly lodgepole pine and Sitka spruce), might offer an answer.[12] The peaty soils, however, could not usually, even with added fertilizer, supply enough nutrients for commercially viable growth. Exposure and late frosts caused further problems. These monocultural forests made little contribution to biodiversity in any case, and in some cases obliterated remnants of specialized bog vegetation communities.[13]

It was another initiative from Brussels, the EU Habitats Directive of 1992, that provided the impetus for a shift in the opposite direction. This directive opened doors toward radical solutions to several key questions: how might Ireland's remaining bogs might be conserved, and how might cutaways and cutovers be restored, or at least rehabilitated as sites of biological diversity related to their bogland origins? The directive specifies habitats and plant and animal species that member states are obliged to protect as Special Areas of Conservation (SACs). It is, along with the Birds Directive of 1979, the mainstay of the EU's conservation policy. It lists active (that is, peat-producing) raised bogs and blanket bogs as priority habitats for protection and, specifically, restoration. This puts a real onus (with fines attached for noncompliance) on the Irish government to honor its many promises to preserve and restore our bogs. It also lists "degraded raised bogs still capable of natural regeneration" as worthy of designation as SACs. A further layer of protection for bogs became available in 2000 when the NPWS acquired the power to designate sites considered important for protection of habitat and wildlife at a national but not necessarily European level as National Heritage Areas (NHAs). Once again restoration was specified as an option, and once again bogs were prioritized.

Restoration or Rehabilitation?

The idea of restoring bogs seemed counterintuitive, and got people thinking. There was a general assumption that bogs, which produce fossil fuel over many centuries, indeed, sometimes over millennia, were not candidates for restoration. The best that could be hoped for a degraded bog, in ecological terms, was surely *rehabilitation* in the best sense, the creation of new wildlife habitats. Spontaneous regeneration on cutaways and cutovers suggested that this could be quite successful, but the plants that flourish on such sites are not classic bog species. Like the primroses and violets celebrated in Kavanagh's poem, they are vagrants from other systems, attracted by the nutrients released by water where the peat has been shaved down close to mineral-rich bedrock. Much better than a black desert, obviously, but offering no secure home to the bogs' unique communities of mosses, ferns, rushes, sedges, heathers, orchids, and asphodels.

In the first years after the directive was issued, Catherine Farrell, a young postgraduate ecologist, noticed something rather more hopeful while studying the vegetation of the cutaway bog at Bellacorick for Bord na Móna.[14] Some of the sites she surveyed showed little sign of natural regeneration of any sort. Where there was regeneration, the most successful colonizer was the soft rush, an aggressive plant that makes little contribution to biodiversity or the renewal of peat-forming processes. On one small site, however, she noticed a really striking regeneration of two key plants for bog formation—Sphagnum (the *S. cuspidatum* species that is one of the bog's first building blocks, though it is not itself peat-forming) and bog cotton. Soft rush was present but not dominant. The apparent source of this new community was almost too obvious to be true. She found that the drains on

PANEL 7 • An Ecologist in Bord Na Móna: Battling for Bits of Ground

When Catherine Farrell was a young ecology student in the early 1990s, she remembers having "an inner urge to stop destruction, to stop people destroying the place, to 'save the planet,'" though she rather cringes at that last phrase today.

And yet, as soon as you had finished your doctorate in 2001, you accepted an invitation to work for a company that has, in fact, destroyed bogs all over the country?

"Well, it has. I was working for a university, I had carried out my PhD research at the Mayo Bord na Móna site. I had met Bord na Móna people there. I saw they weren't intentional destroyers of the universe, they were good people, they all worked with me, understood what I was doing, and had an innate belief that this was the right thing to do.

"For me, working for Bord na Móna was about restoration, a huge opportunity, because no work had been done on restoring Atlantic blanket bog at that stage [March 2001]. That was very much my perspective." Inevitably, she found that some managers had very different plans, including the construction of wind farms and conifer plantations. She had to learn corporate survival skills, and says there were—and are—times when she has to "take up a sword, battling for bits of ground."

She was not at all sure how the company would react to her interpretation of her role, but she found that senior executives were supportive when she made a strong case for restoration: "They said they needed someone like me, who would tell them the truth of the situation really, but that doesn't always make you popular. In a commercial company, even one with strong environmental values, someone in my position can be seen as holding back development. But I have seen significant changes in the culture of the company in my time there. I think the culture outside the company, in our society as a whole, needs to change a great deal more. And that in turn would make it easier to make further changes in corporate culture."

What other changes have you seen in the board's approach, and in national policy?

"There is more money for restoration, industry has started to support it, give it a more substantial platform, it is taken more seriously. A lot of positive moves, from my perspective, from a peatland focus."

"There would have been opposition [from conservationists] to restoration here, a feeling that Bord na Móna would say, well, [if restoration is possible later] 'let's ditch all remaining bogs, because we can put them back again.' But what Bord na Móna has learned is that you *cannot* 'put back the bogs'; what you have should be preserved, but . . . there is a role for restoration and rehabilitation" (all quotations are from an interview with the author, November 2010).

the site had been blocked accidentally. Water levels had risen to a point where soft rush had difficulty regenerating and could not hog all the space, but the Sphagnum and bog cotton were, quite literally, in their element. No one had attempted to restore cutaway Atlantic blanket bog in Ireland before, but it seemed like it really might be as easy as Bill Mitsch's lighthearted mantra for all wetland restoration: "Just add water!"[15]

Over the next few years, Farrell tested her restoration hypothesis at Bellacorick. She blocked drains on an adjacent site by plugging them mechanically with peat—she is not a big fan of the small plastic dams often favored for bog restoration. This site was also surrounded with ridges of peat to further assist rewetting and ponding. Working with dumbfounded digger-operators long accustomed to making drains, not destroying them, she had to put up with a good deal of banter. "How much water do you want to retain, is it just a dash or would you like a little more?" they would ask her, as though they were offering to dilute a glass of whiskey.[16] They must have poured the right measures, as positive results came remarkably quickly. She had also monitored a nearby control site still subject to drainage. Within two years, while there was little change on the control site, the experimental site had been transformed.

Total vegetation cover increased from 40 percent to 93 percent. Sphagnum (mostly *S. cuspidatum*) shot up from 1 percent cover to 31 percent, and bog cotton and rushes (significantly, bulbous rush as well as soft rush) had also increased. Farrell was cautious in her report. She did not claim to have initiated Atlantic blanket bog restoration, but she suggested that she had the germ of a postindustrial management plan for the Bellacorick cutaway, which could reinstate peat-forming processes that will "maximise the ecological potential of the site . . . in a self-sustaining vegetation complex . . . compatible with the surrounding Atlantic blanket bog landscape."[17] It is worth noting one counterintuitive aspect of her experiment, reflected in a number of restoration projects: what we might call the "obvious" biodiversity in the experimental plot declined quite radically, from twenty-nine species to just eight, as the overall vegetative cover increased. But if the aim is to restore peat-forming processes, then this is an enrichment, not an impoverishment. Being species poor doesn't mean that the system is of less value in this case, explains Farrell. "In time more

28. **The right stuff.** Catherine Farrell examines Sphagnum moss before spreading it on a bog to speed up restoration—or at least rehabilitation—at the Oweninny site in the west of Ireland. (Photograph courtesy of Bord na Móna.)

species will colonize, but [the system] needs to be species poor for the initial phase. This is similar to the process in naturally occurring bog pools within more complex bogs."[18]

Seven years after making her first observations, Farrell leads me out onto the experimental site. It is the kind of Mayo day when the waves of rain sweeping in from the ocean give the phrase "Atlantic climate" a particularly vivid meaning. The ground at our feet is even wetter than the air, with a lush mosaic of Sphagnum, its spectrum running from pea green to cranberry red, effortlessly soaking up the surplus moisture. I am rather glad of the remaining soft rush, whose tussocks provide hard mats that stop my sinking up to my thighs in the bog. "Yes, it does help you move through the bog, but I feel *disdain* for that plant," Farrell says airily but emphatically, an unusually strong word from a woman who is generally exceptionally calm, and quietly spoken.[19] In small densities soft rush helps the Sphagnum find shelter to establish, but she is sick of seeing it crowd out other species on sites still subject to drainage. But she welcomes the bulbous rush, which forms little islands on the ponding water, to which the spreading Sphagnum can attach multiple anchors. And she loves the bog cotton, a sedge that she sees as a crucial indicator that the new system is moving in a desirable direction. It's a very familiar plant of the Irish bogs, even to casual observers, in late spring. Its ubiquitous tufting white seed heads can make a sunny bog look as though it has been spangled with snow.

It is distinctive in a different way now, though I would not have recognized it without its headdress. Farrell explains that, in order to thrive in the nutrient-poor bog, it recycles minerals from its leaves to its roots and back again, so that the foliage is sometimes almost emerald green, sometimes a strong wine- or orange-red, sometimes both at once. This variation, along with the patches of Sphagnum, gives a rich texture to the ground cover, an effect intensified by the canopy of gunmetal-gray sky. The whole image is momentarily bound together by a five-second rainbow. The bog can be a very beautiful place at these moments, but Farrell, having drawn my attention briefly to the glory in the sky, directs it back again to the world at our feet. There is great beauty here too, as multiple hummocks of moss capture light like miniature fairy-tale cities. She points out a plant that frequently protrudes from the hummocks, its stiff shoots glistening like a forest of tiny minarets. This is not Sphagnum, but *Polytrichum commune,* or common hair moss. She shows how it provides a kind of climbing frame through which some of the peat-forming Sphagnum mosses can begin to erect their mounds.

"A Mish-Mash That Wants to Be a Bog"

We can already see how these hummock-forming species are coming in, successors to the carpet-forming *S. cuspidatum,* and beginning to swamp out, quite literally, even some of the tallest stalks of the soft rush. The overall effect is a little like a series of miniatures of the roof of London's Millennium Dome.

"In fifty years' time I wonder if the soft rush will still be here at all, at the rate the Sphagnum is growing," she speculates. Caught up on a wave of overheated optimism, I ask her when the black bog rush, emblem and indicator species of Irish Atlantic blanket bog,[20] will put in a first appearance. "I have no expectation of that," she says carefully. "There is absolutely no evidence of it coming back. What you have here is very different to Atlantic blanket bog; what you are looking at is the young bog as it may have been four thousand years ago. The rate of Sphagnum growth here far exceeds what is happening across the road [on the neighboring intact bogs] where it is very, very slow, with 5 to 20 percent cover, and peat formation is just ticking over." Would she not then consider planting black bog rush, to accelerate the system's trajectory toward restoration? "The plant is abundant in the area and if conditions were right for it to establish, it would," she says. "I haven't considered planting it at all. Better to stabilize the site first, and *garden* later, if required . . . and if there is money available to do so."

She also points out that she cannot even be sure that the Atlantic blanket bog system started off from a similar point to her experimentally rewetted plots, because the pollen evidence is not clear—the record does not provide sufficient clarity to offer a historical reference system for bogs in a formative stage. "So I can't restore Atlantic blanket bog here. I certainly can't guarantee that the same climate that made it happen will come back, the whole process is so long. What I can do is very active rehabilitation through management, getting back peat-forming conditions, so that this is like a juvenile bog, a teenager, something that needs to find its own identity. I have a mishmash out here that wants to be a bog but does not necessarily want to be a black bog rush–dominated bog. We are not getting all the Atlantic blanket bog species back in here"—and she lists half a dozen key absentees. So the new bog may turn out to be a something completely different: "a peatland with really high-cover Sphagnum, which is quite rare in the Western world."

We may, then, it seems, be standing not on a restored bog, but on an emerging ecosystem whose nature is still to be determined but which is already enhanced in biodiversity and providing ecosystem services. For example, as we have seen, a degraded bog will acidify adjacent rivers as the rainfall carries huge quantities of exposed and eroding peat down into the watershed. There is also increasing awareness in the region of the threat of massive peat slides from eroded cutaways and cutovers. "The blocking of drains meant that we stopped the erosive power of the water as well—a critical point in the rehabilitation," she adds. She points out, too, that whatever its ultimate ecological category, the rehabilitated bog will function as a useful corridor linking the blanket bogs surrounding Bellacorick and fragmented by the industrial exploitation of the site to each other and to small, relatively intact fragments on the site itself that were never exploited due to their topography or that were saved because of their high biological value.

Among the very few sites in the latter category is Bellacorick Iron Flush, a minerotrophic fen within the bog that was found to harbor several rarities, including two very rare mosses. For one of the latter, however, it seems that rehabilitation has come too late. It could not be found on a recent survey, probably due to changes made to the flush by adjacent drainage and extraction. This was the last redoubt of *Meesia triquetra* in Ireland and Britain, and so it must now be considered extinct in these islands. This is an object lesson in how conservation in isolation often fails, without the joined-up thinking that leads to joined-up landscapes.

Farrell concedes that one reason she is reluctant to use the word "restoration" in this context is that it has legal implications. She has spent the morning before our field trip at a long-running and acrimonious hearing about the environmental impact of a Shell gas pipeline on the region. As the peat board's chief ecologist, she will not make claims that she cannot substantiate in law and in scientific journals. "When you are being held to the barrel of a gun as to whether something is a priority habitat or not, it is wise to take the lower rung when talking about this degraded peat system in Bellacorick. So I describe what we are doing here as "assisting recovery toward a peat-forming condition.'" That in itself would, of course, be very exciting, since her work offers hope that the postindustrial wasteland at Bellacorick can once again flourish as bog, even if it never replicates the original and larger ecosystem in which it is still embedded.

Developments over the last two years, however, have made Farrell a little more cautious about making even that claim. Long before she came to work at the site, Coillte had been experimenting with planting conifers on its cutaways. As we have seen, such plantations had not done well, and when extracted the trees will not be replaced. But they have already left a legacy here that looks rather daunting. Even on the experimental site itself, little lodgepole pines are springing up, and this phenomenon is repeated right across the Bellacorick site, and at Bangor too. I took an optimistic view when I first saw the trees at Bangor, where they are scattered out like a savanna. Might it not be useful, I ask Farrell, to have occasional perching sites for birds on the recovering bog? They could attract finches, pipits and buntings and thus assist the spread of seeds from nearby intact bog. Individual trees could be vantage points for rare raptors like hen harriers, merlins, perhaps even the reintroduced white-tailed and golden eagles that have begun to turn up in the area. True, these trees are aliens, but might they not bring a biodiversity benefit?

By way of an answer, she takes me deep into Bellacorick and I rapidly concede that my savanna idea was hopelessly off the mark. Here the new trees are forming quite dense thickets and already reach well above our heads. They may not be the stuff of commercial timber, but they look quite vigorous at this stage. If they mature, they could dry out the painstakingly rewetted bogs and shade its vegetation, shifting the ecosystem toward monoculture. To make things worse, rhododendrons have also started to sprout, some right in the middle of that first experimental plot.

"Look at that little runt," Farrell exclaims when we find one barely the size of a golf ball rising from a cradle in a piece of half-submerged bog pine, itself a reminder that trees had been here before. "Well, they certainly want to get in on the act now, spoiling our restoration plan, even our rehabilitation plan. Are they going to survive, or can we rewet the site enough so that they will fall over and die? I suppose you just have to wait and see."

The march of the pines, combined with the sudden eruption of rhododendron, certainly casts a shadow over Farrell's initial optimistic 2003 statement: "At Bellacorick, it is evident that, with time, peat-forming conditions can be restored with minimal management and cost."[21] As in so many cases, the price of restoration, or even of effective rehabilitation, is likely to include long-term, periodically intensive, and always adaptive management. And it is not at all clear that funding for such management will be available in recession-ridden Ireland. The spread of lodgepoles is recent and rapid, as is the explosion of rhododendron in the region. Just two years earlier, Farrell thought neither was a serious problem. In any case, bringing back a bog community and bog processes here is obviously going to be a long struggle. At the end of the day, it *is* a lot more complex than just adding water.

Turbary: Private Rights and Public Wrongs

As the tragedy of the commons was played out on the blanket bogs through the folly of overgrazing, a similar drama is still enacted on raised bogs through their exposure to turbary, the traditional right of private individuals and families to cut turf on public or common land. All land rights are deeply emotive issues in Ireland, even today. A long history of colonization, expropriation of indigenous communities, and the eventual return or redistribution of confiscated land still resonates. And turbary rights are no less jealously guarded for being much more vaguely documented than property rights. If a family can cut turf on the same bank of bog over thirty years, without using either secrecy or force, then its turbary rights are established in perpetuity, though they may never be recorded in writing.

As in the case of overgrazing, each individual turf-cutter tends to "think that their own little bit does no harm," in the words of Gerry McNally, land manager of Bord na Móna.[22] Raised bogs are especially vulnerable to turbary because of their hydrological structure. A patch of intact blanket bog can often survive—to a point—alongside turf-cutting operations. But a raised bog, even when its dome (colloquially known as "the high bog") has not been directly drained at all, may quickly dry out and collapse due to turf-cutting along its margins. "Turbary simply bleeds the edges of the bogs," says Catherine Farrell.[23] Once again, many small operators make for a much bigger impact than any one individual acknowledges. But why, you may be wondering, is a turf-cutter permitted to operate on the edge of an SAC at all, and often within it? The answer again lies in the complicated politics of Ireland's rural land.

When the first SAC list was published in 1997, all industrial-scale peat extraction that might impact on them was halted. Private extraction should have ceased as well. However, there was prolonged and strenuous objection from local families with turbary rights, despite offers of compensation and alternative cutting sites. Some families said they were dependent on the bog for affordable domestic fuel; others claimed a traditional cultural practice, an integral part of rural life in the summer, was being destroyed.

The Irish Peat Conservation Council, an NGO that advocates the preservation of bogs, responded creatively to this cultural argument. If people want to spend summer evenings on the bogs, the council proposed, they could just as well spend them blocking drains as cutting them, restoring bogs rather than destroying them. "You *could* get people involved in the work and make it [restoration] a community effort," says Catherine Farrell. "People are not all about money and all about negativity. People like to get out with their children. You could show them what the bog is about. It was the smell of the heather and tea in a bottle that people were dreaming about, the sun and the breeze, that was the romanticism, and they can still have that . . . without the back-breaking work of cutting turf."[24]

But politics is rarely a creative art, and the minister responsible simply caved in and gave a ten-year derogation from the new regulations to private turf-cutters—thus placing the issue neatly beyond her own likely period of electability. Similar derogations were granted when further SACs and NHAs were announced in 2003–2004. The NPWS reported that while most cutting was for domestic purposes, some semicommercial cutting was still taking place on designated sites. The government ignored the service's call for immediate cessation of all these activities. The NPWS has expended no less than €18 million in buying out only 5 percent of turbary rights, and this has had little overall impact.[25] Critics say the parks service failed to communicate with the turf-cutters and with the general public. This is almost certainly true, but how much can an overworked and underresourced organization do, overseen as it is by politicians running scared of losing votes on an emotive issue, and one easily exploited by populist rivals? The Irish Peat Conservation Council estimates that one-third of Ireland's remaining active raised bog has disappeared thanks to these "periods of grace," and the NPWS concurs.[26]

The first set of derogations expired in 2010, but turf-cutting is continuing at many of the sites where it is now supposedly fully outlawed, literally undermining the work of restoration at key sites. A new minister is insisting that EU law must be strictly observed, but enforcement remains feeble. Counterproposals by the turf-cutting lobbyists,[27] while often highly emotive in their headlines, actually suggest that some common ground may exist, but the NPWS is neither resourced nor empowered to engage with them. As Ireland's economic crisis dominates relationships with the EU, the environmental crisis, which may have greater impact in the long run, is largely ignored on all fronts.

A "Crap Bog" Gets a Lot of Tender Loving Care

Killyconny raised bog traverses the border between the counties of Cavan and Monaghan. This is the middle of the "basket of eggs" landscape where the Irish language gave the word "drumlin" to English-language geography.[28] The hollows between these humpy hills fostered many a raised bog, and the "encroaching horizons" of Heaney's "Bogland" frame the landscape. Killyconny has a more fortunate recent history than many SACs, in that the NPWS was at least able to buy half the turbary rights from a single estate in 1999.[29] In addition, relatively little drainage had ever been done on the "high bog" itself. However, there is heavily drained cutover all around its margins, which has caused some subsidence, and the high bog has also been degraded by burning.[30]

In other respects, too, Killyconny is an unlikely SAC. "It's a crap bog, really," announces Jim Ryan, the NPWS raised bog specialist, at the annual meeting of the Irish section of the International Peat Society in October 2010. This seems odd, because its restoration has been chosen as the star theme for the day, both for a seminar presentation and a field trip. Ryan is more than half-serious, however, and the fact that this bog is considered worthy of special conservation status is itself an indicator of how few good raised bogs remain. The basic problem with Killyconny is that it is structured like two lobes, really two distinct small bogs, formed on adjacent ancient lakes. As the bogs rose like (very) slow-motion soufflés above the countryside, they spilled over toward each other, forming a narrow neck of peat above a mineral ridge. A raised bog needs a certain critical mass relative to its immediate surroundings, and Killyconny's split nature reduces that by half. Moreover, the relatively intact dome was severely encroached by cutover on each lobe, and at one point by a Coillte plantation of lodgepole pine on the cutover. The lower level of the cutover, and the presence of trees, have combined to draw water off the high bog, making it much dryer than it should be, so that its characteristic vegetation is struggling to survive.

Given the bog's basic condition, Judit Keleman, the NPWS regional manager, is remarkably undaunted, and has good things to tell us about six years of restoration work. Her restoration principles are delightfully concise and no-nonsense, delivered in an Irish accent inflected by her Hungarian origins:

First principle: you can't manage raised bog without getting it wetter.
Second principle: you can't flood another man's land, only your own.
Third principle: you can't buy land without consent.
Fourth principle: you have to buy all the area before you can flood any of it, because water spreads.
Fifth principle: to get a bog wetter, first control water loss.
Sixth principle: if you want water to dribble off the bog slowly, create a suitable slope.
Seventh principle: if this does not work, add more water.

Practice does not coincide precisely with principle, of course, and the NPWS has not been able to block all the drains on the cutover, because it does not own all of it. And even after turbary rights are withdrawn, the NPWS cannot operate on private land without the agreement of the owners. Having lost their turbary rights, these owners are not necessarily well disposed to cooperating, though the land is of no economic value to them if they can no longer extract peat from it.[31] There is also the risk that agricultural land belonging to neighbors could be flooded if some big drains are blocked, and this blockage is possibly crucial to the success of the restoration. Readings show that the water table is still fluctuating to twenty centimeters below the surface; the ideal range for raised bog is between surface level and ten centimeters below.

Meanwhile, in order to make the high bog wetter, Keleman's team are collecting the water, which pours off it all too rapidly, into three "lagoons" at the facebank, the point of intersection between high bog and cutover. A pump, powered by photovoltaic cells, recirculates this water via pipes to the crown of the high bog. The beauty of this system is that the source of the "added" water is the bog itself, and it therefore has the right acidity levels, and the right propagules, microfauna and microflora, to nurture the system. The pump is not, she says wryly, "intended for use in perpetuity."[32] Without it, however, the dome might already have dried out, cracked, and collapsed, and it may be needed for a very long time.

When she takes us out on the bog under clear blue skies on a crisp autumn afternoon, we rapidly see that some of the best restoration results so far have come on the cutover, where a patchwork of Sphagnum species is gleaming in the sunlight. It is too early to say whether peat formation is taking place but, as she puts it, "the right species are here, and there are no wrong species here." This is all the more impressive as we learn that the cutover was just bare, black peat when turbary ceased here in 1999. Some heath and grassland vegetation came back spontaneously over the next five years, but drain-blocking from 2005 onward precipitated the rapid development of bog flora with good Sphagnum cover including peat-forming species.[33] No artificial propagation was necessary to achieve this. The drains were blocked with peat where possible, which then forms self-vegetating dams, and with plastic dams where the surface was too unstable for machinery to do the digging and plugging. The water flow from the high bog across the cutover then became so heavy that it flooded the road that borders it. Radical action was needed. A ditch two meters deep and sixty centimeters wide was dug along the perimeter, and plastic sheeting half a millimeter thick was inserted into it, so that it blocks the below-surface flow. The peat excavated from the ditch now forms a mound or "bund" that acts as an above-surface barrier.[34]

So far, the bund is working well and has created a "soak" of exposed water running along the inside of much of the bund. Keleman worries, however, that some plant roots could penetrate the plastic over time, and no one in our group can reassure her to the contrary. The heavy water-logging following the drain-blocking has now created "a very dangerous site," she tells us. One false step can find you up to your thighs in bog water, and sinking. The NPWS quickly produced signage to warn people that this is not a place for a careless ramble.

Up on the high bog, however, the surface is dryer than it should be, as often giving your footfall the firm and springy rebound of heathland as the squelchy ankle-suck of good bog. The indications from the vegetation up here are ambiguous. Sphagnum specialists in our group confidently distinguish mosses that look almost identical to the layperson, and soon find that the key peat-forming species are indeed present, though only in smallish patches. There is also a lot of bog asphodel, shrivelled to straw-like stalks after its autumn glory. This is an appropriate plant for raised bog in smallish populations, but its very abundance here could be a sign that the peat is drying out. The attractive "reindeer lichen" (*Cladonia rangiferina*, the one whose finely spun gray webs make convenient foliage for tiny trees on architects' models), is also frequently underfoot. Again, it's a typical species for a raised bog ecosystem, but too much of it indicates that water may be too far below the surface.

"The amount of water leaving the bog is phenomenal," says Keleman, and this has created a problem at the facebank, originally a vertical drop of up to two-and-a-half meters from the uncut zone to the cutover. It became a series of waterfalls in wet weather and began to collapse, eating away at the high bog edges. Keleman and her team found the answer was to slope the bank to a sixty-degree slant, and scatter its bare surface with "scraw," a carpet of living bog vegetation and soil cleared in excavations elsewhere that will hopefully take root and stabilize the bank. The key problem remains, however: that of keeping as much water as possible on the high bog in the first place. In 2009, the team built a four-hundred-fifty meter linear dam on the high bog, well back from the facebank. It's a high-risk strategy, radical surgery at the bog's heart, involving digging a ditch of similar dimensions to the roadside one and packing it with compacted peat and plastic liner. A bund is then built on top from peat excavated nearby and reinforced with plastic dam sections. Some quite severe downsides have emerged. Cracking is occurring in the peat between the bund and the facebank, indicating that this section is becoming *too* dry. The cracks could undermine the dam and raise the prospect of accelerated erosion and collapse. After heavy rainfall, surface water surges in small but powerful torrents along the inner edge of the dam, causing significant erosion.

"I do not expect cracks here," says Keleman, "but the build-up of water pressure may mean that the dam will burst. We will prevent this with cross-pipes to allow for excess water to leave the bog." How serious a threat does the cracking represent? What measurable improvements has the dam created on the high bog? "Cracking is a slow crumbling of the edges where peat cutting banks existed. It is different from site to site. We do not know as yet how the linear dam will work out but water level measurements should inform us," she tells me by e-mail after the trip.

Still on Life Support

Killyconny is a bog on life support, its vital circulation system maintained through pumps and stints and bypasses and intravenous drips, its organic structures propped up with plastic crutches and radical surgery. On Young D. Choi's spectrum from lithe and sensuous limb to prosthetic leg, this is somewhere in the middle.[35] Jim Ryan comments that, given the worryingly dry condition of the high bog, "I think that the majority of the active bog will form on the cutaway, but let's wait and see. This is a very fast response, the Sphagnum now present suggests it will be peat forming, but can we keep it going? Can we keep the water table stable?"[36] It would be a lot easier to do this if the large drains still unblocked could be dammed, but further engineering advice is needed to determine whether this would flood the neighbor's fields as well as the bog.

Fluctuations in the water table are probably also the main reason why a separate but related attempted restoration on one substantial corner of the Killyconny cutaway is also proving problematic. Coillte had forested eleven hectares of the cutover area with lodgepole pines and some downy birch in 1979. Michael Delaney, project manager with the company's EU-LIFE restorations, explains to the seminar that this kind of plantation was part of a policy of "social forestry," often not profitable, but an investment to provide jobs in rural communities—similar, then, to Bord na Móna's peat extraction in Mayo. As shifting ecological and economic thinking brought this policy under critical scrutiny, the company selected it as one of their EU-LIFE raised bog restoration projects. In 2003 they clear-cut and removed the trees (apart from a stand of birch in one perennial dry corner), blocked drains, and wind-rowed lopped-off branches to encourage revegetation.

Initial results were encouraging. The water table rose rapidly from an average of about sixty centimeters below the surface to a very promising zero to ten centimeters. Sphagnum mosses, including peat-forming species, returned quite rapidly. However, as he shows us when he takes us out to the site, the lodgepoles, and especially the birch, are regenerating rapidly. This may be related to the fact that the

Coillte site is right beside the biggest unblocked drains, and monitoring shows that the ground is at some points drying out drastically for some periods in recent years. The presence of trees, of course, is accelerating that process, and so there is now a feedback loop that thwarts the originally intended direction of the project—the restoration of active bog on the cutover.

He is curious as to what we thought should be done about this. He had been experimenting with cutting the birch by hand, and with using herbicides (one or two participants react almost physically against the idea of poison as a restoration tool). In any case, the trees are winning. Perhaps a new tack is needed?

"Let nature take over, trees are part of the past of the bog, and they may be part of its future," one participant volunteers.

"You need to make the site as wet as possible, and then see if they still come back," Catherine Farrell counterproposes. This proposal, of course, depends on their being able to block the remaining drains.

Catherine O'Connell, chief executive of the Irish Peat Conservation Council, insists that peat formation, which can take place in some wet woodlands, should be the bottom line. "With peat formation, the trees are okay. Without it, they are not."[37]

"Wet woodland with Sphagnum is a very rare habitat in Ireland," Jim Ryan reminds us. "There are only 150 hectares in the country. Perhaps we should include it within our definition of active raised bog?"

Michael Delaney himself recalls that a major focus of the Coillte restoration was to support the hydrology of the remaining high bog, and that a wet woodland on his part of the site might do that as well as a more typical bog plant community.

This is a quintessential restoration debate, covering the familiar issues of historical reference system and trajectory, pragmatism and principle. As Catherine Farrell said earlier, the problem with having an idealized concept of restoration is that it may make you miss the good things happening at the edges of your project, unintended but rich in biodiversity.

The restoration project at Killyconny bog is also fortunate in having strong support from the local community. Jim Smith, who once cut turf there himself, is now a passionate advocate of its restoration. He has argued in articles in the *Anglo-Celt,* he tells me, that we should value the bog as much as if it were a fragment of rain forest, contributing as it does to biodiversity and carbon sequestration. He stresses that it can be used for teaching "biology, ecology, chemistry, history, geography and folklore." He calls the bog "this friend in our midst."

And perhaps that is part of the answer to the kind of dilemma posed by the persistence of birch trees on the cutover: if the bog is our friend, there may be times when the most friendly attitude is to let it go its own way.

Irish Raised Bogs: World Headquarters or Brink of Extinction?

"We have the headquarters of this ecosystem type in the world," Farrell reminds a group of volunteers working on another raised bog, during a peat society outing the following year. "We should cherish it."

But do we? And can we?

The answers to these questions, and especially the second one, await the judgment of time, but are both perilously close to negative at the time of writing.

Since the warning bells for Irish bog ecosystems rang out in the 1980s, tremendous work has been done to save and restore our bogs by NGOs, by individual scientists, and by researchers and rangers in the NPWS. More recently, there has been some innovative engagement with peatland restoration by the two commercial semistate companies that between them own most of Ireland's peatlands, Coillte and Bord na Móna. There has been a remarkable and welcome shift in their business cultures. But the number of healthy remaining sites continues to decline in all our categories of peatland. None of them is fully intact, and many of them become more degraded with every passing year. Irish public opinion is still barely aware of an issue affecting almost 20 percent of our national landmass, with significant value not only for its rare and biodiverse communities, but also for ecosystem services like flood protection and climate stabilization. Many of the few who are really familiar with these landscapes today are hostile to preservation and restoration; private turf-cutters paradoxically cherish peat extraction more than they cherish the peatlands on which, of course, sustainable extraction itself depends.

The restoration of "active" peatlands presents special challenges, because peat-forming processes require highly specialized hydrology, especially for raised bogs. This often depends on control of conditions well outside the site boundaries, at least as they are currently drawn under the EU Habitats directive.

"Restoration is only successful when the whole [raised] bog is treated as a unit. If this is not done, one is at best simply slowing down the rate of loss of active bog," Jim Ryan told a conference at the conclusion of the first Coillte LIFE Restoring Raised Bog Project in 2008.[38] Since no raised bog is currently so treated, the scenario is grim, and made grimmer by the persistence of turbary.

Climate change also presents special challenges for the restoration of raised bog, indeed for bogs generally, not only in the sense of future shifts, but because bog formation was often strongly influenced by radical climate fluctuations in the past, which are most unlikely to recur in similar rhythms in the future.

"What is a raised bog?" Farrell asks provocatively. "Is it something that happened in the past, is it something that will happen in the future? We don't know if it can happen in the future, because of climate. We know it happened over the last ten thousand years because of climate and how it oscillated, drying out periods and wet periods. But is it something for the future?"[39]

There are many problems too, associated with afteruses of cutaway bog, even those which come with green labels attached, like afforestation and wind farms (see panel 8).

The prospect for restoring, in any full sense of that phrase, more than a sparse handful of active raised bogs now seems remote. It is worth persisting on the most promising ones because at least some valuable plant and animal communities may survive in these habitats. And, where restoration of the peat-forming processes does prove possible over a long period, our children and grandchildren may see some of these domed shapes, once so characteristic of the Irish midlands, growing very slowly toward the sky once again.

Worse but Better: Ruined Peatlands Blossom

Fragile as the prospects are for raised bog restoration as such, the restoration-based research in recent years on all Irish bog types has borne remarkable and somewhat unexpected fruit. In its first report to the European Commission on the conservation status of

PANEL 8 • Greenhouse Gases Down on the Bog: The Triple Whammy

> Irish peatlands are a huge carbon store, containing more than 75% of the soil organic carbon in Ireland. Natural peatlands . . . play an important role in the regulation of the global climate by actively removing carbon from the atmosphere but this important function is reversed (i.e., there is a net release of carbon) when the peatland is damaged. (Renou-Wilson et al. 2011, xiii)

Since the last Ice Age, peat bogs worldwide have been storing very substantial quantities of the greenhouse gas carbon dioxide (CO_2), equivalent to about half the CO_2 in the atmosphere. Bogs are therefore one of the world's most effective "carbon sinks." At deeper levels, peatlands may also store very significant amounts of methane (CH_4), which is twenty-three times more potent than CO_2 as a greenhouse gas. However, while tundra permafrost locks CH_4 down, the much milder Irish conditions allow it to escape quite easily.

Under natural conditions, bogs do release CO_2 into the atmosphere in times of drought, when the water table drops, but store it again as the water table rises. Conversely, CH_4 emissions decrease in times of drought and increase when the water table rises. There is, therefore, a strong element of swings and roundabouts in the greenhouse gas budget for Irish bogs—they are sources of emissions as well as carbon sinks.

It is clear that the extraction of peat, especially when it is mechanized, creates a kind of triple whammy in the acceleration of such emissions:

1. The destruction of peat-forming processes and the removal of layers of peat prevent the bogs from providing the services of a carbon sink into the future.
2. The drainage of the water table and the removal of surface peat release CO_2. (Conversely, drainage tends to lock down CH_4 as the peat dries out, though it will still escape from drainage ditches.)
3. The burning of peat fuels, domestically and especially in power stations, unlocks still more carbon, as does the combustion of the oil and diesel which drives the extraction machinery.

After-uses of Irish cutaway bogs, including some "green" uses, may further increase emissions. The surface disturbance created by afforestation, and the fossil fuels consumed by machinery in the process, may outweigh any new storage benefits from the plantation trees. Similarly, the construction of wind farms may cause carbon losses greater than the clean energy gains.

Can peatland restoration return the Irish bogs to their role as a carbon sink? Experience in Finland and Canada suggests it can, as does recent

PANEL 8 *(continued)*

research by David Wilson that shows that the rehabilitated peatland at Oweninny is now a net CO_2 sink. But if a warming climate dries out the Irish landscape in the future, then peat-forming processes may be choked off. The jury is still out, as restoration may often require some surface engineering disturbance, and the fossil fuel emissions associated with machinery obviously also have a negative impact.

Rewetting and natural regeneration in general should both assist in the storage of carbon from the atmosphere and prevent new releases by blanketing exposed peat. But recent research in Ireland indicates that rewetting will also create hotspots of methane emissions (Wilson et al. 2009). If slowing down climate change rates in the short term were the main criterion for after-use of Irish peatlands, reafforestation would be a significantly better short-term option than restoration through rewetting and natural regeneration; but in the long term, restoration may ensure carbon uptake for thousands of years, whereas tree coverage would oxidize the peat and ultimately lead to more emissions.

These are complex equations with many variables—specialists insist that a great deal more research is required before the carbon budgets of bogs in varying stages of degradation and restoration can be confidently accounted for. It also appears that there may be quite drastic variations, between different bogs, and within the same bog, for reasons that are not yet well understood.

Most specialists agree that, while exploitation of the bogs is undoubtedly accelerating climate change, it is too early to claim significant climate change mitigation benefits result from Irish peatland restoration. The gains are certainly likely to be much more modest than those from the restoration of tropical peatlands, especially in Indonesia, where clearances are causing vast greenhouse gas emissions. (For much of the information in this panel, I am particularly indebted to the work and comments of David Wilson of University College, Dublin, and Earthy Matters Environmental Consultancy. See especially Wilson 2006 and Wilson et al. 2009.)

EU-designated protected habitats in Ireland, the NPWS bleakly categorized the condition of active raised bog, and of degraded raised bog capable of restoration, as "unfavourable/bad."[40] But it also stated that "a third category, Secondary degraded raised bog was identified during this assessment. This includes highly drained high bog devoid of vegetation (including the majority of Bord na Móna sites), cutaway, cutover bog and occasionally reclaimed agriculture land with peaty soils."[41] On these apparently ruined peatlands, the report continues, "it may be possible and may even be more feasible to restore to Active bog than in some areas of Degraded bog."

I found myself reading these highly counterintuitive sentences three or four times, then again. How exactly could it be more feasible to restore more degraded categories of peatland to active raised bog, than to restore less degraded categories? I turned to Catherine Farrell for an interpretation. "That makes perfect sense to me," she replied. "Take a raised bog, and drain it. In some there is still some active raised bog, and you keep it in situ, conserve it, and if you own the whole site you might be able to restore it, even enhance it. But some degraded raised bogs are completely knackered, they have lost active raised bog, though they still have a dome and are technically still capable of restoration. But the costs are very high, there are very deep ditches, it is completely drained and dried out. . . . It's gone beyond point of redemption.

"Now the third category the NPWS are talking about would be much more degraded again, on the face of it. Maybe you had complete removal of the vegetative layer, maybe you had a midlands bog with a couple feet taken off but you still had the Sphagnum layer. Well, the potential to restore peat-forming vegetation there is far greater than on one of these dried out old skeletons of a raised bog. You can breathe more life into a more damaged system, which does not seem to make sense, but it brings it back. You saw it on Killyconny, you were getting better Sphagnum establishment on the cutover than on the raised bog dome."

Warming to her theme, she went on to raise blunt questions about attempting to restore a system that evolved under very specific conditions that are most unlikely to be repeated. She firmly rejects the kind of purism in restoration thinking which "puts chains on a system, and puts blinkers on our outlook. It's all good," she said of a more open-minded approach, "as long as you are not draining and destroying."[42]

Of course this also fits in well with Farrell's own thinking on the management of degraded Atlantic blanket bog, where she generally eschews the word "restoration." The plant communities that emerge as she rewets the bog at Oweninny manifest many differences from the classic Atlantic blanket bog communities that define the historical reference system—the unexploited bog across the road. But they do produce a biodiverse system, much of it peat-forming, and these are much, much better outcomes than an eroding "black desert," or a monocultural and inappropriate forestry plantation.

Florence Renou-Wilson, lead author of the *BOGLAND* report,[43] is a lot more cautious initially about the 2007 NPWS discovery of this "worse but better" category of degraded raised bog. "We have to be careful when this sort of statement is produced," she writes. "Indeed, it could be used by Bord na Móna for their advantage!"[44] But then she adds, "It is my belief also that industrial cutaway peatlands may be in some cases a "better" start for a restoration programme than a degraded [raised] bog which has been cut on the margins only. It is easier to raise the water table in an industrial cutaway bog where meters of peat have been taken off and therefore lie well below the water table. By comparison the degraded bog with marginal cutting is a more difficult task, with an unknown number of 'leaks.'" She points out that, nevertheless, water table fluctuation may be a problem, but given the size of many cutaway sites she believes there will usually be plenty of "pockets" of very wet land where the peat-forming process can take off.

Restoration: Changing Mindsets, Revealing Options

Reviewing progress in peatland research and conservation since his own groundbreaking work in the 1980s, Matthijs Schouten contrasted the achievements "on paper"—a vast increase in the amount of bogland officially protected—to the reality of continuing degradation due to turbary around key sites. He cited the NPWS ranking of the condition of these designated habitats "at the lowest level possible—i.e. *bad*."[45] He continued with a sobering assessment of the prospects for Irish peatland restoration at ecosystem level.

> Rehabilitation of cutover, cutaway and afforested peatlands seems promising, but it cannot be a compensation for the deterioration of the still relatively intact peatland sites. Restoration measures in formerly afforested blanket bogs or on cutaway blanket peat may locally bring back peatland communities . . . but whether the ecosystem can be restored remains unclear. In cutover and cutaway raised bogs, ecosystem restoration is even more difficult. Dutch projects on cutaway peatlands have shown that, usually at high expenses [*sic*], Sphagnum

> growth can be re-initiated and re-colonisation by bog communities realized; restoration of ecosystem functions, however, appears to be very difficult and—if at all possible—will take a very long time. That is not to say that the vast area of cutaway raised bog in Ireland does not offer an enormous potential for wetland creation.[46]

Restoration on Irish bogland, then, seems a slim figure fighting a slide toward the extinction of very scarce and precious ecosystems. And what progress it has made is constantly undermined by continued private cutting on the margins of supposedly protected sites. It is hard even to hope that this situation will change before it is too late, given the economic and political maelstrom into which the country has sunk in the brief period since Schouten's address. Nevertheless, restoration efforts have greatly increased our knowledge of how bogs work, and restoration thinking has been influential in changing the mindsets of powerful corporations like Coillte and Bord na Móna. And where the extent of degradation or other factors make restoration impossible, at least in the short term, restoration research has indeed demonstrated options for remarkably successful rehabilitation into other types of wetland.

After the conference in 2010 that launched Bord na Móna's aforementioned Biodiversity Action Plan, Catherine Farrell led us on a field trip to Ballycon. This is—or rather was—a raised bog more or less leveled by peat extraction. It was therefore completely destroyed as a raised bog. Its characteristic hydrology cannot be recreated by any currently available technology. However, the construction of a shallow peat dam across the area has created large ponds, fringed by reed mace and bog cotton, gleaming red and gold in brilliant November sunshine. This recently rehabilitated bog has begun to attract rare breeding birds, like lapwing and ringed plover. Last year, one very rare wanderer to Ireland, a crane, spent a couple of weeks there. On the field trip, we are greeted by the musical calls of a flock of ninety whooper swans, scarce winter visitors from Iceland. They are partial to the bulbs of the marsh arrow grass that spring up, ironically enough, on the remaining areas of bare peat.

A Bord na Móna railway line runs along the edge of the bog, still feeding the nearby Edenderry power station with milled peat from other areas. The company staff insist that the railway is "sacrosanct"—

29. **A railway line through rehabilitation.** Ballycon bog, where the mined areas have been rehabilitated into wetlands (*right*), but an industrial railway line remains, as a trade-off between business and environmental interests. The visitors are from the Intergovernmental Panel on Climate Change. (Photograph courtesy of Bord na Móna.)

and acknowledge that its presence has limited the options for rewetting the entire area, and therefore the extent of the rehabilitation. This is one of the compromises between the board's "three bottom lines," announced in 2008 by its CEO Gabriel D'Arcy: it must continue to make a profit, and to support local human communities, as well as honor its newfound mission to protect and enhance the environment. The biodiversity plan stresses that there will be multiple uses for Bord na Móna bogs, post-extraction. Their "afterlife" will accommodate windfarms and other industrial initiatives, and general recreational facilities, alongside the promotion of biological recovery. "All have to work together," says D'Arcy, "none can be to the exclusion of the others."[47]

This will obviously not be easy, and conflicts of interest will arise. But curiously, in this case, the railway line's gravel has itself become home to some relatively plant rare species, like blue fleabane, which would never have found a home on the original bog. So once again a human intervention closes off one option, but opens up another.

There are a lot of odd swings and roundabouts in the business of biodiversity.

On the same occasion, Bord na Móna land manager Gerry Ryan remarked that some economists in Ireland are at last beginning to take seriously the idea that biodiversity itself has monetary as well as other values. As Bord na Móna's corporate vision shifts from the exploitation of bogs to their multiple afteruses, the case for a natural capital and ecosystem services evaluation of the company's assets becomes more and more compelling. But it has yet to take place.

One man at the conference could personally put very direct monetary value on the company's plans. Ray Stapleton, general manager of the nearby Lullymore Heritage Centre, describes the launch of the Biodiversity Action Plan as a "historic day. We have seen in Britain how wildlife and wetland centers attract large numbers of visitors. It will be a massive boost to local communities if Bord na Móna is serious about all this. It gives me hope." At a time when hope is a very scarce commodity, and the bankers and politicians are telling us we are bankrupt, it is good to be reminded of the great natural wealth that this country still holds. Growing that wealth through helping diverse flora and fauna flourish on the cutaway bogs would be a very positive response, in many ways, to Ireland's current crisis, and will certainly attract ecotourism, among many other benefits.

The rehabilitation of Ballycon demonstrates dramatically how quickly new and highly biodiverse systems can be created on bogs that have been reduced to the status of black deserts. The flowers are often different, for sure, but at least they are blooming again.

This good work is no substitute, however, for the protection of the remaining intact bogs, nor for the fullest possible restoration of bog ecosystems where this is still ecologically feasible. It is vital that Bord na Móna, national and local policymakers and, above all, the Irish public, recognize this. For the handful of bogs that have not been irremediably degraded, the options of preservation, assisted by restoration and rehabilitation, still offer a kind of redemption. If I may put it this way, they offer a last-ditch opportunity to avoid a corporate and national disgrace: the extinction of unique European ecosystems in an advanced country in the twenty-first century. For additional material on bog restoration, see http://www.press.uchicago.edu/sites/woodworth/.

14 *Walk Like a Chameleon: Three Trends, One Story*

We shall not cease from exploration
And the end of all our exploring
Will be to arrive where we started
And know the place for the first time.
T. S. ELIOT, "Quartet No. 4: Little Gidding"

Sometimes criticized as an exercise in nostalgia, [restoration] is, at its best and most deliberate, quite the opposite of that—not a daydream of the past but a troubling encounter with time and change. * BILL JORDAN

Do I contradict myself? Very well, then, I contradict myself? I contain multitudes. * WALT WHITMAN

[Climate change and novel ecosystems are] the big question of the age for conservationists and restorationists; just as they are getting on their feet, all the rules are changing. * RICHARD HOBBS

Is the word restoration still relevant? Is restoration still a useful concept? * RICHARD HOBBS

Our planet's future may depend on the maturation of the young discipline of ecological restoration. . . . In its short life it has assumed a major role in sustainable development efforts across the globe. * LESLIE ROBERTS, RICHARD STONE, AND ANDREW SUGDEN, editorial in *Science*

Do we prefer the original or the transformed ecosystem? Which version favors our ecological values, our economic wellbeing, our cultural fulfillment, and our aesthetic preferences? * JAMES ARONSON AND ANDRE CLEWELL, *Ecological Restoration*

The degree to which a restored ecosystem resembles its reference cannot be discounted, but it is ultimately of secondary importance relative to the value of that ecosystem to people. * JAMES ARONSON AND ANDRE CLEWELL, *Ecological Restoration*

The restorationist is, like a chameleon, always looking backwards, with one eye, while looking steadfastly forward with the other one! * JELTE VAN ANDEL AND JAMES ARONSON, adaptation of a Malagasy proverb in *Restoration Ecology*

* *

The story of ecological restoration seems to take us full circle. Restorationists started off wanting to hand back restored ecosystems to the play of natural forces and have ended up learning that they have to manage them, actively or passively, forever. But it is not quite that simple.

The early American restorationists saw prairies, wetlands, and forests disappearing, drastically modified by human domination. They dreamed of restoring a pristine past, of a nature "uncontaminated" by humanity. They were influenced by a fusion of the Judeo-Christian vision of the world before the Fall and the ecological vision of a climax ecosystem in eternal equilibrium (for which read also the biological abundance of the Americas before 1492).

Their dream was born of a nightmare that we often trace to the Industrial Revolution, if we come from European and North American cultures. But this nightmare has been a recurring one in other and older human histories. It must have ridden on the tormented sleep of the inhabitants of Easter Island and was no doubt familiar to the citizens of many other vanished civilizations.[1] The expansion of human populations, coupled with our compulsion to consume much more than we need to survive, has eaten bare many (relatively) small worlds in the past.

The founders of the modern American environmental movement experienced these forces of expansion and consumption at new scales and new speeds. A surging population that pursued unprecedented economic growth was eating away vast swathes of the "wild," territory that had so recently seemed unfathomable, infinite. The "wilderness" that fed their cities in more ways than most of them understood was shrinking before their waking eyes.

"In Wildness is the preservation of the world," Henry David Thoreau told his audience at the Concord Lyceum on April 23, 1851,[2] and the phrase resonates, albeit uneasily, today. It is easy to mock Thoreau a little now, when we know that he went back to a comfortable home-cooked meal most nights from Walden Pond. The "wildness" of Walden was indeed of a very tame, almost suburban, variety. And when Thoreau did encounter wilderness in the raw, on the summit of Mount Katahdin, a landscape that truly seemed to bear no human trace, he was more alienated than exhilarated.[3]

Nevertheless, in the paragraph that couches his "wildness" dictum, Thoreau reminds us of two vital insights about our place in the world: that the basis of all civilization is access to nature's resources (what we might now call natural capital and ecosystem goods and services); and that, when an intimate personal relationship to the natural world ceases to be the norm in a culture, that culture is close to collapse.

Thoreau's language may seem overblown today. But he is worth reading and rereading, not least for his insight that a domesticated nature is as much in need of the wild as we are: "Every tree sends its fibres forth in search of the Wild. The cities import it at any price. Men plow and sail for it. From the forest and wilderness come the tonics and barks which brace mankind."[4]

A century later, human expansion was impacting the entire planet. An emerging conservation movement in the United States had won significant national and even international victories—including the 1916 National Park Service Act and the 1918 Migratory Bird Treaty Act—but had no strategy for winning a war it seemed doomed to lose. In this new context, Aldo Leopold began to articulate a similar dual vision to Thoreau's: we need wild nature both for our material and existential well-being. In the last analysis, we need it for our survival.

Almost a century later, we coined the term "anthropocene" to describe the era where human impact is evident everywhere—from the poles to the equator to the stratosphere. Under our heedless pressure, the natural world is approaching a tipping point where many species, resources, and ecosystems are on the verge of extinction, exhaustion, or unprecedented change. Meanwhile, despite heroic efforts by conservationists and some educators, the relationship of our globalized culture to the natural world is for the most part dim, and getting dimmer by the minute.[5] This is a very dangerous combination.

If you have read this far, you are very likely part of the minority that is already painfully aware of this crisis. So please forgive me for rehearsing its elements here one more time. This book is mainly about the positive, if challenging, environmental scenarios that an exploration of restoration opens up. But we would be guilty of a glib, Pollyanna optimism if we ever allowed ourselves to forget that today's restoration movement lives in the shadow of environmental catastrophe.

The exponential expansion of the human population, coupled with a globally dominant ideology dedicated to ever-increasing consumerism, has turned the entire planet into an Easter Island. What we used to call "materialism" has somehow blinded us to a central material fact about our existence on Earth: our material resources are finite. We are certainly running out of wilderness. We are also running out of raw materials, and ultimately we are running out of space. Peter Raven points out that if the peoples of developing countries achieve the levels of consumption that is the current norm in the developed world, we will soon need the resources of several more Earth-sized planets to support our species.[6] Unless we make radical changes in our population control, our economic and cultural paradigms, and ultimately in our personal lives, we seem doomed to consume the last of what sustains us.

Meanwhile, our culture has put so many screens between our daily lives and their roots in the soil of wild nature that most of us have no direct perception of the source and scale of the crisis. When children think that milk comes from cartons and not from cows, there is little chance that they will understand that cows need pastures; that pastures need fallowing; and that fallow lands need nurturing from nearby wildlands to maintain a vital cycle as old as agriculture itself.

This is a harsh, indeed terrifying, reality of our twenty-first-century existence, and it is not within the scope of this book to propose any grand solutions. Perhaps the gargantuan follies of our financial institutions in recent years and the increasingly naked subservience of all currently significant political parties to "market forces" will spark popular movements to take democratic control of our economies. Perhaps our cultures will tire of the tyranny of moronic consumerism. Without such paradigm shifts in global culture, it may well be that all restorationists are doing is rearranging blocks of seminatural real estate on the deck of planet *Titanic*, their backs to the coming iceberg.

But we should not entirely despair at the chronically alienated relationship between our species and the rest of nature today. We may derive at least a little legitimate hope from historical precedents. We have learned a great deal in recent decades, sometimes precisely through attempts at ecological restoration, about human influence on landscapes that we had previously thought of as "purely natural." The knowledge that forests, savannas, prairies, and wetlands, once dreamed of as pristine, often bear deep if superficially invisible human footprints, may be a romantic's nightmare. But it reminds realists that we have always been part of a web of interconnections with the natural world.

As we have seen repeatedly in our journey through restoration projects, these interconnections are by no means universally negative. Again and again, we have found that biodiversity benefits can blossom from human interventions in ecosystems. Native American burning regimes fostered the flora and fauna of the Midwest, while preindustrial agriculture opened up scores of significant ecological niches right across Eurasia and through much of Africa.

First Get Your Garden

Wise cultures knew these interactions had their limits; that the cultivation or manipulation of landscapes must alternate with at least a degree of wilderness if, as RNC advocates put it today, natural capital stocks were to remain buoyant and ecosystems were to continue to supply the goods and services vital to human societies. The Maori, as we saw in chapter 11, learned through bitter experience that species disappear forever if harvested too heavily. True, that knowledge came too late to save the moa. But the Maori learned the lesson, and adapted their management of nature to include the concept of *tapu* (taboo). This was a ban, imposed by local elders, on hunting particular species at particular times, depending on local conditions. Such bans probably saved species like the *kereru* (New Zealand wood pigeon) from the moa's fate. It pioneered a new relationship between hunting and conservation for the Maori and prefigures the concept of sustainability. It is a pity that there was no way for the Maori to transfer this concept to North American settlers who, centuries later, failed to grasp in time that even the vast populations of the passenger pigeon had a finite limit.

The Lacandon agricultural system is close to an ideal paradigm of how human beings can live sustainably. It is an unbroken cycle, starting with the clearance of a strictly limited patch of forest, followed by a flexible repertoire of cultivation, manipulation and exploitation, of animals as well as of vegetation, and finally moving toward the restoration of the forest. And then a further cycle may follow, beginning with another phase of clearance.

Indeed, whether the word "restoration" is totally appropriate to the Lacandon system is debatable, since it is an abstraction of one element from a whole that they understand as seamless. But their methods certainly offer a solid toolbox of specific restoration techniques to neighboring peoples, and some illuminating restoration principles to anyone, anywhere, who cares to pay attention.

Of course, very few if any cultures are likely to replicate the Lacandon way of living even in broad outline. Most of us have already modified our environments in far too many ways and far too many times over to even conceive of such an option in its fullest sense. But of all the lessons that this remarkable culture offers us, at least one is surely universal: restoration is only fully possible when an intact "historical reference system" still exists, in close proximity to the cultivated or degraded site. The presence of primary, or at least old-growth, forest alongside the Lacandon *milpas* and *acahuales* is not only an open and living text book on the kind of soil, plant, and animal communities that they want to restore; it is also an invaluable source of flora and fauna to repopulate these cultivated and managed sites with species that have diminished or disappeared on them.

So where we are trying to restore to a state that existed on a site in the past, and still exists in the present elsewhere, the level of success will increase with the proximity of a thriving historical reference system.

And that leads us an absolutely fundamental point about the relationship of conservation and restoration. Conservation is not only always more desirable (and much less costly) than restoration from the point of view of maintaining biodiversity; restoration itself depends on some degree of successful conservation in order to achieve its own fullest expression. The identification, acquisition and protection of mature, biodiverse ecosystems with intact processes—and these processes may include appropriate sustainable agriculture—remain urgent priorities. This is both because they are valuable in themselves and because they are the *sine qua non* of comprehensive restoration of those systems on appropriate sites. First, as Dan Janzen and Winnie Hallwachs might say, get your garden. (And yes, I can hear them add, make sure at once that you have your society's permission to let wild things flourish in it!)

But such islands of public (and private) conservation can only be one part of a holistic environmental strategy in a world with ever-increasing pressure on land from the human population. The question of what happens on the vast ranges of agricultural land that generally surround such wild islands as we have managed to conserve is also a crucial one. We have "redesigned the landscape," to borrow Keith Bradby's phrase, for too long without any thought for the impact of such reconfigurations on biodiversity, and also on ecosystem processes that affect us even more directly, such as flood control, water purification, and the renewal of fertile earth.

The future may be best secured by the fostering of a willingness on all sides to blur, or better, blend, the rigid concepts of "conserved" and "productive" land. Nature reserves will be much more acceptable to poor rural people and to human economies in general, if people can carry out productive activities within their boundaries on a sustainable basis. Depending on circumstances—it will always be case-specific—the limited extraction of timber, or the hunting and fishing of abundant species, or managed grazing, may do no harm to a park's biodiversity, and will often do some good. And, as Janzen and Hallwachs repeatedly point out, such activities will certainly help secure the sociological underpinning of the park, without which all attempts at conservation will founder.

Conversely, if numerous pockets of biodiversity are allowed to flourish again on productive land, as they often used to before the introduction of industrialized agriculture, the quality of rural life will be enriched in many material and cultural ways. Such developments can also, where appropriate, create a network of corridors, on micro and macro scales, that will facilitate the flow of genes within—and between—species. It is, of course, also essential that productive land is exploited sustainably, and does not leave a legacy of impoverished "old fields." Many of the major restoration projects we have visited—Working for Water, the Cinque Terre National Park, the Gondwana Link, the Área de Conservación Guanacaste, Irish woodlands and bog restorations, the Lacandon *milpa* cycle—involve diverse mixtures of agricultural production and conservation. And this kind of mix has been cogently advocated in New Zealand by Bruce Clarkson and John Craig, among others. The same principles can be extended right into the heart of urban areas, alongside the exciting new developments we are seeing in urban agriculture, in countries as different as Cuba, Canada, and Ireland.

These principles have been discussed in conservation circles for decades now, but they have not made much impression on public discourse, and less on public policy. They violate a variety of taboos—all a lot less useful than the Maori *tapu*—about conservation principles, on the one hand, and about the rights of private property on the other.

Restoration Reconciles Opposites

It seems to me that the practice of ecological restoration is exceptionally well equipped to interface with many of the challenges that such a revamped global strategy for conservation will encounter. There is something about the way in which the attempt to restore forces us to *engage* with both nature and human society that can lead, I believe, to the reconciliation between the traditionally opposed imperatives of the conservation of ecosystems, on the one hand, and their exploitation for production and consumption, on the other.

It is still too early to accurately describe the ecological restoration movement as having a "global strategy," but the players to develop such a strategy have been lining up strongly over the past decade. Ecological restoration is now recognized as a frontline weapon of choice by many of the leading national and international state and nonprofit conservation agencies and campaigns across world. Advocacy by SER has played a leading role in raising the profile of restoration, and the SER primer is recognized as a key guiding document by, among others, the International Union for the Conservation of Nature, Parks

Canada, the TEEB reports, Nature Conservation Foundation India, the Food and Agriculture Organization, the United Nations Environment Programme, Kew Royal Botanic Gardens.[7]

SER has therefore been deeply influential on the rapid development of restoration theory and practice internationally in recent years, but it has no monopoly on either. Many of the largest restoration projects in the United States, including the Comprehensive Everglades Restoration Plan, the Chesapeake Bay Program, the crossborder Great Lakes Restoration Initiative, and the San Francisco South Bay Salt Pond Restoration Project, are only loosely grouped together, for example, in the National Conference on Ecosystem Restoration. They have never been affiliated with SER, though there are moves afoot at the time of writing to invite them within the SER umbrella. But the very existence of such huge projects shows how much restoration has become part of US mainstream thinking.

Meanwhile, countries as diverse as Cuba and New Zealand, each more or less isolated from SER in their own ways for many years, have autonomously developed their own restoration guidelines. These have turned out to be remarkably similar to the primer, and are thus are good endorsements of its wide applicability.[8]

The mainstreaming of restoration is also increasingly evident in much broader fields of public interest, in particular in the creation of jobs and in legislation. The very ambitious Atlantic Rainforest Restoration Pact, which aims to restore fifteen million hectares of native forest in Brazil by 2050, estimates that it will generate at least three million jobs in that period.[9] Meanwhile, the Food and Agriculture Organization's chief forestry economist has envisaged the creation of ten million jobs in the establishment, restoration, and improvement of woodlands worldwide in the short term.[10] These are remarkable figures from reputable sources; even if they are overestimated it remains clear that restoration is already playing an increasingly high-profile role in national economies. For that role alone, but also for many other reasons, specific legislation and regulation of restoration procedures and targets is increasingly required. Just how specific that legislation should be, however, has become the subject for an illuminating debate among restorationists in Brazil, one of the countries where restoration legislation is most advanced.[11] Meanwhile, restoration's rise on the radars of the broad conservation movement, and of economic and political policymakers, has been accompanied by endorsement by peers in the scientific community.

Our Planet's Future May Depend . . .

This was epitomized by a prestigious and substantial special section of *Science* magazine on restoration ecology in July 2009. "Our planet's future may depend on the maturation of the young discipline of ecological restoration," the section editors lead off in a glowing opening article in this normally sober journal, "in its short life it has assumed a major role in sustainable development efforts across the globe."[12]

One article in particular underpinned that upbeat assessment. This was a meta-analysis of eighty-nine published studies of restoration projects, across all continents except Antarctica and across a wide range of ecosystems. It showed that they had, on average, increased biodiversity by 44 percent and ecosystem services by 25 percent. It also showed a positive relationship between increases in biodiversity and increases in ecosystems services. This indicates that biodiversity is not a luxury, but a condition for success in the more pragmatic pursuit of natural capital and ecosystem goods and services. Especially given the short timescale of many of the projects, these outcomes are remarkably positive. The article quite rightly entered the standard caveat that restoration of a degraded system remains a poor substitute for conservation of an intact one. And it also stressed the urgency of developing improved restoration techniques, and of monitoring restoration outcomes.[13]

At an early stage in restoration ecology's history, Anthony Bradshaw rightly highlighted the unique research potential of ecological restoration, as an "acid test" of ecological theory.[14] If you think you know how something works, there is no better test for that assumption than to attempt to build it from the ground up. In an innovative attempt to develop this insight, Kew Royal Botanic Gardens are embarking on a major long-term project that will take an ecosystem apart very carefully, and then attempt to reassemble it. Bruce Pavlik, Kew's first head of restoration ecology, has very high ambitions for his brainchild. He says it is "the restoration equivalent of the Human Genome Project—a search for information, interactions, regulatory mechanisms, and hidden wonders."[15] If it gets off the ground—and it will need a lot of funds to do so—Pavlik's project may represent an unprecedented leap in scale and sophistication for restoration ecology, though it will take many years to produce definitive outcomes.

Restoration's greatly enhanced profile and prospects brings with it greatly increased challenges. Stepping out into the forefront of public affairs, the ideas that underlie it will be tested as never before. In order to gauge whether they can take the strain, we need to explore some of the main trends of thinking in the restoration movement. Not coincidentally, they are in ferment just as these challenges are looming larger than ever before.

This movement is, as we have seen, a broad church, which is no bad thing. And, like all dynamic movements, it is sometimes given to energetic disputes about what direction it should take. I have observed, with fascination—though occasionally with dismay—the robust arguments that have developed between three broad trends in restoration thinking over the last five years. These debates are the fruits of encounters with the limits of restoration practice, and the theories that develop out of such encounters. Yet the principles that are emerging from them sometimes seem mutually exclusive. One recent set of polemics, in particular, is proving deeply divisive and potentially damaging to the clear articulation of restoration policy, just at the very moment when mainstream policymakers are becoming aware of the rich potential of restoration. Wisely conducted within a spirit of pluralism and open-mindedness, these debates could prove mutually enriching and ultimately contribute toward a new contract between our species and our world.

However, our adversarial and highly competitive culture can channel such differences into intellectual, professional, and quasipolitical factionalism. The scientific method eschews dogma in its pure form, but scientists can adhere as dogmatically as any priest to theories they happen to find personally congenial. The study and practice of restoration confers no special impunity against the distortions generated by ambitious egos and alpha (fe)male posturing, nor against our deeply ingrained social tendency to form a like-minded communities ("us") and demonize those who hold different views as a hostile "them."

I suggested in chapter 8 that the contending arguments within the restoration movement are as likely to be expressed as internal conflicts within individuals as any one of them is likely to be embraced entirely by any one person. But things have moved on since the 2007 SER conference in San José, and the tendency of individuals to identify with a particular trend of thinking has, I believe, intensified. In broad terms, three such trends can be fairly easily distinguished, though each is complex in itself.

First, there are restorationists who, like Bill Jordan, insist that the essence of restoration is the recovery of *historical* or, as he now puts it, *classic* ecosystems. Their overriding concern is to bring into being, on a degraded site, a new ecological community that replicates, in as many respects as possible, an ecosystem that existed there in the past. Within this trend, there is usually a strong emphasis on the value of nature in itself, on what Jordan provocatively describes as "a studied disregard for human interests." Jordan fears that factoring human needs or tastes into environmental equations endangers the integrity of nature, leading to the loss of unique but economically unproductive—or aesthetically unattractive—ecosystems. At times he seems to see only a slippery slope between the high ground of respect for conserving nature-in-itself, that hard-won value espoused by traditional environmentalism, and the consumerism that is gobbling up resources and landscapes on a planetary scale. Anyone who abandons this principle, he suggests, will slide to the bottom, regardless of her or his good intentions, and the restoration enterprise will be fatally compromised. I will call this trend of thinking the *ecocentric restoration* trend, borrowing a recent phrase of Jordan. He also describes it as "other-centered restoration," stressing a focus on restoring ecosystems for their own sakes, as it were, and not for ours.[16]

Second, there are those who take what appears to be the opposite approach, and, like James Aronson, focus on the human interface with ecosystems, and on the dependence of human societies on natural capital and the ecosystem goods and services it generates. They call for a new fusion of economic and environmental analysis and strategies, to halt the decline in natural capital stocks, and ultimately to augment them, through ecological restoration. They do not reject the target of restoring to historical reference systems—in fact, Aronson is a strong advocate of this principle—but they prioritize the benefits to humans when assessing a range of such references. They are particularly driven by awareness that most of the world's remaining biodiversity hot spots are concentrated in regions of chronic human poverty. They argue that conservation strategies will never work if local people see them as an obstacle to decent livelihoods. They believe that poverty alleviation through development is compatible with the restoration of biodiversity, but only if the vital importance of restoring natural capital is recognized, both in the developed and developing worlds. I call this the *RNC trend.*

Third, there are those restorationists like Richard Hobbs who stress that a combination of new, human-generated environmental factors (especially climate change, land degradation, and alien invasive species) are leading to the emergence of "novel ecosystems." They argue that this development makes the restoration of many or most ecosystems to their historical range of variation extremely difficult, impossibly expensive, or simply impossible. They say that the species communities in novel ecosystems have no historical precedent, and will probably frustrate all attempts to restore them to a historically based condition. They also stress that new findings about the complexity of interactions between soil, plant, and animal communities indicate that straightforward restoration to a historical reference system may not be a realistic goal under the best of circumstances. Ergo, scarce scientific and economic resources dictate that the target of restoring historical ecological communities must often be abandoned; the best that can be achieved in many cases today is the restoration of some ecosystem functions and services. Consequently, they make proposals that many restorationists consider a kind of heresy, for example, the acceptance of persistent alien species where the cost or difficulty of removing them is (in their view) prohibitive, and especially where the aliens may perform similar ecosystem functions as the displaced natives did. I will call this trend the *novel ecosystem trend.*

Irreconcilable Differences?

There are clearly overlaps, as well as sharp breaks, between these trends. Few who advocate ecocentric restoration are entirely insensible to human needs, the value of natural capital, or the awkward fact that the historical target is very often more aspirational than realistic, and more so than ever in our anthropocene era. Those who advocate RNC recognize the value of classic landscapes, and often have a pragmatic attitude to the challenges posed by truly novel ecosystems. And those who now contemplate the abandonment of historical references and the embrace of novel ecosystems often do so most reluctantly, well aware that their analysis is itself a grim comment on the degraded ecological state of our world.

One can often identify cultural value judgments in the debates between these trends. Those most closely associated with the ecocentric trend argue that the emphasis on human needs in the RNC trend will inevitably compromise what they understand as the ecological integrity of restoration. Meanwhile, both ecocentric and RNC trends fear that the acceptance, and sometimes the embrace, of "no-analog" ecological communities by the novel ecosystem trend reflects defeatism in the face of the difficulties that were always evident in the restoration enterprise. They also suspect the novel ecosystem trend of hubris. They point out that Hobbs and his colleagues warn of the limitations of restoration, but then, rather paradoxically, appear to revive the old human delusion that we are the masters of our universe, and can engineer and redesign novel systems according to our whims. Both ecocentric and RNC trends sometimes wonder aloud whether the proposals of the novel ecosystem trend merit the label "restoration" at all. The novel ecosystem trend has, in my view, given its critics hostages to fortune over the last few years, through creating a polemical narrative that overstates some arguments and understates others.

Meanwhile, the harder-hitting proponents of the novel ecosystem analysis, having made their break, often painful personally, with historical ecosystems, sometimes accuse those who still adhere to traditional restoration of romantic and irresponsible nostalgia, of squandering resources on unattainable goals. They warn repeatedly, and wisely, of the "moral hazard" run by restorationists who make promises that they cannot fulfill.

I believe that these differences, while real, are not as irreconcilable as they sometimes appear. Hopefully, they represent the growing pains of the young discipline of restoration ecology as it approaches its first adulthood, rather than any definitive rupture.[17] Each trend shares the restoration vision of a very particular and very radical kind of engagement with nature. Despite, and sometimes because of, its frequent frustrations and occasional gross failures, this restoration engagement can be uniquely illuminating, from a variety of perspectives: ecological, conservationist, cultural, economic, and philosophical.

I would like to conclude our journey through the world of restoration, then, by engaging with the ideas of these three influential restorationists: Bill Jordan, James Aronson, and Richard Hobbs, each of whom has thought long and deep about these matters, and each of whom I have identified as, broadly speaking, representing one of the trends outlined above. I must stress, however, that this division is based on my own observations and not on the existence of any organized "factions." Others might well subdivide the thinking within restoration differently. But I believe that the identification of these trends highlights the range of options—and challenges—faced by the movement.

"Ecocentric" Restoration

Bill Jordan has thought longer—and in my view deeper—about ecological restoration than most of his colleagues. His quirky but erudite philosophical bent leads him to ask questions about restoration that most ecologists do not even consider, and tease out how the concepts and practices in restoration both reflect our relationship with nature and lead us to rethink that relationship. I am not always convinced by his conclusions, but I think his work is well worth exploring, because it throws much light on the key issue of historical reference systems and opens up a fruitful discussion about the deeper motives that lie behind restoration. Jordan's path has led him to question the direction in which the movement he helped establish appears to be heading, though he interrogates his own most dearly held positions with ferocious honesty. He is very tall, angular and rangy, with a prominent forehead and balding crown. He engages in conversation with a penetrating stare that can morph, though almost imperceptibly, to an ironic twinkle. He knows that he bears more than a passing resemblance to the male figure in Grant Wood's *American Gothic*—at least I assume he does because, when I visited his home in Woodstock, Illinois, he asked me to photograph him in his garden with a pumpkin and a pitchfork.

Jordan arrived at the heart of the reemergent Midwestern restora-

30. **Ecocentric philosopher.** Bill Jordan argues that we should restore "classic landscapes . . . with a studied disregard for human interests." (Photograph courtesy of Barbara Jordan.)

tion movement, Madison Arboretum, three decades ago. He became the first editor of *Management & Restoration Notes* in 1981 and remained at the helm for two decades. Under his editorship it became *Ecological Restoration,* the major journal for practitioners, at least in the United States. So he was the first to cast a critical eye over many influential articles in the field, by authors ranging from Andre Clewell to Joy Zedler, and from Stephen Packard to Eric Katz and Jon Mendelson. He may well have coined the phrase "ecological restoration," as no one seems to be able to find an earlier reference than one under his byline in 1985. But he is too modest to push this claim himself.[18] He moved to the Chicago region in 2001, where he founded and is the director of the New Academy for Nature and Culture, and where his collaborators include Liam Heneghan. Its mission is "to develop restoration as the basis for a gracious, sustainable relationship between people and natural ecosystems, and as a key element in a new environmentalism."[19]

I first met Jordan at the 2005 St. Louis workshop on restoration of natural capital. He was clearly out of sympathy with the emphasis of the sessions, led by James Aronson, and, it sometimes seemed, with their essence. He made an arcane presentation on restoration as humanity's gift to nature, which I suspect went over the heads of many present; it certainly went over mine. This was not a happy encounter for him, and he departed early. But he left me with a copy of his provocative reflections on restoration, *The Sunflower Forest*.[20] If anyone else has peered at our subject from so many angles, through so many lenses, I haven't found them yet. I have never reread a book so many times and have rarely been more richly rewarded, though also at times more baffled and irritated. When I heard him give a presentation at the Perth SER conference in 2009, I could see that some in the audience were as confused as I had been in St. Louis. But I realized that listening to Jordan is like hearing an improvised jazz solo. It helps a lot to know the basic score—his writings—but even then he is unpredictable. Sometimes he will soar with inspirational clarity, and sometimes he will crash. But always from a height.

Jordan insists that a key principle of restoration is to keep fidelity with a "classic landscape . . . that is, one that has a relatively long history, and has value in part because it has passed some test of survival over time."[21] His other criteria for classic landscapes include "some kind of distinctive ecological value," recognized by experts in the field. That sounds straightforward, but Jordan's thinking is subtle and leads us to places that are often challenging, though occasionally, I fear, willfully obscure. He is sometimes read as an exponent of unreconstructed, old-fashioned, Madison Arboretum–type restoration orthodoxy, but while his roots are deep in those prairies, the branches of his thinking have spread out in a permanent process of radical self-revision. A 1988 conference in England triggered his evolution away from the use of terms like "natural" and "native" as targets for restoration. He saw that his European colleagues had no single point like "1491" in their ecopolitical calendar to which they could ascribe a pristine condition. Instead, they tended to pick and choose different points in their histories—medieval mowing meadows, for example, or beech forests cultivated during the Roman Empire—as targets for restoration. At the same time, he was coming to understand that the "1491" that had inspired so much early American restoration was built on an illusionary view of a virgin "new" world, which was really just as "old" and often just as manhandled as Europe. America's landscapes, just like European mowing meadows and beech forests, reflected changing human as well as changing "natural" ecological influences over many, many centuries. So he began to use the term "historic" for landscapes that were appropriate targets for restoration, and by that he meant simply landscapes that had existed at some time in the past. But he found that concept much too broad, so he added the designation "classic" for historic landscapes that had "qualities not found in any other kind of system."

Most Recent, Most Radical

In *The Sunflower Forest* Jordan draws on anthropology, literature, ethics, philosophy, and religion as well as ecology to wrestle with the implications—and motives—for restoring to classic landscapes. In doing so, he ultimately articulates aspirations that are probably too refined, demanding, and altruistic to attract many followers in practice. But they can function as a useful benchmark for one particular type of restoration and raise flags that are often relevant to other points on the spectrum of restoration options, too. Jordan's most recent expression of his position is also his most radical one, though its outline was already visible in *The Sunflower Forest*. He argued there that we should restore classic ecosystems for their own sakes, using that provocative criterion, a "studied disregard for human interests"; he insisted that the aim of the restorationist is to "act in such a way as to disappear from the landscape"; that he should "let nature work through him [*sic*] on its own terms."[22]

Since then he has extended these arguments under the rubric of "ecocentric"—"other-centered"—restoration, in a fascinating if characteristically idiosyncratic history of the field in the United States and Australia.[23] He defines "ecocentric restoration" as "the form of management that entails minimum management, and that [is] undertaken strictly for the sake of the system, as a tribute to nature as given and its inherent value."[24] The only aim of such restoration should be fidelity to a classic landscape, and "most certainly not [a landscape] more useful for us, more stable, more beautiful or anything like that."[25] Such a high-minded aim, in its purest form, breaches the frontiers of feasibility in ecology and indeed in ethics, as he himself acknowledges. We are almost always incapable of restoring historical ecosystems with anything approaching total fidelity. Moreover, Jordan fully recognizes the poignant irony that the only way such a system can be even partially restored is by compromising its "otherness" through managing it, albeit if only to compensate for other anthropogenic influences. On the ethical front, he is well aware that the cultivation of selflessness may be ultimately self-interested, a conundrum that has kept moral philosophers spinning on the heads of pins for millennia.

But Jordan says that the impossibility of what he proposes is precisely his point. Ecocentric restoration confronts us with our limits, with the tragic (and comic) elements in life that technocratic and consumerist cultures try to ignore, but that are the stuff of religious myths and high art. He is unabashed in using terms like "sacramental" to describe such restoration. He says that he regards the practice of ecocentric restoration as "deeply impractical—and important for just this reason—the same reason [the] Sabbath is important, especially for a society that has lost touch with it and is committed to a value system based on . . . 'absolute work.'" He argues that restoration along these lines constitutes a unique recognition of the "otherness" of nature, and that, contrary to what you might think, it is not romantic or idealistic, but a repudiation of the "sentimentality" of a tradition in environmentalism—stretching back to and beyond Thoreau—that stresses a rather cozy and highly selective human "oneness" with nature.

I have to part company with the pure version of his theory for several reasons. His insistence on the "otherness" of nature, without which ecocentric restoration makes no sense, would surely force us to seek as references only ecosystems on which humans had made no mark in the past, and these are rarer than hen's teeth. As will be obvious by now, I take the opposite position: it is one of the happy insights of restoration that we are inextricably involved with nature, and that restoration demands human presence, not human absence, in ecosystems. We are, in many senses (and not sentimental ones at all), more a part of nature than we are usually aware of, and we lose sight of this relationship at our peril. And in looking for ecosystems without the mark of human influence as the most appropriate targets for restoration, I cannot help but suspect that Jordan remains more influenced by the pristine vision of 1491 than he is willing to recognize.

Yet he has a telling point in reminding us that nature does not exist solely to satisfy human needs, material and spiritual. This is what the "sentimental" version of oneness with nature often suggests, albeit unconsciously. True, nature is often benign to us, and we are indeed its creatures. But it may also annihilate us with earthquakes, tsunamis, epidemics and glaciations—what TEEB describes as "ecosystem *disservices*."[26] Like so many of evolution's creations before us, *Homo sapiens* may become one day just another creature that nature finds surplus to its requirements. Our current rapacious relationship with the environment recklessly speeds that day. But even if we do learn to live "in harmony" with nature, that day may still come.

So I can see ecocentric restoration as a bracing philosophical proposal, and (for some) as a significant personal environmental practice. It is the restoration equivalent of a retreat to a monastery, which will do no harm and may bring great insights. But I cannot imagine that it will, in the foreseeable future, attract more than a small minority of adherents. And they will, for the most part, be restricted to regions that have sufficient economic surpluses to take time off for an environmental Sabbath. For most of the human world, restoration will have to find a convergence between our self-interest and whatever ideological commitment to the conservation of biodiversity our cultures may have—or it simply will not happen at all. This is the realm of the RNC trend, where restoration takes place in the context of what we might call "*a studied regard* for human interest."

But the compartments of the restoration movement are not watertight. Surprising as it may seem, we can find some very clear and moving expressions of the importance of this synergy between ecological restoration and economics in Jordan's own work. Like Walt Whitman's mind, this man's thinking contains multitudes, and he is not afraid to boldly contradict himself. Or rather, he develops his thinking by embracing and exploring contradictions. In the very paragraph in *The Sunflower Forest* in which he declares that "the aim of the restorationist is to remove the mark of his own kind from the landscape," he continues, "Yet through the process of restoration he enters into a peculiarly profound and intimate relationship with it."[27] True, in this case he is not talking about an economic relationship, but throughout the book he lauds restoration-linked activities that benefit us very directly. He talks about the pressing need to "find ways in which the ecosystems can contribute to, or in some way intermesh with, the economy of the human community." He hopes for a day when ecology and economy will "cease to be in opposition" and, instead, "pull each other along."[28] He even praises the "beautiful intimacy" of agriculture,[29] and predicts that, in the near future, landscape-scale restored prairies could be rotated with cultivated croplands—to the benefit of both.[30] Like his forerunner Ralph Waldo Emerson, it may be that he considers rigid consistency to be a hobgoblin of little minds. It is certainly important to stress that ecocentric restoration is not his sole concern. Indeed, his own summation of the significance of restoration amply recognizes its economic function: "Restoration is important for a number of reasons. It is important because [A,] it is a way of returning classic ecosystems to the landscape . . . [B,] because it provides a unique context for a positive, active relationship with the classic landscape . . . [C,] *because it offers a way of rebuilding the ecological capital of soil, air, water [etc] . . . that is the basis of all human economies.*"[31]

This is not, however, what interests him most about restoration as a philosopher, and before we move on to the RNC trend it is worth considering one of his most cherished but difficult ideas, which has more implications for restoration practice than is at first evident. This is his concept of the close relationship between restoration and "shame," which we first encountered briefly in chapter 5.[32] He continues, "But it [restoration] is also important for exactly the reasons that four generations of environmentalists have been skeptical about it: because it is at every point an encounter with shame."[33]

Jordan's use of the word "shame" is problematic, and even his colleagues in the New Academy for Nature and Culture struggle to come to terms with what he means by it. But he prefers its robust Anglo-Saxon force to abstractions like "existential anxiety," though he concedes that it has a very similar shade of meaning. In the following passage he articulates a familiar sense of human unease in the world, and suggests how restoration enables us to recognize it rather than denying it in the warm but dulling comfort of a sentimental pantheism: "Restoration is shameful because it involves killing and a measure of hegemony over the land; because the restoration effort is never fully successful and never complete; because it dramatizes not only our troubling dependence on the natural landscape, but—equally troubling—its dependence on us."[34]

For those who find such ruminations too "ethereal," awkward, or uncomfortable, it is important to point out that he gives it a very practical application, which, as we have seen, offers one path to a resolution of the painful conflict at the heart of the Chicago North Branch Restoration Project controversy.[35] He argues that what we might call "occasions of shame" in restoration—Jordan was brought up as a Catholic—should prompt the creation of rituals that could explore and interpret the meaning of restoration projects both to participants and to the wider community. This is not just a matter of having a congenial picnic after an IAP clearance or a midnight dance after a prairie burn, though such traditions are excellent for bonding participants and reaching out to neighbors. No, Jordan insists on grasping and brandishing the painful aspects of restoration that the North Branch movement in Chicago tried to ignore, or to disguise with cosmetic measures, to its own great cost. Restoration often involves measures that are troubling—"shameful" as Jordan would have it—both to many conservationists and to "developed" cultures influenced by traditional (and sentimental!) environmentalist thinking: culling deer and extirpating possums, cutting down trees and spreading herbicide on shrubs. Our ancestors in many cultures recognized that disturbing existential dilemmas are inherent in practices we find necessary for our survival. Hunting involves killing other species; Promethean fire makes us violators of landscapes; cultivation, and later mining, both require us to cut open Mother Earth. They elaborated many rituals to resolve the anxieties they felt about these activities. It would be presumptuous to prescribe what rituals restorationists should use in any given situation, but it is a question that each project should perhaps consider—though our antennae would need to be alert to the high potential for a cringe factor. As well as their inherent value, such rituals could provide opportunities to interface with local artists, performers, and religious leaders.

Jordan also argues that restoration offers a kind of long-distance benefit to those wilderness areas that remain more or less intact. Restoration work encourages individuals and communities to engage with nature right in their own neighborhoods. Such engagement feeds a hunger for an intimate experience and knowledge of wild things and wild processes and teaches us that they are much more accessible than we often think. We do not have to go to Yosemite, the Kruger, or the Amazon to experience our relationship with ecosystems. And these relatively wild areas would suffer fewer of our footprints—which in turn make them less wild—if more people had more of their encounters with nature closer to home. We can be intimately engaged with nature locally, even on our high street. And we do not need to turn the wilderness into a high street in order to satisfy the needs of an expanding population for contact with nature.

The RNC Trend: An International Perspective

James Aronson, the leading figure in the RNC trend of ecological restoration whom we met first in chapter 3, obviously comes at many of these questions from a very different, even opposed, perspective, and temperament, to Jordan's. The human factor is always part of the RNC equation. The encounter between the two men at the St. Louis RNC workshop was not exactly a meeting of minds. And yet, the more I have talked and corresponded with both men since those first meetings, the more I have become convinced that their positions on restoration are more complementary than contradictory. Their undoubtedly different orientations must be due in part to the fact that Jordan views restoration through a philosopher's lens, and Aronson through an ecologist's. But I suspect that another key element in the development of Aronson's position is his extensive international experience and vision, whereas Jordan has focused intensely—and until recently almost exclusively—on the experience of restorationists in the United States.

Aronson left his hometown of St. Louis in the late 1970s for Israel, where he studied applied and economic botany and agroforestry systems. Ironically, for someone who would become keenly aware of the threat posed by IAPs, he worked for a period selecting drought- and salt-tolerant alien plants that might become productive in the Negev Desert. But he came to the conclusion that Israeli agricultural policy was unrealistic, because it required unsustainable transfers of water and energy to make the desert a breadbasket. Simultaneously, he was

31. **Natural capital advocate.** "The degree to which a restored ecosystem resembles its reference cannot be discounted, but it is ultimately of secondary importance relative to the value of that ecosystem to people."—James Aronson (Photograph courtesy of Thibaud Aronson.)

becoming increasingly attracted to the then brand-new discipline of restoration ecology, drawing him toward a much more appropriate match of native plants and native ecosystems.

After spending some time in Latin America studying arid ecosystems, in 1992 he became the first restoration ecologist to be hired by the Centre d'Ecologie Fonctionnelle et Evolutive in Montpelier, France, part of a prestigious national research network, the CNRS. His home for the last two decades has been a Languedoc *mas,* or farmhouse. From here, he collaborates on a stream of journal articles and books. In his scarce spare time, he gardens and, with his family, produces a small quantity of good olive oil. But he also restlessly travels the world, especially the developing world, in search of projects that marry ecological and social aspirations.

In 1993, he was the lead author on a pair of articles in the just-established journal *Restoration Ecology*.[36] In these he marked out two zones of the restoration territory he was to explore zealously over the next two decades. The first was a geographic and socioeconomic orientation toward what he calls the global "South." By this he means those developing countries that have generally been very poor in terms of income per capita but are often still very rich in biodiversity. His work, he wrote, was based on the view that the conservation of well-functioning ecosystems was best achieved through making them economically useful to local people.[37] From the outset, Aronson viewed ecological and social restoration as joined at the hip.

The other zone he maps out in this same article is a dogged determination to define the terms of restoration ecology, which, even today, and even in peer-reviewed journals, often remain remarkably vague. Mainstream ecology has attempted to achieve greater precision through an embrace of mathematics over the past fifty years, and Aronson himself has written of the need for the study of ecosystem health to move "from metaphor to . . . measurement."[38] But, like Jordan, Aronson is a creature of fruitful contradictions. In some ways, his own bent is often more metaphorical than mathematical. He is as likely to focus on broad, effective practical strokes as on fine-detail academic brushwork. His instinct is probably right here. Restoration ecology is an activist science, like conservation biology, working against an ecological clock accelerated exponentially by human impacts. If we wait to restore a system until all its structural and functional equations had been demonstrated, it will very likely have crossed several more thresholds of degradation, perhaps irreversibly, before we even begin to restore it.[39] So there is a constant tension in Aronson's writing between an awareness of the need to apply rigorous scientific principles in restoration ecology, and the pragmatic recognition that ecological restoration projects are bound to be "messy,"[40] requiring the use of intuitive, even artistic, skills as much as strictly rational ones: "Restoration is not akin to civil engineering, where each task has a highly predictable outcome, nor is it like gardening or farming, where local conditions can be tightly controlled until crops are harvested. . . . [It] is more like raising children . . . unexpected circumstances arise that necessitate prompt resolution using ingenuity and assertiveness."[41]

Despite this recognition that high levels of precision are very difficult or impossible to achieve in restoration science, Aronson repeatedly insists that an agreed terminology is essential if effective guidelines for restoration are to be laid down and followed. This is an even more pressing question today, when bold attempts are being made to legislate for restoration on a grand scale. In these 1993 articles, Aronson and his coauthors ventured definitions of key terms that would be fiercely debated within SER for the next decade, and sometimes still are. These included "restoration" itself, as opposed to related but distinct options for post-degradation land use like "rehabilitation" or "reallocation." The core element in restoration strictly speaking, they argued, citing an early SER definition, lies in the intention to return a degraded site to an appropriate "historic ecosystem," despite the difficulties that this almost always entails.

Rehabilitation, in their view, is much less ambitious and does not put the same stress on the most complete possible restoration of native ecological communities. It may simply aim at the recovery of functions and services—usually directly beneficial to humans—that have become blocked or lost due to degradation. Reallocation is very different again: the option to accept entirely new plant and animal communities on the site, and therefore quite probably new functions and services, where land is very badly degraded. This option sounds very similar to the embrace of a "novel ecosystems," though that term had not yet been coined when they wrote the article. But even at this early stage, Aronson saw reallocation—rightly, I believe—as completely distinct from restoration. The blurring of these categories—especially by industrial interests eager to put a greenwashed spin on their cheap fixes to damaged landscapes—has done much damage to the reputation of restoration among conservationists. To some extent, it still does.

The authors also listed eighteen "vital ecosystem attributes," nine related to structure and nine to function, that could act as check lists in planning restoration projects. And they offered ten hypotheses for restoration ecologists to test, many of which still require a lot of experimental work nearly two decades later.[42] While these articles only reach toward the clearer definitions that would emerge in later work (by Aronson, and by many others), they are still worth revisiting to tease out the basic agenda for restoration ecology.

The central insistence that restoration targets should be based on a historical reference system has been a constant of Aronson's work, though it has also been constantly refined. Two years later he led a succinct and cogent response to an editorial in the same journal by S. T. A. Pickett and V. Thomas Parker, who argued that the use of reference systems was a "trap" for restorationists. They said the concept was a hangover from the obsolete ecological model of climax systems in eternal equilibrium; it was worse than useless in the light of the new ecological paradigm of a flux of nature.[43] While embracing the flux paradigm, Aronson and his colleagues argued that a reference system, "*even if arbitrary and imperfect*," remains essential if there is to be any "control" for the experiment inherent in every worthwhile restoration project—essential, that is, for the experiment to be evaluated in any meaningful way.[44]

There is certainly a significant shift here from the attachment of early American restorationists to the idea that there is always one ideal past state to be used as a blueprint. Aronson acknowledges that restorationists are always making a choice between a whole series of states in the past as their restoration target—and are unlikely to be able to replicate any of them completely. He argues, however, that choosing one of these states as a reference is necessary "to clarify research goals and methodology," and for "purposes of project design and evaluation."[45] When we last discussed these ideas, he stressed that that "I've moved more towards a notion of sequential references, since it is often not easy or possible to pick one single reference for a 100-year-long project."[46] The freedom to choose multiple possible references from the past—a forested system from four centuries ago, an agricultural/scrub mosaic from two centuries ago, and so on—is matched here by recognition that any restoration has several possible trajectories into the future (see fig. 32). Some of these will be determined by management methods, and more are likely to be determined by unknown elements within the system, or by the impact of events in the "matrix landscape" that surrounds it. Indeed, management will probably have to be repeatedly modified to cope with the latter—"adaptive management," though they do not use the phrase here.

And then, in a sentence that is almost an aside, Aronson adds,

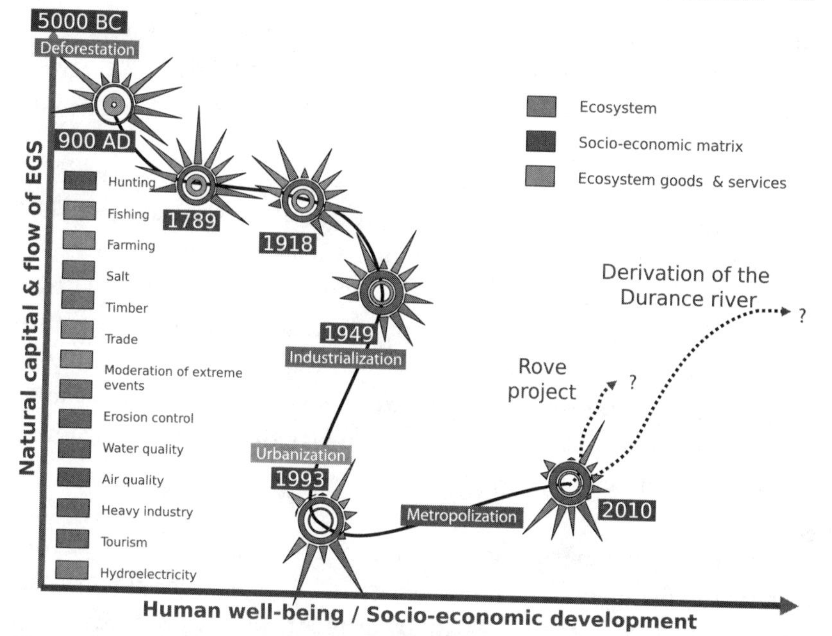

32. **Multiple historical references, multiple futures.** A sequential reference model applied to the landscape around the Berre Lagoon in France, showing six different points in its historical range of variation related to natural capital and ecosystem goods and services, and an indication of its future trajectories under the impact of proposed major infrastructural changes: a tunnel, a channel, and river engineering. (Figure courtesy of James Aronson and Michael Weinstein; originally published in M. P. Weinstein and R.E. Turner, *Sustainability Science: The Emerging Paradigm and the Urban Environment* [New York: Springer, 2012].)

"The reference system, particularly if it is an indigenous, historic one, should provide the essential oil of motivation and inspiration to keep the project rolling."[47]

Not a Pitfall, a Useful Tool

In an interview in 2011, he gave great emphasis to this additional motivational benefit in using historical reference systems as restoration targets.[48] Such a reference, he argued, is a very powerful tool for gripping the imagination of stakeholders. The process of choosing one reference from several points in the past is an adventure that can restore relationships within and between human communities, an antidote to our habitual contemporary alienation from both nature and our neighbors. Establishing a clear target based on a historical reference system, even if it has to be modified repeatedly over time, gives communities a coherent and cohesive restoration narrative:

> The reference gives you orientation, and a goal, and a yardstick or a benchmark or a compass to help you evaluate where you are going, to plan where you want to go, to monitor what you are doing against a benchmark, if you have been lucky enough to set up a control within your experiment, and to inspire your co-workers, and neighbors, and stake-holders through the metaphorical use of the terminology. And the word restoration applied to ecosystems is metaphorical, because you can't turn back the clock . . . if you play the tape of life again it comes out differently.[49]

The conclusion of the 1995 article contains a succinct summary of the requirements of successful restoration: "Projects should, insofar as possible, be undertaken as an experiment, whereby something is defined as a reference, relevant indicators or 'vital ecosystem attributes' are identified, copious baseline data are gathered at the outset, regular sessions of data collection punctuate the project, and enough flexibility is maintained throughout so that the original 'ecosystem of reference' does not become a pitfall but rather remains a useful tool."[50]

The exploration of the significance and role of "ecosystems of reference" underpins much of Aronson's subsequent collaboration with Andre F. Clewell, an influential founder member of SER. Clewell, an elegant and energetic Floridian now in his late seventies, is unusual in that he combines in his long life experience the professions of restoration ecologist and ecological restoration entrepreneur. The collaboration found its first expression in the 2002 definitive version of the SER primer, over which animated and often heated discussions had taken place in the society over the previous ten years.[51] The 2004 edition remains a touchstone of restoration internationally today, though it is again under considerable pressure for further revision.

Its core ideas are unbundled and discussed in detail in the most significant Clewell-Aronson collaboration to date, *Ecological Restoration: Principles, Values and Structures of an Emerging Profession*,[52] one of a major series on restoration science and practice published by Island Press (and edited by Aronson). Like Jordan, the authors make a strong defense of the historical reference system approach. But they part company from him in their clearly human-centered view of restoration. Not for them the desire to restore to a historical system "for its own sake." They recognize that, at least in the case of cultural landscapes—which today means most if not all of our landscapes—multiple choices always exist; and that the target ecosystem for a restoration will always ultimately be chosen for human-centered reasons of one sort or another. "The question of desirability [of a restoration target] is reduced to human values. Do we prefer the original or the transformed ecosystem? Which version favors our ecological values, our economic wellbeing, our cultural fulfillment, and our aesthetic preferences? . . . The selection of the preferred alternative state in restoration planning is critical, because stakeholders and also the members of the broader community will have to live with the results."And again: "The degree to which a restored ecosystem resembles its reference cannot be discounted, but it is ultimately of secondary importance relative to the value of that ecosystem to people."[53]

This human-centered approach to restoration found its clearest practical expression in *Restoring Natural Capital: Science, Business and Practice*, the edited volume workshopped at the St. Louis symposium in 2005.[54] Propagating an RNC strategy, especially in the developing world, became and remains one of the main thrusts of Aronson's activity over the intervening years. But he has also been deeply and passionately involved in the debate about novel ecosystems that has opened up a sharp rift among restorationists in the past five years. This debate sometimes seems in danger of destabilizing a movement that is only just finding its feet. Ironically, Aronson was one of the coauthors of two of the seminal articles around these issues in 2006.[55] Yet he soon became one of the leading critics of the path taken by some of his fellow authors, as they developed a root-and-branch revision of restoration fundamentals in the context of accelerating global environmental change.

Novel Ecosystems Trend: Dark Side or Shining Light?

Since 2006, this revision has been carried forward through an energetic series of thought-provoking and provocative articles, workshops, and conference presentations by Richard Hobbs, James Harris, Eric Higgs, Young D. Choi, Cara Nelson, Lauren Hallett, Stephen Jackson, and others. Strong passions bubble below the surface good manners of scientific discourse in these discussions. The image of alpha males marking out territory, on all sides, occasionally springs to mind. Hobbs, Harris, and Higgs (sometimes described by their critics as "H-cubed") lament only half-jokingly that they have made themselves "sitting ducks" for "target practice" by those in SER who accuse them of abandoning historical reference ecosystems, tearing up the rulebook for restoration and "generally going over to the dark side."[56] Young D. Choi told the 2011 SER conference in Mérida, Mexico, that when he proposes the retention of some invasive species in novel ecosystems, he is labeled as suffering from "historiophobia" and "future mania." Choi is a very mild-mannered and scrupulously courteous man, so it was a surprise when he continued with some visible emotion: "These people spit at me. And I will wipe that spit from my face."[57]

Higgs is also explicit about the angst he has suffered while making his still reluctant transition from ardent defender of historical reference system positions to "official worrier"—his phrase—for the novel ecosystem trend. "It hurts my brain to think about it," he confessed in Mérida. "Nothing in my career has proved so conceptually and ethically challenging as novel ecosystems."[58] As well as this internal conflict, he is clearly pained by the "overblown rhetoric" he says that some of their critics have brought to the debate.

There might, however, be rather less rhetoric blowing around today if Hobbs and his colleagues had framed their own positions with fewer polemical declarations and with a little more sensitivity to the legitimate concerns of those who have engaged them in that criticism. There is also that curious wobble, already noted, in the way they advance their arguments. They stress—I would argue that they overstress—the limitations of restoration science and the consequent uncertainties involved in ecological restoration to historical trajectories, even under the best circumstances. But then they make sweeping references to the potential for "engineering" and even "designing" novel ecosystems, surely underestimating the great dangers this involves. It is in this field that our knowledge is, by definition, most limited, and our experience almost virgin. They do repeatedly acknowledge that they may be guilty of hubris; yet they then make grand statements that can only be described as, well, hubristic. Hobbs's own erudition and serious intent sometimes make awkward bedfellows with his wickedly irreverent wit and sharp eye for the catchy phrase. His excitement about a mold-breaking theory sometimes seems to overwhelm his awareness—keen though this is—of the value of what may be being broken by his breakthroughs. For example, the subtitle of his 2006 article on novel ecosystems is "Theoretical and Management Aspects of the *New Ecological World order*."[59] This phrase does not sound so much like a warning of difficult new times ahead, but—at least in my frame of reference—it echoes the optimistic triumphalism of the rhetoric of George H. W. Bush.[60] No doubt irony was intended, and respect and affection for the old order is indeed periodically expressed within the text. But that is not what remains with the reader. And the phrase "novel ecosystem" itself is very problematic: it sounds bright, shiny, and positive—much more attractive than "irreversibly degraded ecosystem," for example. Hobbs originally used "emerging ecosystem," but his editors did not find it catchy enough. Dumbing down is not a monopoly of the tabloid press—science journals play that game too, as Hobbs himself has commented.[61]

Words and their meanings can be every bit as tricky in science as they are in literature. The phrases scientists choose to portray natural phenomena can shape public perceptions of the world and affect crucial decisions made by policymakers. Since he adopted this phrase in 2006, there has been confusion and controversy over whether Hobbs is simply describing the existence and implications of a new ecological phenomenon or embracing it with excitement and enthusiasm.

Normal Service Will Not Be Resumed

When I interviewed him three years later, in his hometown of Fremantle, I felt him to be more energized than depressed by his exploration of apparently irreversible changes to historical ecosystems: "The emergence of novel ecosystems," he says, "makes it a fascinating time to be involved with restoration ecology. A number of people are still in strong denial. People used to wait, when their TV sets went wonky, to see if normal service would be restored. Well, normal service will *not* be restored now," he concludes, with a tone that seems to register as much gung-ho satisfaction as regret. There is a striking contrast between the ringing certainty of such declarations about the future, and the deep strain of doubt he quite properly expresses about our knowledge of past and present ecosystems. Ecosystems, he insists, are not nearly as easy to manipulate as had once been thought. This is a recurrent theme in his critique of attempts to restore historical reference systems, especially in a rapidly changing global system:

> We are really just at the start of a learning curve . . . we are only just getting our fingernails into understanding the above-ground processes, it is so amazingly complex. Half the time it is difficult even to know what you are measuring and whether it even makes sense or not, we are a long way from making sense of it.[62]

This is a remarkable (but not untypical) statement for a leading figure in a cutting-edge field of ecology, and the editor of its leading journal. It is, of course, part of the job of scientists to draw our attention to what we don't know, and there has always been a refreshing, iconoclastic honesty in Hobbs's positions. But the view he expresses here does seem excessively bleak. Is the glass of restoration ecology not at least one-quarter full? And is it not getting fuller all the time, with successes like Alcoa, Kings Park, and Justin Jonson's old fields at Peniup on the Gondwana Link, to name just a few projects close to Hobbs's current home territory around the University of Western Australia? It is hard to reconcile such a downbeat assessment of the knowledge base of restoration ecology with his very positive comments on these projects.[63]

Certainly, Steve Hopper took the view that Hobbs and his collaborators had been "overly pessimistic" about the prospects for restoring old fields in southwest Australia to historical ecosystems.[64] In a response, Hobbs (and Rachel Standish) concluded, "The incipient threat of ongoing degradation and climate change demands more radical approaches to restoration than ever before."[65] Passionate concern about the urgency of responding to this threat of unprecedented and accelerating global change is the driver of Hobbs's new thinking. He can be caustic about those who do not sign up 100 percent to his mission. He is simply an unblinkered realist, "playing the hand in front of us," he told me in response to my critical comments on his articles. Such realism, he added, is "obviously causing angst to people, you included,

33. **Novel contribution.** Richard Hobbs at work on one of his experimental plots on Jasper Ridge at Stanford University: "The emergence of novel ecosystems makes it a fascinating time to be involved with restoration ecology." (Photograph courtesy of Lauren Hallett.)

who would prefer the world to be a more slowly changing and less complicated place."[66] The allegation that his critics have their heads in the sand is a constant theme of his presentations on the issue.

In 2009, Hobbs published a new article on novel ecosystems with Harris and Higgs that made clear just how radical their new approach would be. They stated at the outset that novel ecosystems "will require significant revision of conservation and restoration norms and practices *away from the traditional place-based focus on existing or historical assemblages [of species]*."[67] There is an unfortunate ambiguity in this sentence, which conflates two issues. Are the authors simply saying that "novel ecosystems" are a reality, and that they need a different management approach than historical ecosystems do? Many restorationists would accept that position without too much difficulty. Or are they really saying that, given the speed at which our world is changing, restoration must abandon historical ecosystems entirely, or almost entirely, as references for restoration? This is a much more contentious position.

Two things would assist greatly in dispelling this ambiguity: a clearer definition of novel ecosystems and a convincing estimate of how prevalent such systems now are and are likely to become. And without the first, it is impossible to quantify the second. There was an advance toward a definition in the 2009 article, especially through a figure that shows the trajectories of "historical" ecosystems moving across thresholds of change in their biotic and/or abiotic conditions into "hybrid" ecosystems, and then into "novel" ecosystems. Restoration from hybrid to historical systems is "feasible," the authors say. But returning across the threshold from novel systems back toward hybrid and historical systems is "difficult and/or impossible," unless there is "significant management input."[68]

We are approaching a clear and useful definition here, but for some reason the authors themselves hold back from crossing a different type of threshold,[69] one of meaning that would, once and for all, distinguish novel ecosystems from hybrid and historic ones and make explicit their significance for restoration. The definition remains blurred because *all* degraded ecosystems are "difficult" to restore, and *all* require at least some "significant management input" for successful restoration.

A clearer set of definitions for this figure would surely be:

historical ecosystems continue to evolve along their historical trajectory, within their historical range of variation;
hybrid ecosystems have been modified so that they are now evolving outside the historical range of variation, but they can still be feasibly restored to some point within that range;
novel ecosystems have been so radically modified that it is impossible to restore them to any point within their historical range of variation.

There are still (at least) two crucial words that beg further questions within this alternative set of definitions. What do we mean when say that the restoration of a hybrid system is "feasible"? And what precisely do we mean when we say that the restoration of a novel ecosystem is "impossible"? Obviously, there are two (intimately related) aspects in each case. A restoration is feasible if (a) we have sufficient ecological knowledge to attempt it in reasonable expectation of success, and (b) we have sufficient resources to carry out such an experiment. A restoration is indeed impossible where we lack either such knowledge, or such resources, or both. Evidently, advances in ecological knowledge also depend to a considerable degree on the allocation of resources for research. Most crucially of all, the decision as to whether we allocate such resources to restoration depends directly on our cultural values, economic values, and political will. Whether or not we designate an ecosystem as novel, therefore, may be more a question of public policy than of ecological analysis. Here, I believe, we are reaching the nub of question and can begin to see in sharp focus why the issue of novel ecosystems became so deeply contentious. The concept is very susceptible to cooption by those with interests in avoiding the costs of restoration. And it is also very attractive to those who are seduced by the vision that we can engineer and design the biosphere as we please.

Concerns that this is the direction in which Hobbs and his colleagues want to take restoration were further fed by points like these:

- "Removing the requirement to aim for a historic ecosystem increases the range of options available and could enable *reduced investment of effort and resources still to achieve valuable outcomes.*
- "It [restoration] will probably *change its focus from damage control to ecosystem engineering or 'designer ecosystems'*"
- "In this regard, cultural norms of nature will evolve alongside changing ecosystems, and *it is likely that our present beliefs require significant adjustment.*"[70]

An Expression of Defeatism?

Unsurprisingly, critics have been moved to describe this thinking as an expression of defeatism—or worse—"going over the dark side," to use Hobbs's own tongue-in-cheek self-description. Why, one could well ask, should restorationists change their "cultural norms of nature" in response to the degradation of ecosystems? On the contrary, should they not continue to attempt to reverse that degradation, inspired by existing cultural norms of the value of historical ecosystems? This passage put me in mind of an exchange I had with Jim Harris on the field trip to Peniup.[71] "SER must learn to speak in a language that engineers will understand," he said, "if we are to make a significant impact on policy worldwide." I could not help but wonder whether it would not be more appropriate for SER, while of course listening to engineers, and to experts from many other disciplines, to make a reciprocal demand: other disciplines also need to learn the language of restoration.

In e-mail correspondence with James Aronson over this period, I noted his rising level of irritation with the tendency in these articles to reject the need for historical reference systems in exchange for, as he sees it, a poorly defined new paradigm: "Simply put, if there is no longer any concern or regard for historical reference systems . . . then it is not restoration we're talking about anymore and let's not use the word. Let's say we're talking about designer systems or eco-engineering or 'intervention ecology' . . . If we use the words "ecological restoration," let's insure [*sic*] a minimum of actual content for the phrase and avoid letting 'restoration' be used to describe anything and everything that involves manipulation of biota or ecosystems . . . there is a real danger in that regard!"[72]

The most comprehensive critique of the novel ecosystem trend of restoration thinking to date is an unpublished paper by James Aronson and his colleagues. This article accuses Hobbs and company of sowing confusion, above all because they have failed to offer facts and figures to substantiate their claims: "There is no documentation of empirical thresholds defining historical, hybrid, or novel ecosystems, and there is no way to ascertain whether the hybrid or novel state is transient or stable. . . . To our knowledge, there are as yet no quantitative criteria to ascertain these points." These authors argue that truly novel ecosystems are probably much rarer than implied in the articles about them; that native species are likely to much more resilient to climate change than is suggested; they reject "the caricature that ecological restoration is a romantic quest to relive the past." They warn of the danger that novel ecosystems theory, as currently presented, could gravely undermine conservation efforts. The arrival of a single invasive plant, they say, could be enough for a policy maker to designate an ecosystem as "novel" and abandon restoration and conservation efforts there.[73]

Similar concerns are shared by other senior scientists. Steve Hopper wrote to me:

I worry about the more extreme views published on the concept of "novel ecosystems." Proponents claim that we are currently in such an unprecedented period of human impact and global change that the past is no longer a prescriptive guide for what might happen in the future. This is convenient academically, allowing an abandonment of the need to research and understand the historical and evolutionary literature, and to focus on modeling and conjuring up future scenarios full of assumptions about biological pattern and process about which we know very little at present.

Some advocates of the "novel ecosystems" concept have made extraordinary, speculative, published statements suggesting, for instance, that all tropical cloud forest or the Greater Cape's fynbos may well all be doomed. I get very nervous when I read scientists claiming absolutes like this. They don't read like serious scientific hypotheses. Moreover, the practical interpretation by misguided or gullible conservation managers of such claims by novel ecosystem proponents has a potentially negative impact on vitally needed and urgent efforts for biodiversity conservation. There seems to be an inability by some academics to resist appearing controversial in challenging a false straw man. This is disturbing to watch from the sidelines.

Meanwhile the real and difficult scientific challenges, tinged with a spark of hope for biodiversity, have not gone away and should not be deflected by such diversions. We live in a time when scientific humility is likely to pay much better dividends than hubris, and when enhanced collaboration between evolutionary and ecological scientists has great promise to make genuinely new strides in theory and problem-solving pertinent to all humanity at this moment.

In fairness, the passages quoted above from the Hobbs and colleagues' 2009 article are interpolated by the authors' own self-questioning: "caution is required: will we be capable of understanding what is best in a rapidly changing world? Will such activities be restoration or evolve into new types of intervention that respond to the rise of novel ecosystems?"[74] However, no satisfactory answers to these questions are attempted, and these are not the points that linger in the mind. Their concluding sentence seemed to me to set out the tasks that needed to be completed before the theoretical discussion can be usefully taken much further:

More detailed ongoing examination of novel ecosystems, and how to recognize, quantify and manage them, is required.[75]

This was not forthcoming, however, in Hobbs's other 2009 article on the subject, which appeared in a context where one might have expected him to celebrate restoration's successes to date. Instead, he focuses on the novel challenges that our rapidly changing biosphere poses to restoration's barely established principles. A brave choice, undoubtedly, but one that again fanned increasingly familiar concerns among his critics. This context was the aforementioned special section of *Science* magazine, which had given restoration ecology such a ringing imprimatur in an editorial comment.

Hobbs's contribution, however, may well have made some readers wonder why the field deserved such accolades at all. "For many ecosystems," Hobbs writes with paleoecologist Stephen T. Jackson, "restoration to a historic standard is anachronistic." The authors list the difficulties—long familiar from restoration literature—of identifying appropriate historical reference systems for restoration projects, and they conclude: "These observations have the potential for setting restoration ecology adrift from its moorings in notions of objectively identifiable natural states of ecosystems. . . . Ecological restoration finds new moorings in emphasizing restoration of ecosystem function, goods, and services."[76]

Almost at the stroke of a pen, then, restoration within a historical range of variations, hitherto the *sine qua non* of the discipline, appears to be relegated to the sidelines. It is replaced as the anchor of restoration by "ecosystem function, goods, and services." The authors do concede some space to restoring historical systems that may survive in the "new world order," but this concern seems subsidiary here to a new goal: "developing our capacity to engineer ecosystems successfully."[77]

The rich, finely drawn meshes of biodiverse communities must yield, it seemed to many readers, to the broad strokes of those aspects of ecosystems that are likely to directly benefit human beings. This sounds like a fast route to the bottom of the slippery slope Bill Jordan warns us about: the inevitable consequence, with the best will in the world, if we lose our grasp on the firm ground of historical ecological references. The RNC trend should be happy enough to see Hobbs's recognition of the importance of ecological goods and services—though they wonder at the omission from his work of the natural capital that produces them. But James Aronson and Andre Clewell nevertheless argue that such a reductionist reading of restoration removes vital dimensions from the big picture. Hobbs, however, has a valid challenge in return: "Is it better to refrain from recognizing and discussing these issues for fear of precipitating a slide, or should we embark on the discussion and at least aim to stand at the top of the slide with our eyes open?"[78]

He made a further major contribution to the debate in June 2011, in an article with Stanford graduate student Lauren M. Hallett and two of that university's stellar ecologists, Paul R. Erlich and Harold A. Mooney.[79] This article sought to place restoration ecology, along with fields like conservation biology, under a new umbrella, "intervention ecology." There is much to commend this initiative, and there are many points of convergence that the other restoration trends of thought would surely welcome: the need for an inter- and transdisciplinary approach to conservation, and the recognition that negotiating changes in environmental policy at global, national and local levels is key to success.

Regarding restoration, however, the general view in the piece seems clear: its "traditional goals are unachievable" due to "escalating global change" and it is time to "move on" from restoration, time indeed for restoration science "to metamorphose into something related but different."[80] Again, one has to salute Hobbs's courage in proposing the wholesale reinvention of such a young discipline in the face of unprecedented environmental turmoil. What is a little disturbing, however, is that some of the arguments marshaled here against the "moral hazards" of "traditional" restoration seem to set up straw men, in assailing positions that most restorationists have abandoned long ago. How many serious restorationists today, for example, promise the public that it is really possible to restore to a "previous static state"?[81] Most restorationists today recognize that ecosystems are dynamic, not static, and do their best to restore toward a *future* trajectory, guided by a reference system within a historical range of *variation*.[82] These principles lie at the heart of the SER primer definition of restoration and seem to hold up pretty well, in many cases, despite the acceleration of global change. Another apparent straw man set up here and in the earlier Hobbs articles about novel ecosystems revives one of the oldest charges laid against ecological restoration: that false promises of restoration tomorrow—usually as part of a mitigation deal by extractive or infrastructural businesses—may be used to justify degradation and destruction today.[83] While unscrupulous corporations undoubtedly deploy this tactic all too often, this is not a critique of restoration at all, but rather of failures to rigorously insist on restoration principles.

Despite these concerns, "Intervention Ecology" is an important article, and its authors recognize the weight of a number of counterarguments. In particular, they acknowledge that abandonment of the goal of restoration to a historical reference system could be seen as "lowering the bar" for restoration standards.[84] And they continue to highlight the danger of hubris in their own aspiration to "take hold of the steering wheel of life," no less, "and becoming intelligent, compassionate, and otherwise mindful managers of the planet."[85]

There is no question but that the issues so energetically raised by Hobbs and his colleagues are very pressing for the restoration movement. If the biosphere really were in such flux that familiar place-based historical ecological communities and systems are going to vanish from our planet in short order, then attempting to conserve them, much less restore them, would be futile. Worse, it would be a grossly sentimental and irresponsible waste of time and resources at a moment of extreme crisis. If they are right about the extent and depth of this crisis, we should focus instead on learning about rapidly evolving novel ecosystems, and manage them as best we can to ensure that they continue to supply the goods and services that are vital to our lives. We should leave other species to survive this period of rapid and unpredictable evolution as best they can. Of course we should also, as they themselves conclude, do all we can to slow down the pace of all human-induced changes. But they do not think that such changes can be slowed down very much, or for very long. Just enough, perhaps, to learn the new management skills required by the new ecological world order with a better chance of success, but even about this prospect they are far from sanguine.[86]

But we should, however, think very hard before we accept these conclusions unreservedly, because the assumptions underlying them remain open to question. The speed and impact of climate change globally remains extremely difficult to predict with any real precision, and many argue that it will in any case be patchy, with many historical ecosystems persisting in situ, within their historical range of variation. Their resilience may also be greater than the more pessimistic forecasts imagine. Such persistence and resilience would make continued efforts to restore degraded sites within historical ranges of variation much more feasible than Hobbs and his colleagues suggest here. It still seems too early to base restoration science priorities, and restoration policy generally, on the most radical scenarios for global change.

In the hope of clarifying some of these issues, or at least of hearing them well argued out in debate, I decided to travel to the aforementioned SER 2011 conference in Mexico. Hobbs was leading a two-session symposium under the title "Moving Targets, Shifting Sands," and Eric Higgs was leading a subsequent session on novel ecosystems. It seemed legitimate to hope that an entire day with a long list of distinguished speakers might go some way toward getting closer to the heart of this issue. As so often at SER conferences, however, the intellectual passions expressed on these issues at informal encounters in bars and restaurants did not carry over into the formal meetings, which broadly conformed to the published views of their organizers.

I was able to clarify one point to my own satisfaction during one of the brief pauses for questions that did occur. In several presentations, Hobbs had again stressed the important question of moral hazard: the public, and policymakers, are becoming enthusiastic about restoration, because they understand it to mean, naturally enough, that degraded ecosystems can be restored to something like their former glory. But that is a promise, Hobbs argued, on which restoration rarely comes close to delivering. He cited numerous instances of failed or very inadequate restoration projects to underline his point. But *why* had these projects failed, I asked him. Was it because the sites involved had become novel ecosystems that would resist any restoration efforts? Or was it simply because these projects were badly executed, or inadequately resourced, or both? He replied without hesitation that, in most cases, the failures were due to poor restoration work and/or poor resources. And at a later point in the sessions he volunteered the view that "nothing is irreversible if you throw enough resources and money and everything at it."

This straightforward exchange takes us back, it seems to me, to the core problem with the way the novel ecosystem trend is often presented. Its proponents have certainly identified a significant ecological development. Unprecedented scales of human impact on the biosphere are the defining characteristic of our anthropocene period. Climate change, land degradation, and introduced invasive species are indeed creating ecosystems that have no analogs in the past. Most of these systems will probably be even more difficult to restore to their historical trajectories than ecosystems where the historical elements are still dominant, or at least present nearby. In some cases, novel ecosystems may indeed totally resist restoration, in any meaningful sense of the word, as Hobbs and his colleagues argue. However, in many and perhaps most cases, the question of whether such degraded and altered systems are restorable to their historical range of variation may not come down to scientific or technical capacity, as the novel ecosystem trend suggests—or suggested until very recently. The real barrier to restoring a degraded system to its historical range of variation may often be economic, not ecological. And that makes the core issue not one of restoration science, but one of priorities, social as well as ecological. In other words, a matter of galvanizing the political will to dedicate the necessary resources to restoration.

We Haven't Even Started . . .

That, at least, is the conclusion reached by Kingsley Dixon. His job as director of science at Kings Park and Botanic Garden puts him at a fertile point of intersection between theory and practice. Novel ecosystems theory, he argues, is dangerously detached from developments on the ground. Worse, it is undermining the rationale for investment in restoration technology just as that technology is beginning to deliver spectacular results. During a coffee break at the Mérida conference, he put forward the case against this trend, as currently articulated, with his characteristic panache.

Dixon does not mince words: "The term 'novel ecosystem' is a fashionable concept, a cute term," he began. "It has gained currency with a whole lot of people who are saying, 'Oh, *right*, I can now define my weed-infested patch as a novel ecosystem.' It gives legitimacy to the illegitimate."

Wait a minute, I counter, isn't that is a rather malign reading of the term? Surely Hobbs and his colleagues have a major and significant point, that new systems are rapidly emerging that are indeed impossible to restore? Elsewhere, might the theory not help us establish priorities where restoration would cost so much that resources would be better spent elsewhere?

Dixon gives a good impression of a barely controlled explosion in the glitzy and artificial surroundings of the Fiesta Americana cafe. "*There!* You have hit the real nail on the head with that big word, *resources*. Globally, what are we spending on a real effort to put back ecosystem functions, using your local palette of indigenous species? It is minute, totally minute. We haven't even started. People think that restored biodiversity and ecosystems have nothing to do with the crises in water and soil quality and so on. Well, they have *everything* to do with it."

The issue, he argues, is not whether we can afford to invest in restoration, but whether we can any longer afford *not* to invest in it. And now, just as the arguments for restoration are gaining leverage among policy makers, the novel ecosystems theory seems to be knocking the blocks from under them.

> That is why I am getting so anxious, upset, and, yes, *angry* about this talk of novel ecosystems. It *is* dropping the bar. We have not even started this restoration journey. When I hear this discussion, all I get is a vision of the corrupted [eco]systems getting worse and worse with no government support. My state, Western Australia, is really, really rich. We make from iron ore alone, not from gold or anything else, AU$64 billion annually. But if I look at government investment in ecological restoration, how much is there? You guessed it, nothing, zero. The support we are getting [for restoration] at Kings Park is coming from the resources sector, private companies.
>
> Hobbs is really after principles and theories, and he has done some great stuff. But now, when we go to government and say let's have .0001 of our state budget put into restoration, and that is all we are

> asking for, he is jettisoning that case. They can say, "Look, they are just weedy systems, get used to it."

He gives the example of a volunteer citizen coastal restoration group he started eleven years ago. "We made terrible mistakes, we thought we could never do it, at first. Now we are looking at 100 percent species reinstatement."[87]

Dixon has direct experience of remarkably successful restorations in very different but difficult circumstances, like Alcoa's jarrah forest, and at Kings Park's Eliza Escarpment. He firmly believes that it is far too early to be saying we should "move on from restoration"—just as restoration technology is taking off. He believes Kings Park and Botanic Garden is entering into relationships with innovative restoration managers like Justin Jonson at Peniup on the Gondwana Link, which are only beginning to yield results:

> Kings Park is developing a whole lot of technology toolboxes for restoration, but they are all in their infancy. We have a five-year time frame; potentially we have a $30 million budget, all about technology. We will impact on even the very good work Jonson is already doing, so that he not only gets a lovely canopy and some nice shrubs, he gets 80 percent of the species he wants to get in there. In some of these systems Hobbs and company are calling "novel," we may be able to do a lot more, we are unpacking these systems, and saying gee, it is really interesting, but this is what we are going to have to do to these soils . . . and no one has done any of this stuff.
>
> Imagine if we started investing in the science of how to put them back. Making decisions about "novel ecosystems" now, when we haven't even started, is crazy.

A Fork in the Road or a New Convergence?

Richard Hobbs spoke at the conference about restoration ecology approaching a "fork in the road," which suggests an "either/or" scenario. After seven years of researching the restoration movement, I personally found the growing polarization on both sides of this debate disturbing and also, oddly, sadly familiar. If I may digress for a moment, let me explain that I grew up during the worst years of the Irish conflict, and then spent much of my professional life reporting on the Basque conflict. In both cases, people were under enormous pressure to identify with one camp, and demonize the other one. Worse, most people—and I include myself at certain moments—were often very comfortable with that black-and-white, good-and-evil scenario.

I'm not for a moment suggesting that restoration ecologists are about to start shooting each other—though Hobbs jokingly portrays himself, Harris, and Higgs as literally under gunfire in one Powerpoint slide. But there seems to be an inherent human tendency to think in binary terms, to need to believe that those who are not totally with us on any issue must be totally against us. Science is no more immune from this lethal tendency than politics is. Hobbs and his colleagues often caricature their critics as traditionalists fixated on a vanished past, incapable of looking a challenging future in the eye. Those critics often caricature the "novel ecosystem" advocates as sacrificing vital conservation principles for the high excitement—or high profile—of propounding a new theory.

Yet the more I read both sides, the more I become convinced that there is very significant common ground between them, indeed, between all three trends I have identified. There is a solid commitment to the conservation of biodiversity, for starters. There is a shared recognition that ecosystems are dynamic, and that restoration to a static past is not an option. There is a shared understanding that if human beings restore ecosystems, they will always do so in what they perceive as human interests, however altruistic they claim to be. A deeper understanding of our species' complex enmeshment with natural systems throughout our history is a universal lesson of restoration, properly understood.

So when James Aronson invited me to chair a mini symposium entitled "Novel Ecosystems: New Normal or Red Herring," at the 2012 SER Europe conference in the Czech Republic, it seemed a positive, if challenging, opportunity to see if that common ground really existed.

A broad range of positions, from Hobbs's to Aronson's and Dixon's (the latter virtually, via Powerpoint), were represented on the platform. There were no surprises in what they said, except perhaps for those who seemed to expect them to physically assault each other. ("Do you have a good strong helmet?" one eminent scientist had asked me when she heard I was chairing the session). The event was hardly a love-in, but, for the first time, the broad range of positions on the utility of the novel ecosystems concept among SER members was presented within the same four walls. I'm not aware that any of the leading players significantly shifted their positions as a result of their exchanges, but the common commitment to restoring biodiversity was repeatedly underlined. And Hobbs moved some ways to reassure those who feared he was advocating any blanket abandonment of restoration to historical reference systems: "Don't give up standard approaches," he said, "But we do need to augment and adapt them."

Perhaps the most important contributions came from the floor. The conference director, Karel Prach, had generously allocated an unprecedented three hours across two sessions to ensure maximum discussion, and people flocked to both of them. While some of those present said they found the novel ecosystem concept very useful, especially in contexts like post-mining restoration, more said that it was already having negative, and in some cases potentially disastrous, impacts on public policy. One American, who asked not to be named, claimed that one federal agency had already dropped the hard-won criterion that native seed should always be supplied to conservation projects, because the novel ecosystem theory suggested that alien plants could often be equally or more appropriate. Kris de Kleer, a member of the SER Europe board, warned that EU bureaucrats might behave similarly, and close down multimillion-euro restoration projects. "This 'novel ecosystem' phrase gives an alibi to those who do not want to invest in restoration," he said. "It is playing with fire. We need scientists to focus on restoration, on the successful eradication of invasives, not on this theory."

To which Jim Harris replied robustly, "I understand fear of abuse of the term, but I'm disturbed by the notion that we shouldn't be talking about it. I can't understand that, as scientists, we should not be publishing what we observe." He has a strong point. One cannot self-censor investigation because policy-makers might hijack it for negative purposes. But the question of how that investigation should be responsibly communicated remains contentious.

Following the conference, I had a detailed email conversation around these questions with Hobbs. He would not concede any ground at all to my critique of the language used in his articles. He insisted that the problem was not his choice of words, but the misreading of them by those who could not bring themselves to accept that novel situations require novel analysis and novel solutions. If it was not exactly a dialogue of the deaf, we certainly were not hearing each other very well.

However, at the same time, he sent me some chapters from the forthcoming book he has edited on the subject with Higgs and Carol M. Hall, *Novel Ecosystems: Intervening in the New Ecological World Order*.[88] The title led me to expect more of the same, but I found that that expectation was, indeed, to judge a book by its cover. On the basis of these chapters, which include the synthesis and conclusions, the book takes a much cooler, more measured and nuanced approach to the issues than anything I had seen on the subject previously. Two points strike me as particularly important in addressing the most serious concerns that have been raised about the concept.

The definition of novel ecosystems presented early in the book introduces a new element that explicitly acknowledges that we are not talking about a purely scientific category here: "Novel ecosystems are distinguished from hybrid ecosystems by practical limitations (a combination of ecological, environmental *and social* thresholds) on the recovery of historical qualities."[89] And this acknowledgment that the designation of novelty is conditioned not by ecological factors alone, but also by "social drivers," is spelled out very clearly in the same chapter. These include "deeply embedded economic constraints, political intransigence, or cultural values." While addressing these constraints will "not be easy," the editors continue, it can—and should—be done: "Thresholds matter, but they should not be trotted out as an excuse to avoid resolute action."[90] It is good to see this recognition that calling an ecosystem novel can indeed be used as an excuse—or alibi—for not investing in restoration made head-on and upfront, so early in the book. And in the conclusions, the related issue of abandoning goals related to historical systems is saluted as "a real concern . . . that should not be dismissed lightly."[91]

The second point is that, in the material I have seen at least, the hubristic tone is gone, replaced by exemplary scientific humility. The identification and measurement of even purely ecological thresholds is admitted to be extremely difficult, and often far beyond our current scientific capacities, and the authors also concede that the concept of irreversibility "might not hold water in all situations . . . with enough effort, even massive changes can be to some extent reversed."[92]

At the same time, Lauren Hallett sent me her own draft chapter from the book, "Towards a Conceptual Framework for Novel Ecosystems," and it confirmed my impression that, if the rest of the book meets these high standards it will clarify many of the issues that have been contentious and obscure in the debate so far.[93] While she is cautious about overstating claims, she is particularly strong in giving examples of how alien species may be critically useful in maintaining ecosystem functions, and even saving native species from extinction, in specific novel ecosystem contexts.

So perhaps it is not a question of "embracing" novel ecosystems and "abandoning" historical restoration, but of treating each ecosystem on a case-specific basis? A more pluralistic approach to the future of restoration, ideally synthesizing the three current trends, should be the aim of all concerned.

Only the restoration of hybrid systems toward their historical range of variability is ecological restoration in the fullest sense, certainly in the sense understood by the ecocentric ecosystems trend. Nevertheless, especially if we succeed in slowing down current rates of global change, there is still likely to be much scope for this particularly happy form of ecological intervention in the coming century. The recognition of the limits of restoration in managing novel ecosystems, and some hybrid ones, should satisfy the most compelling argument from the novel ecosystems trend: restorationists must scrupulously avoid the moral hazard involved in promising policymakers and the public outcomes that restoration cannot deliver. And we should certainly be willing to consider novel solutions, including the retention of alien species where appropriate, in the interests of restoring the greatest possible diversity. Meanwhile, the recognition that natural capital and ecosystem goods and services are vital elements to evaluate in all three scenarios should meet the concerns of the RNC trend.

Managing ecosystems to stay within their historical range of variation, whether that management is active or passive, may seem like conservation in its simplest sense. But in these scenarios, too, restoration thinking brings something fresh and important to the table.

It sees beyond the blanket imperative to preserve "wilderness" as we happen to find it in the present. It contributes a methodology based on exploring a site's sequence of historical reference systems as a guide to present and future management. Like the chameleon in the Malagasy proverb, "the restorationist is . . . always looking backwards, with one eye, while looking steadfastly forward with the other one."[94] One might add that the restoration also requires a third eye, to keep a sharp focus on all the ecological interactions that are taking place in the present.

Within this framework, I believe and hope, there is room for consensus between all the restoration trends, while respecting, and exploring, real differences.[95] Clearly, there is much to be done on all these fronts, but also much reason to believe that much can be achieved. And perhaps the best place to bring this discussion to an end is to remind ourselves, in one of Kingsley Dixon's favorite phrases: the journey of restoration has "hardly even started yet." But the preparations made for the trip over the last few decades are impressive, and it promises to be a deeply rewarding—and constantly challenging—environmental voyage.

15 *Conclusions: Why Restore?*

Over the last seven years, I have been looking at ecological restoration projects and talking to restoration volunteers, land managers, conservation professionals, restoration practitioners, and restoration ecologists. Essentially, I have been asking them all a single question: why restore?

It is time I tried to answer that question myself.

We should restore because there is something badly off-kilter in our relationship with the rest of the natural world, and ecological restoration offers unique and refreshing perspectives for setting that relationship on a better course.

Since humanity embarked on the great adventure of agriculture, our unprecedented dominance as a single species on this planet has long imposed extensive changes on its ecosystems, and on its awe-inspiring biodiversity. Some of these changes have been positive. We opened up forests into savannas. We turned relatively uniform systems into patchworks and mosaics that offered new niches in which many species could flourish and indeed evolve. Often, however, we exceeded the sustainable carrying capacity of particular ecosystems, landscapes, and even regions. We tipped them over into degradation, drastically impairing ecological functions and impoverishing their variety of life forms.

The surge in our productive powers during the industrial and technological revolutions, and the surge in population that has followed these engines of growth, has globalized human impacts on the entire biosphere. Our consumption has already outstripped the carrying capacity of the planet; we are altering its macrodynamics—climate, carbon cycles, nutrient flows—with outcomes that could ultimately repeat the tragedy of Easter Island on a planetary scale.

In this threatening scenario, ecological restoration offers a suite of options that have been widely, though as yet unsystematically, tested in diverse global contexts, and from which a wide range of encouraging lessons can be learned.

The first lesson is a psychological one, that of crossing a threshold from denial and despair to hope. The environmental movement has generally tended to focus on catastrophic scenarios, on narratives of loss and desolation. The impact of these narratives is often more numbing than energizing. Moreover, when some of the more apocalyptic narratives prove to be false—or at least premature—our very human response is to eat, drink, and be merry while there is still food on the table. We mock the messengers who tell us that the crops are failing, that the cupboards will soon be bare, that the sea levels are rising.

Ecological restoration, in contrast, offers us a narrative that has the capacity to galvanize individuals, communities, and societies to action. Not that this is an easy task. We have been too accustomed to viewing the human relationship with nature as inherently destructive and dysfunctional. The very idea that we can also have a constructive relationship with the world around us—that indeed this may be the more "natural" relationship—is hard to grasp at first. But once grasped, it can be electrifying.

Change you can believe in, if you like, but not as a phrase coined by a political spin doctor: this is change you can believe in because you can, on a small scale, make it happen with your own hands, your own sweat, and your own skills, and you can see the results with your own eyes. Restoration is an exceptionally good fit for that rather tired but still wise environmentalist slogan, "think globally, act locally."

If this book has demonstrated one thing, I hope it is that the best restoration is not based on nostalgia for the past, but on an engagement with the present, which of course contains the past; and that this engagement can weave from the present a richer future. The backward look is certainly part of restoration, because knowledge of the past of a degraded system enables us to imagine its future possibilities. The engagement essential to restoration deepens a knowledge of the sequence of past ecological states that have led to the present system; it deepens our knowledge of the roles biotic communities and abiotic processes are playing in the present, and of what sequence of states we can reasonably imagine, and work towards, in the future.

I think these simple principles apply, no matter which of the broad trends in the restoration movement sketched out in our last chapter one chooses to follow. Even ecosystems that are truly novel will benefit from this approach. They have not dropped down from the skies without a past attached to them, and we can learn from that past, however painful its lessons may be, and steer them towards a better future with judicious use of that knowledge.

We need ecocentric restorationists whose love for classic landscapes leads them to attempt to restore them, with as much fidelity as possible, under present conditions. They will produce work of great value, bringing back and maintaining, however imperfectly, natural communities and functions, whose vast range of secrets and values we will fully unlock and appreciate only over many generations. Ironically, among the secrets of classic landscapes, there are likely to be many more revelations of hitherto unsuspected marks of human agency in the past.

We also need natural capital restorationists with a sharp focus on human roles and human needs in ecosystems, and who prioritize the restoration of natural capital, and the ecosystem goods and services that flow from it. Without this trend's contribution, it is unlikely that the future will hold biodiverse ecosystems to restore; it is also unlikely that there will be humans around to restore them. But there will be pressures to reduce the RNC trend to become nothing more than the purveyors of an ever-increasing breadbasket. The global ideology that urges some of us to eat too much, while many others have to make do with much too little, will not easily surrender its present tight if sickly grip on our lives. The RNC trend should never lose sight of the goal that ecosystem goods and services should be distributed democratically, and that involves engagement with politics. And it should always remember that societies do not live by bread alone but also require the full flow of cultural, aesthetic and spiritual services that only healthy ecosystems and landscapes can supply.

And we need restorationists who explore the very challenging world of novel ecosystems and whose engagement with them may help us steer a course through an ecological ocean of no-analog

34. **Restored to Irish skies.** A red kite soars over Wicklow. (Photograph courtesy of Marc Ruddock.)

futures. But they too must resist pressure, in their case, from the siren calls of an engineering tradition that suggests we can turn nature into a designer shopping mall of human-orientated functions and services. Restoration's unique engagement with nature should always warn them against the hubris of thinking that we are the lords of all we survey. We are really only—but what a big only!—uniquely conscious participants in the mysteries of one small but infinitely diverse and ever-evolving planet.

For restoration to fulfill any of the potential outlined above, it needs to mainstream its stories into our globalized but still, happily, very diverse cultures. If this book has made any contribution to that process, I will be very happy.

I look out our door at Cois Abhainn, overlooking the Avonbeg River in Ireland's Wicklow Mountains. I see the fine, remnant native oak woods of Ballinacor Estate struggling to regenerate as invasive rhododendron strides through the understory. No native flowers blossom where the thick exotic canopies hold sway. The gamekeepers across the river are struggling to turn back the tide, and that battle remains in the balance. But now we know, from restoration experience with rhododendron in Kerry and Donegal, that it is winnable.

If I am lucky, a red kite may sail overhead. Reintroduced after persecution had exterminated them on our island several centuries back, these exquisitely graceful birds of prey are doing well after just a few years. When someone shot one, quite possibly by accident, on a nearby farm, the farmer was so upset that he banned all shooting on his land. Unlike his forbears, he sees no threat from this elegant creature. And he is prepared to pay a price for its presence. The kite itself does no damage, but he has now lost the benefit—pest control—offered by casual hunters on his farm. A similar reintroduction scheme with sea eagles in Kerry has run into trouble, however, as some farmers still see the restoration of these huge birds, which feed on fish, as a threat to their lambs, and poison them illegally. Good communication remains essential to restoration success.

If I am very lucky indeed—it hasn't happened yet—I may hear a woodpecker hammering nearby. Our only native species, the great spotted woodpecker, also vanished centuries ago; we are not sure why. Since I began writing this book, they have begun to return of their own accord. It seems they have expanded beyond the carrying capacity of British woodlands—and their more recent habitat of suburban gardens with their feeders—and are seeking a new frontier. The pioneer woodpeckers have chosen to build their nests, in most cases, in our few remaining old-growth native woods. In the valley across the hills from our house, they are nesting in the restored woods at Clara and Ballygannon.

I could hardly have imagined any of this even twenty years ago. Some things are getting better. But there is still a lot of very hard work to be done.

Acknowledgments

A book written over ten years, on and off, incurs incalculable debts of many kinds.

I cannot adequately express my thanks to the restoration community, scientists, professional practitioners, and citizen volunteers, who have welcomed me into their worlds and been so generous with their knowledge, their time, and often with their warm hospitality and other resources. To all of those who appear in these pages, though I know you will not always be happy with what you find here, I am enormously grateful. To all those who gave me of their time and knowledge and who did not find their way into the final cut of this book, I offer my thanks in equal measure. The expertise and commitment I encountered in so many places was always instructive and humbling, and I wish—though the reader may not!—that I could have written a book twice the size and included everyone.

It is regrettably impossible in this space to mention all those who helped me, but there are some who must get special mention. From the moment we met in Zaragoza seven years ago, James Aronson has mentored my research with tremendous generosity, sustaining and patient faith, and indefatigable energy. I am grateful for everything, including and perhaps especially for passionate disagreements, a mark of true friendship. This project would have fallen at several fences without his support. *Kadima*, James!

In Ireland, Declan Little not only unlocked arcane secrets of woodland restoration ecology, he also opened doors to a series of invaluable funding sources, starting with his own NGO, Woodlands of Ireland, at a moment when my own resources were faltering badly. Catherine

Farrell was also exceptionally generous with her time, wise advice, and warm encouragement. Richard Hobbs gave me a much-needed morale boost through his support at an early stage, and has been unfailingly helpful and good-humored—and occasionally appropriately robust!—in our many exchanges over the intervening years. All four of the above also made insightful comments on specific chapters, and in Aronson's case, on the entire first draft.

A number of other people gave critical assistance by reading individual chapters, improving them in all cases. As usual, any remaining errors, and all opinions expressed, remain my own. Thanks to Bill Allen, Keith Bradby, John Cross, Winnie Hallwachs, Liam Heneghan, Dan Janzen, Christo Marais, Colin Miskelly, Riccardo Nardelli, Sandro and Erika Pignatti, Guy Preston, Samuel Levy Tacher, and Lynne Westphal, many of whom also assisted me in many other ways. I am also grateful to my brother-in-law, Pa Duhig, who made perceptive comments on the whole raw manuscript. And also to an anonymous reviewer for University of Chicago Press, who went above and beyond the call of duty in making perceptive and enriching criticisms and suggestions.

I would also like to acknowledge a number of other individuals for their encouragement, time and support, and my apologies to anyone I may have omitted: Sasha Alexander, George Archibald, John and Jane Balaban, Luis Balaguer, Robert Baldock, Domingo Ballarte, Jeb Barzen, Rich Beilfuss, Aaron Bennett, Sandra Berrios, James Blignaut, Petra Blix, Tom Blinkhorn, Steve Bosak, Keith Bowers, Orla Brady, Nick Brandt, Mary Carmel Burke, Jonathan Cobb, Andre Clewell, Ed Collins, Kate Conley, Steve Cornelius, John Craig, Barbara Dean, Kris de Cleer, Kingsley Dixon, Hank and Dottie Edmondson, Dave Egan, Don Falk, Andy Friedland, Paul Gobster, Jennifer Gouza, Lauren Hallett, Janet, Doug, and Spencer Hardy, Jim Harris, Brooke Hecht, Steve Hopper, Jonathan Hughes, Bill and Buffy Jordan, Amanda Keesing, Brian and Mary Kennedy, mk LeFevour, Michele Lilley, Roy Lubke, Jan Marshall, Anabel Martín, Jesus Matos Maderos, Betsy McGean, Maud McKee, Stephen and Teri Meyer, Richard Nairn, Marlene Izquierdo Navarro, Claire O'Neill, Terry Osborne, Bruce Pavlik, Edurne Portela, Mike Powell, Karel Prach, George Rabb, Peter Raven, Heather Ray, Karen Rodriguez, Laurel Ross, Jack Shepherd, Richard Stamelman, Sharon Tribou St. Martin, Steve Packard, Mikel Valladares, Michael Viney,

Ross Virginia, Elizabeth and Jason Wallers, Steve Whisenant, Peter and Diane Wyse-Jackson.

To Ciaran Benson for the prolonged loan of a laptop and constant encouragement, and to Bill Roxby for invaluable help and assistance in selecting and editing images, I also offer my thanks, as well as to all who generously assisted me in supplying photographs and figures, who are individually acknowledged in the captions.

I could not have finished the book without the financial support of a number of institutions. The whole enterprise began to roll with a William B. Quarton fellowship on the International Writing Program at the University of Iowa in 2003. I thank especially the director, Christopher Merrill, who introduced me to the idea of ecological restoration. Gregory Norminton's unique contribution is acknowledged in the preface. A further fellowship at the Dickey Center for International Understanding at Dartmouth College, New Hampshire, in 2008 gave me invaluable time for reflection, research access, discussion, networking, and writing, thanks to the kind and ongoing support of the director, Ken Yalowitz and the staff at the Center. I could not have undertaken my New Zealand research without a travel grant from the University of Waikato, kindly brokered by Bruce Clarkson. In Ireland, six bodies provided generous—and admirably disinterested—financial research grants. My thanks to Bord na Móna (Catherine Farrell and Gabriel D'Arcy); Coillte (Tim Crowley and Gerard Murphy); the Heritage Council (Cliona O'Brien and Michael Starrett); Kate Bowe PR (Kate Bowe); the National Parks and Wildlife Service (John Cross and Ciaran O'Keeffe); and Woodlands of Ireland (Declan Little).

In terms of developing the book proposal, I owe a special debt of gratitude to Ivan Mulcahy, now of Mulcahy Conway Associates, for his professional advice. Various media outlets also assisted me in developing ideas, and sometimes sections of articles, that have found their way into the text. Thanks especially to Karl Meyer and Linda Wrigley of the *World Policy Journal*; Ivan Oransky at *The Scientist*; Mrill Ingrams at *Ecological Restoration*; Sheila Wayman, Conor Goodman, and Dick Ahlstrom at the *Irish Times*; and Jack Power at the *Irish Examiner*.

Writers dream, usually in vain, of finding an empathetic editor. I cherish the day I approached Christie Henry, Editorial Director of Sciences and Social Sciences, at the University of Chicago Press booth at a conference in 2007. She immediately responded to the project with the warmest interest, and has been constantly supportive, motivational, insightful, and incisive through many vicissitudes and, I must confess, too many delays on my part, ever since. At the University of Chicago Press I must also thank Abby Collier for prompt, patient, and ever-helpful responses to an endless stream of editorial queries, and Levi Stahl for his professionalism and supportive commitment to promoting this book. I also thank Mary Gehl for her courteous and good-humored editing assistance, and Dominic Carroll for his meticulous work on the index.

I don't know what it takes to live with a writer whose projects go on forever, periodically hemorrhage money, and sometimes create exceptional stress. Whatever it is, I am enormously fortunate that my wife, Trish Long, has it in spades. Through her boundless emotional and material generosity and, in a word, her love, she has sustained me unstintingly through three books now. *Go raibh mile maith agat, a stór agus a ghrá mo chroí.*

Glossary

This short glossary includes technical words and phrases that may be unfamiliar to the general reader, as well as familiar words with distinct meanings in an ecological context. Generally I have not included phrases here whose meaning is the subject of substantial discussion in the text, like "ecological restoration" or "novel ecosystem." For many definitions here, I am indebted to the much more comprehensive and very useful glossary in *Ecological Restoration* by Andre F. Clewell and James Aronson.

abiotic: describes the "nonliving" elements of an ecosystem, such as minerals, geology, topography, hydrology, climate. Compare with *biota*

alien invasive plant (AIP): nonnative species that displace natives from their historical ecological territory—alien mammals and birds may also impact natives.

anthropocene: the contemporary geological period in which most or all ecosystems are impacted by humans

anthropogenic: shaped by humans

biodiversity: the diversity of life in genes, species, communities of species, ecosystems and biomes

biome: a regional ecological unit, usually defined by vegetation, such as the coniferous forest biome of northern Europe

biota: the living (biotic) elements of an ecosystem

bryophyte: nonvascular plants, including mosses and liverworts

climax concept, climax community: see *natural succession*

cultural ecosystems: ecosystems where the influence of humans is the key conditioning factor

ecology: the study of interactions among living organisms and between organisms and their environment

ecosystem: a distinctive complex of living organisms and the abiotic environment with which they interact at a specific location

ecosystem goods and services: the benefits nature supplies to humans

ecotone: a transition zone between distinct ecosystems

endemic : of species, unique to a particular region

epiphyte: a plant that lives nonparasitically on another plant, such as many ferns, orchids, mosses, and bromeliads.

habitat: a place where conditions are suitable for a particular species

historical range of variation: the series of relatively stable states through which an ecosystem has moved in the past, any one of which may be selected as the appropriate historical reference system (see below) for a restoration project

historical reference system (or model): the ecosystem state chosen, from research on appropriately similar local systems in the present or past, as the target for the outcome of the restoration of a degraded system in the future

hydrology: the input, retention, output, and recycling of water

indigenous: native to a given location

mitigation: compensation for ecological damage or loss by infrastructural development in one place through ecological restoration, rehabilitation, etc, somewhere else

multivariate analysis: statistical technique for analyzing data with many simultaneous variables

mycorrhiza: mutual association between a plant root and a soil fungus that facilitates the plant's mineral uptake

natural capital: natural resource stocks that may be renewable (ecosystems, organisms), nonrenewable (oil, coal, minerals, etc.), replenishable (the atmosphere, potable water, fertile soils), or cultivated (crops, forest plantations, etc.); the source of ecosystem goods and services

natural succession: the sequence of stages, usually defined by the dominant plant community at each stage, that takes place as an ecosystem develops or recovers from disturbance. Early ecological theory suggested that each system reaches a permanently stable state, a "climax community." More recent research reveals that systems are always in a dynamic state of flux, though relatively stable states may persist for long periods.

pristine: entirely uninfluenced by people

rehabilitation: as opposed to restoration, the recovery of some ecosystem processes after damage, without recovery of the historical biota

resilience: the capacity of an ecosystem to persist on its trajectory despite disturbances

rhizome: underground plant stem producing both belowground roots and aboveground shoots

riparian: pertaining to rivers

salinization: increase of salt level in soil at root zone, usually caused by changes in human land use

seed bank: dormant seeds in the soil that can replenish vegetation following disturbance

threshold: a "tipping point" at which an ecosystem, under pressure from novel environmental or human impacts, switches into a radically different state, outside its historical range of variation

trajectory: the path (through changing states) that an ecosystem is expected to follow within its historical range of variability

trophic: related to levels of food supply. The removal of a top predator (like a wolf) generally results in a "trophic cascade," altering predator-prey abundance and relationships throughout a food chain.

FURTHER ECOLOGICAL INFORMATION ONLINE:

The Encyclopedia of Earth offers many articles on ecological topics and theories at http://www.eoearth.org/.

Readers might also enjoy the occasionally eccentric and irreverent glossary at http://www.terrapsych.com/ecology.html.

Notes

CHAPTER 1

1. Peter Matthiessen and Robert Bateman, *The Birds of Heaven: Travels with Cranes*, (New York: North Point Press, 2001).
2. Jeb Barzen, Alison Duff, and Rich Beilfuss, "The Prairie Ecosystem and How it Changes" (paper presented at the Grassland Restoration Network Workshop, Madison Arboretum, August 25–27, 2008).
3. Aldo Leopold and Luna B. Leopold, *Round River: From the Journals of Aldo Leopold*, (New York: Oxford University Press, 1993), 146. The quotation was later popularized by the Stanford biologist Paul Ehrlich as "the first rule of intelligent tinkering is to save all the parts."
4. In a shift that showed how the zeitgeist was turning, Leopold had his department renamed as "wildlife management" at the end of the decade. (Curt Meine, *Correction Lines: Essays on Land, Leopold, and Conservation* [Washington DC: Island Press, 2004], 127). But he remained the kind of conservationist who always carried a gun, and liked to hunt for the pot.
5. Aldo Leopold, *A Sand County Almanac, and Sketches here and there* (London: Oxford University Press, 1968), 129–30.
6. Ibid., 95–101.
7. Ibid., 101.
8. His experience as a land manager, both for the government and on the beloved few acres around his shack, produces many nuggets with restoration resonance. He is astute about the multiple factors that can frustrate a manager's attempt to steer an ecosystem in one direction or another. He argues the case for removing some birch to benefit a favorite stand of pine, but then concludes: "Again, if a drouthy summer follows my removal of the birch's shade, the hotter soil may offset the lesser competition for water, and my pine be none the better for my bias" (ibid., 70.)
9. He tends to reject economic arguments for conservation, because he acknowledges that many members of the "land community" have no de-

monstrable economic value. Taking what we might today call a deep ecology position, he defends that entire community's "biotic right" to exist, "regardless of the presence or absence of economic advantage to us" (ibid., 211). However, he often uses metaphors we would now associate with "restoration of natural capital" (see chapter 3, p. 21) thinking: agriculture takes out "overdrafts on the soil," causing erosion (ibid., 217). And when herons, otters, and merganser are extirpated to increase game fish populations, such management is paying "dividends to one citizen out of capital stock belonging to all" (ibid., 170).

10. Ibid., 199–200.
11. William R. Jordan III, *Our First 50 Years: The University of Wisconsin–Madison Arboretum 1934–1984*, http://digicoll.library.wisc.edu/cgi-bin/EcoNatRes/EcoNatRes-idx?type=turn&entity=EcoNatRes.ArbFirstYrs.p0005&id=EcoNatRes.ArbFirstYrs&isize=M.
12. There are several earlier candidates for the first ecological restoration project in the United States. For a full account, see William R. Jordan and George M. Lubick, *Making Nature Whole: A History of Ecological Restoration* (Washington DC: Island Press, 2011).
13. Jordan, *Our First 50 Years.*
14. Paddy Woodworth, "Saving Cranes and Helping People," *Village*, December 18–22, 2004, 32.
15. For updates on the progress of the Lower Zambezi Valley Delta Program, see http://www.savingcranes.org/lower-zambezi-valley-and-delta-program.html.

CHAPTER 2

1. This chapter is based on an article I wrote for the *Irish Times*, and all quotations, except where otherwise stated, are from a visit to the ICF in October 2003 and a road trip with Operation Migration the following month (Paddy Woodworth, "A Wing and a Prayer," *Irish Times*, October 1, 2005).
2. Aldo Leopold, *A Sand County Almanac, and Sketches Here and There* (London: Oxford University Press, 1968), 96.
3. "Whooping Crane Update, 4 February–5 March 2013," Whooping Crane Eastern Partnership, http://www.bringbackthecranes.org/technicaldatabase/projectupdates/2013/4FebTO5March2013.html
4. The Texas-Canada flock suffered a particularly bad winter in 2011–2012, falling from 279 to 245 birds. See Aransas Project, "State of the Whooping Crane Flock 2011–2012," Whooping Crane Conservation Association, http://whoopingcrane.com/state-of-the-whooping-crane-flock-2011–2012/
5. Joe Duff, e-mail to author, March 19, 2013.
6. For updates on the progress of Operation Migration, see *In the Field with Operation Migration* (blog), http://operationmigration.org/InTheField/.

For information on the Whooping Crane Eastern Partnership, see their website at http://www.bringbackthecranes.org/. For information on crane conservation in general, see the International Crane Foundation website at http://www.savingcranes.org/.

CHAPTER 3

1. Robert L. Beschta and William J. Ripple, "Large Predators and Trophic Cascades in Terrestrial Ecosystems of the Western United States," *Biological Conservation* 142, no. 11(2009): 2401–14, doi:10.1016/j.biocon.2009.06.015.
2. See "Flyaway Conservation," International Crane Foundation, http://www.savingcranes.org/flyway-conservation.html.
3. The journal was first launched by Bill Jordan, a pioneer of restoration writing, as *Restoration & Management Notes* in 1981.
4. Blaine Harden, "Montana Town's Boys Are Its Last Gasp of Hope," *Washington Post,* November 17, 2003.
5. See also Joy Zedler, "Restoration Targets are Changing," www.uwarboretum.org/publications/leaflets/PDF/Leaflet21.pdf. See chapters 8, 14 and 15 for further discussion of novel ecosystems.
6. Mitsch and Costanza subsequently produced papers proposing radical alternatives to federal and state plans for rebuilding New Orleans. See Robert Costanza, William J. Mitsch, and John W. Day Jr., "A New Vision for New Orleans and the Mississippi Delta: Applying Ecological Economics and Ecological Engineering," *Frontiers in Ecology and the Environment* 4, no. 9 (2006): 465–72; idem., "Creating a Sustainable and Desirable New Orleans," *Ecological Engineering* 26, no. 4 (2006): 317–20; idem., "Rebuilding New Orleans: Applying Ecological Economics and Ecological Engineering," *Encyclopedia of the Earth,* August 3, 2007, http://www.eoearth.org/article/Rebuilding_New_Orleans%3A_applying_ecological_economics_and_ecological_engineering. Their plans have not been implemented: see Susanna Murley, "Another Katrina? Not Funny," *Huffington Post,* August 22, 2011, http://www.huffingtonpost.com/susanna-murley/another-katrina-not-funny_b_933273.html. For a recent report coauthored by Costanza on the future of the Mississippi delta, see: D. P. Batker, I. de la Torre, R. Costanza, P. Swedeen, J. Day, R. Boumans, and K. Bagstad, "Gaining Ground: Wetlands, Hurricanes, and the Economy: The Value of Restoring the Mississippi River Delta," *Environmental Law Reporter News and Analysis* 40, no. 11 (2010): 106–10.
7. Robert Costanza et al., "The Value of the World's Ecosystem Services and Natural Capital," *Nature* 387 (1997): 253–60.
8. Richard Hobbs, "Setting Restoration Goals: Mixing Ecological Theory with Social Values" (paper presented at the SER World Conference on Ecological Restoration, Zaragoza, Spain, September 14, 2005).
9. See chapters 8, 14, and 15.

10. Keith Bowers, chairperson's address (presented at the SER World Conference on Ecological Restoration, Zaragoza, Spain, September 14, 2005).
11. Paddy Woodworth, "Restoring Our Wetlands," *Irish Times*, October 1, 2005.
12. James Aronson, Suzanne J. Milton, and James N. Blignaut, *Restoration of Natural Capital: Science, Business and Practice*, (Washington DC: Island Press, 2007). For articles and comment from an RNC perspective, see http://www.rncalliance.org.
13. Aronson et al., *Restoration of Natural Capital*, 208–15.
14. See the Millennium Ecosystem Assessment at http://www.maweb.org/en/index.aspx.
15. See Aronson et al., *Restoration of Natural Capital*, 325.
16. See *Ecosystems and Human Well-Being: A Framework for Assessment* (Washington DC: Millennium Ecosystem Assessment/Island Press, 2003), 56, http://www.maweb.org/en/Framework.aspx.
17. Ibid.
18. See Rudolf De Groot et al., "Integrating the Ecological and Economic Dimensions in Biodiversity and Ecosystem Service Valuation," in *The Economics of Ecosystems and Biodiversity: Ecological and Economic Foundations (TEEB Do)*, ed. Pushpam Kumar (London: Earthscan, 2010), 25, 39–40. A draft version is available on the TEEB website: http://www.teebweb.org/LinkCjlick.aspx?fileticket=4yFN-LAMGI4%3d&tabid=1018&language=en-US.
19. See also: Gretchen C. Daily, ed., *Nature's Services: Societal Dependence on Natural Ecosystems* (Washington DC: Island Press, 1997); Gretchen C. Daily and Katherine Ellison, *The New Economy of Nature: The Quest to make Conservation Profitable* (Washington DC: Island Press, 2002); Rudolf De Groot, M. A. Wilson and R. M. J. Boumans, "A Typology for the Classification, Description and Valuation of Ecosystem Functions, Goods and Services," *Ecological Economics* 41 (2002): 393–408; Millennium Ecosystem Assessment, see http://www.millenniumassessment.org/en/Synthesis.html; De Groot et al., "Integrating Ecological and Economic Dimensions"; Gretchen C. Daily, S. Polasky, J. Goldstein, P. M. Kareiva, H. A. Mooney, L., T. H. Ricketts, J. Salzman, and R. Shallenberger, "Ecosystem Services in Decision-Making: Time to Deliver," *Frontiers in Ecology and the Environment* 7, no. 1(2009): 21–28.
20. Millennium Ecosystem Assessment, "Living Beyond Our Means: Natural Assets and Human Well-Being. Statement from the Board," March 2005, 21, http://www.unep.org/maweb/en/boardstatement.aspx.
21. For a nuanced critique of neoclassical economics from the TEEB and RNC perspectives, see De Groot et al., "Integrating Ecological and Economic Dimensions," 20–21, or http://www.teebweb.org/LinkClick.aspx?fileticket=4yFN-LAMGI4%3d&tabid=1018&language=en-US. For the views of several other economists, including Nobel laureate Kenneth Arrow, see Paddy Woodworth, "What Price Ecological Restoration?" *Scientist*, April 2006, 38–45.

22. For a recent discussion of natural capital, population and sustainability, see Paul R. Ehrlich, Peter M. Kareiva, and Gretchen C. Daily, "Securing Natural Capital and Expanding Equity to Rescale Civilization," *Nature* 486 (2012): 68–73. doi:10.1038/nature11157.
23. Pavan Sukhdev, author interview, Merida, Mexico, August 2011.
24. See http://www.teebweb.org
25. De Groot et al., "Integrating Ecological and Economic Dimensions."
26. See http://www.teeb4me.com.
27. "The Banker Trying to Put a Value on Nature," online interview of Pavan Sukhdev with Jo Confino, *Guardian Sustainable Business*, May 16, 2011, http://www.guardian.co.uk/sustainable-business/pavan-sukhdev-valuing-biodiversity-ecosystem-services.
28. Sue Milton, e-mail to author, November 26, 2005.
29. Michael J. Stevenson, "Problems with Natural Capital: A Response to Clewell," *Restoration Ecology* 8, no. 3 (2000): 212.
30. William R. Jordan III, e-mail to author, January 13, 2006.
31. The phrase is an adaptation of the title of E. F. Schumacher's influential book, *Small Is Beautiful: Economics as if People Mattered* (New York: Harper & Row, 1973).

CHAPTER 4

1. Guy Preston, author interview, January 2006.
2. The total 2010 budget for WfW was $98 million, and this does not include a further $28 million for the related program Working on Fire (Guy Preston and Ahmed Khan, e-mails to author, August 25, 2011).
3. A. J. Lamb, ed., *Jonkershoek Forestry Research Centre, Pamphlet 384*, (Pretoria, South Africa: Forestry Branch of the Dept of Environmental Affairs, 1985), 6.
4. Ronald Good, *The Geography of the Flowering Plants*, 3rd ed. (New York: J. Wiley, 1964). For comparison, Good's Holarctic Floristic Kingdom includes virtually all of North America and Eurasia. Some experts now contend that the Cape should be designated rather less grandly, as a Floral Region. See Barry Cox, "The Biogeographic Regions Reconsidered," *Journal of Biogeography* 28, no. 4 (2001): 511–23. Regardless, it remains exceptionally rich in plants.
5. "UWC's Enviro-Facts Guide to Botany," University of Western Cape, http://www.botany.uwc.ac.za/envfacts/fynbos/.
6. "Cape Floral Region Protected Areas," UNESCO World Heritage Convention, http://whc.unesco.org/en/list/1007.
7. "The Biodiversity Hotspots," Conservation International, http://www.biodiversityhotspots.org/xp/hotspots/hotspotsscience/Pages/hotspots_defined.aspx.
8. Jane Turpie, Barry J. Heydenrych, and Stephen J. Lamberth, "Economic

Value of Terrestrial and Marine Biodiversity in the Cape Floristic Region: Implications for Defining Effective and Socially Optimal Conservation Strategies," *Biological Conservation* 112 (2003): 233.

9. David C.Le Maitre et al., "Invasive Plants and Water Resources in the Western Cape Province, South Africa: Modelling the Consequences of a Lack of Management," *Journal of Applied Ecology* 33 (1996): 161.
10. Richard Cowling, author interview, January 2006.
11. Kader Asmal, author interview, January 2006.
12. Guy Preston and Ahmed Khan, e-mails to author, August 25, 2011.
13. Ferial Haffajee, "South Africa: Water for Everyone," *UNESCO Courier* 52, no. 2 (1999): 26. See also Michael Burns, Michelle Audouin, and Alex Weaver, "Advancing Sustainability Science in South Africa," *South African Journal of Science* 102 (September/October 2006): 382.
14. Asmal, author interview, January 2006.
15. Mike Powell, author interview, January 2006.
16. A. J. Mills and R. M. Cowling, "Rate of Carbon Sequestration at Two Thicket Restoration Sites in the Eastern Cape, South Africa." *Restoration Ecology* 14 (2006): 38–49.
17. Cedric Kleinbooi and Tersia Noubous, author interview, January 2006.
18. She herself is of mixed race, that large grouping still generally classified as "Colored" in the Cape.
19. Abbey-gail Lukas, author interview, January 2006.
20. Ronnie Kasrils, "Launch of Operation Vuselela," Department of Water Affairs, Republic of South Africa, http://www.dwaf.gov.za/Communications/MinisterSpeeches/Kasrils/2003/Launch%20of%20Operation%20Vuselela%2027%20March%2003.pdf
21. Author interviews with WfW employees, January 2006.
22. Christo Marais, author interview, January 2006.
23. Asmal, author interview, January 2006.
24. Richard Cowling, author interview, January 2006.
25. Department of Water Affairs and Forestry, *Working for Water Annual Report*. Pretoria, South Africa: DWAF, 2003–2004.
26. Preston, author interview, January 2006.
27. See Christo Marais and Brian van Wilgen, "The Clearing of Invasive Alien Plants in South Africa: A Preliminary Assessment of Costs and Progress," *South African Journal of Science*, 100 (January/February 2004): 97–103.
28. Patricia Holmes, author interview, January 2006.
29. Christo Marais, e-mail to author, February 1, 2007.
30. Patricia Holmes, David M. Richardson, and Christo Marais, "Costs and Benefits of Restoring Natural Capital Following Alien Plant Invasions in Fynbos Ecosystems in South Africa," in *Restoring Natural Capital*, ed. James Aronson, Suzanne J. Milton, J. N. Blignaut, and the Society for Ecological Restoration International, 188–98 (Washington DC: Island Press, 2007).
31. Asmal, author interview, January 2006.

32. Preston, author interview, January 2006.
33. Marais, author interview, January 2006.
34. This audit, the Common Ground Evaluation, was conducted by an independent team of specialists. It is unpublished, but the main findings are given in Working for Water's 2002–2003 annual report.
35. Powell, author interview, January 2006.
36. Keynote papers from this symposium were published, together with other research largely funded by WfW, in a special edition of the *South African Journal of Science* 100 (January/February 2004).
37. Ian A. W. Macdonald, "A Review of the Inaugural Research Symposium of the Working for Water Programme," *South African Journal of Science* 100 (January/February 2004): 22, 24.
38. Brian W. van Wilgen, "Scientific Challenges in the Field of Invasive Alien Plant Management," *South African Journal of Science* 100 (January/February 2004): 19.
39. David Le Maitre, author interview, January 2006.
40. Preston, author interview, January 2006.
41. Powell, author interview, January 2006.
42. Stephen Hosking, author interview, January 2006.
43. M. Du Preez, S. Tessendorf, and S.G. Hosking, "Application of the Contingent Valuation Method to Estimate the Willingness-to-Pay for Restoring Indigenous Vegetation in Underberg, Kwazulu-Natal, South Africa," *South African Journal of Economic and Management Sciences* 13, no. 2 (2010): 135–57. Hosking subsequently made the clarification that their study was conducted on the assumption that the cost of restoring indigenous vegetation was included in WfW budgets. He now understands that the budgets were just for clearing IAPs, and that the additional costs of active restoration, where necessary, would probably push the price beyond most people's willingness-to-pay (e-mail to author, September 19, 2011).
44. Beatrice Conradie, author interview, January 2006.
45. Marais, author interview, January 2006.
46. Macdonald, "Review of the Inaugural Research Symposium," 27.
47. Barbara Schreiner, author interview, January 2006.
48. Cowling, author interview, April 2008.
49. James Blignaut, Christo Marais, and Jane Turpie, "Determining a Charge for the Clearing of Invasive Alien Plant Species (IAPs) to Augment Water Supply in South Africa," *Water SA* 33, no. 1 (2007): 30.
50. J. D. S. Cullis, A. H. M. Görgens, and C. Marais, "A Strategic Study of the Impact of Invasive Alien Plants in the High Rainfall Catchments and Riparian Zones of South Africa on Total Surface Water Yield," *Water SA* 33, no 1 (2007): 41.
51. C. Marais and A. Wannenberg, "Restoration of Water Resources (Natural Capital) through the Clearing of Invasive Alien Plants from Riparian

Areas in South Africa—Costs and Water Benefits," *South African Journal of Botany* 74, no 3 (July 2008): 526–37.

52. Ibid., 536.
53. J. K. Turpie, C. Marais, and J. N. Blignaut, "The Working for Water Programme: Evolution of a Payments for Ecosystem Services Mechanism that Addresses both Poverty and Ecosystem Service Delivery in South Africa," *Ecological Economics* 65 (2008): 788–98.
54. The sister organizations referred to are, respectively, Working for Woodlands, Working for Wetlands, and Working on Fire.
55. Preston, author interview, January 2006.
56. Holmes, author interview, January 2006.
57. See the discussion of OCBILs in chapter 9 (pp. 246–255).
58. Preston, author interview, January 2006.
59. Marais et al., "Restoration of Water Resources," 536.
60. Jennifer Gouza, author interview, January 2006.
61. Preston, author interview, January 2006.
62. Guy Preston and Lucas Williams, "Working for Water" (paper presented at the Service Delivery Learning Academy Conference, Durban, KwaZulu-Natal, South Africa, July 2003).
63. Marais, author interview, January 2006.
64. Holmes et al., "Costs and Benefits of Restoring Natural Capital," 193.
65. Cowling, author interview, April 2008.
66. Preston, author interview, January 2006.
67. Christo Marais, e-mail to author, November 2008.
68. Nosipho Jezile, author interview, January 2006.
69. See Brian W. van Wilgen et al., "Costs and Benefits of Biological Control of Invasive Alien Plants: Case Studies from South Africa," *South African Journal of Science* 100 (January/February 2004): 113–23.
70. Paddy Woodworth, "What Price Ecological Restoration?" *Scientist* 20, no. 4 (2006), 38–45; Paddy Woodworth, "Working for Water in South Africa: Saving the World on a Single Budget?" *World Policy Journal* 23, no. 2 (2006): 31–44.
71. Preston, author interview, May 2011.
72. Ibid.
73. All the "Working for/Working on" programs are now grouped together under the National Resources Management Programmes.
74. Preston, author interview, May 2011.
75. Ibid.
76. "Management of Invasive Alien Plants," Department of Water Affairs, Republic of South Africa, http://www.dwaf.gov.za/wfw/Control/. One of the most remarkable successes for biocontrol has been in the battle against the notorious IAP *Chromolaena odorata* (triffid weed) in Hluhluwe–iMfolozi National Park (Guy Preston, e-mail to author, October 11, 2001).
77. Preston, author interview, May 2011.

78. Guy Preston, e-mail to author, April 29, 2011.
79. Ibid. See also W. J. de Lange and B. W. van Wilgen, "An Economic Assessment of the Contribution of Biological Control to the Management of Invasive Alien Plants and to the Protection of Ecosystem Services in South Africa," *Biological Invasions* 12, no. 12 (2010): 4113–24.
80. Preston, author interview, May 2011.
81. Patricia Holmes, e-mail to author, May 24, 2011 (original emphasis).
82. Karen J. Esler et al., ed., "Riparian Vegetation Management in Landscapes Invaded by Alien Plants: Insights from South Africa," special issue, *South African Journal of Botany* 74 (2008): 397–554.
83. Karen Esler, e-mail to author, June 6, 2011.
84. See chapter 3, note 31.

CHAPTER 5

1. William R. Jordan, *The Sunflower Forest: Ecological Restoration and the New Communion with Nature* (Berkeley: University of California Press, 2003): 135–36.
2. Liam Heneghan, author interview, May 2010.
3. Much of this information is based on the History section of the Forest Preserve District of Cook County website, itself a précis of a 1949 thesis by William P Hayes: http://fpdcc.com/about/history (emphasis added).
4. Laurel Ross, author interview, May 2010; see also the Chicago Wildnerness website at http://www.chicagowilderness.org/places.php.
5. "The Early History of the Forest Preserve District of Cook County, 1869–1922," Forest Preserve District of Cook County, http://fpdcc.com/about/history/ (emphasis added).
6. Steve Packard, author interview, September 2007.
7. Steve Packard, "Just a Few Oddball Species: Restoration and the Rediscovery of Tallgrass Savanna," *Restoration & Management Notes* 6, no. 1 (Summer 1988): 14–15.
8. Stephen Packard and Cornelia Mutel, *The Tallgrass Restoration Handbook For Prairies, Savannas, and Woodlands* (Washington DC: Island Press, 1997).
9. Packard, author interview, September 2007.
10. William K. Stevens, *Miracle under the Oaks: The Revival of Nature in America* (New York: Pocket Books, 1995); Peter Friederici, *Nature's Restoration: People and Places on the Front Lines of Conservation* (Washington DC: Island Press, 2006).
11. "Science" is not, of course, uncolored by ideological and cultural biases. And cultural and aesthetic preferences are indeed a valid basis for restoration decisions. But scientific jargon should not be used to mask other motivations for restoration, which should be transparent.
12. See chapter 14, p. 395.
13. Packard, "Just a Few Oddball Species," 13.

14. Ibid., 17.
15. Ibid., 13.
16. Ibid., 18.
17. A. D. Bradshaw, "Restoration: The Acid Test for Ecology," in *Restoration Ecology: A Synthetic Approach to Ecological Research*, ed. William R. Jordan, Michael E. Gilpin, and John D. Aber, 53–74 (New York: Cambridge University Press, 1987), 54.
18. Heneghan, author interview, May 2010.
19. Laurel M. Ross, "Illinois' Volunteer Corps: A Model Program with Deep Roots in the Prairie," *Restoration & Management Notes* 12, no. 1 (Summer 1994): 57–59.
20. Ibid., 59.
21. Packard, author interview, September 2007.
22. Packard, "Just a Few Oddball Species," 14.
23. Stevens, *Miracle under the Oaks*, 65.
24. Packard, "Just a Few Oddball Species," 22.
25. Packard, author interview, September 2007. By "presettlement," Packard is referring to the period prior to European colonization.
26. Stephen Packard, "Restoring Oak Ecosystems," *Restoration & Management Notes* 11, no. 1 (Summer 1993): 14.
27. Stephen Packard, "Scything and Gentians (or How To Apologize To A Plant)," September 2007, http://www.sommepreserve.org/insights.html #insightsgentians.
28. Paul Gobster, author interview, March 2008.
29. Ross, author interview, May 2010.
30. J. Mendelson, S. P. Aultz, and J. D. Mendelson, "Carving up the Woods: Savanna Restoration in North-Eastern Illinois," *Restoration & Management Notes* 10, no. 2 (Winter 1992): 127–31.
31. Stevens, *Miracle under the Oaks*, 280.
32. Mendelson, "Carving up the Woods," 17.
33. Eric Katz, "Restoration and Redesign: The Ethical Significance of Human Intervention in Nature," *Restoration & Management Notes* 9, no. 2 (Winter 1991): 90–96.
34. Packard, "Restoring Oak Ecosystems."
35. Ibid., 6. Robert Elliot gives a thoughtfully nuanced case for the traditional position in *Faking Nature: The Ethics of Environmental Restoration* (London: Routledge, 1997). One of its most trenchant advocates is Eric Katz, whose position is summed up in this remarkable sentence: "When we consider the authenticity of a natural system, the presence or absence of human intentionality is the key determining factor of its value and ontological character" (Eric Katz, "Another Look at Restoration: Technology and Artificial Nature," in *Restoring Nature*, ed. Paul H. Gobster and R. Bruce Hull [Washington DC: Island Press, 2000], 41).

36. Ibid., 7.
37. Ibid.
38. Packard, author interview, May 2010.
39. Ibid. (emphasis added).
40. Steve Packard, "No End to Nature," *Restoration & Management Notes* 8, no. 2 (Winter 1990): 72.
41. Packard, "Restoring Oak Ecosystems," 15.
42. Laurel Ross, "The Chicago Wilderness: A Coalition for Urban Conservation," *Restoration & Management Notes* 15, no. 1 (Summer 1997): 24.
43. Ibid, 17.
44. Bathsheeba Birman, author interview, March 2008.
45. Ross, author interview, May 2010 (original emphasis).
46. Ibid.
47. "Bunker Hill Savanna and Sid Yates Flatwoods," North Branch Restoration Project, http://www.northbranchrestoration.org/sites_bunker.html.
48. Stevens, *Miracle under the Oaks*, 66–67.
49. Ross, author interview, May 2010.
50. Gobster and Hull, *Restoring Nature*, 6.
51. Gobster, author interview, May 2010.
52. Jordan, *Sunflower Forest*, 189. For further discussion on Jordan's understanding of "shame" in relationship to restoration, see chapter 14, pp. 404–407.
53. Packard, "Just a Few Oddball Species," 22.
54. Gobster, author interview, March 2008.
55. Paul Gobster, "The Other Side: A Survey of the Arguments," *Restoration & Management Notes* 15, no. 1 (Summer 1997), 32.
56. Gobster and Hull, *Restoring Nature*.
57. Bill Jordan, author interview, September 2007.
58. Jackie Boland, author interview, May 2010.
59. Carol Nelson, author interview, May 2010.
60. Petra Blix, author interview, May 2010.
61. Ross, "The Chicago Wilderness"; author interview, May 2010.
62. Ross, author interview, May 2010.
63. The three programs are:

MOLA: Managing Invasive Buckthorn
This program has its origins in Liam Heneghan's aforementioned efforts, temporarily aborted, to evaluate the effect of various soil treatments on restoration after Buckthorn clearance.

100 Sites for 100 Years: Developing a Long-Term Research Program to Assess Biodiversity Outcomes on Managed Lands
Roughly a hundred sites across the region are being selected for their range of management and restoration histories in woodland, savanna and prairie. Each site will be assessed in terms of current ecological com-

munities and processes, and changes in these baselines will be monitored into the future, to "test hypotheses in conservation and restoration ecology, reveal novel patterns, and provide practical assistance to land-management practitioners." (Liam Heneghan, personal communication, June 2010).

R.E.S.T.O.R.E.—Rethinking Ecological and Social Trends of Restoration Ecology

Social science researchers, research ecologists, and land managers will work together to "(1) understand in more depth the diversity and dynamics of the planning processes that are employed in the restoration of woodlands across Chicago Wilderness; (2) understand whether or not different planning processes lead to predictably different (not necessarily better or more desirable) biodiversity and ecological outcomes; and (3) understand the impact of differences in biodiversity outcomes and different planning processes on support for restoration and conservation of biodiversity by constituents not directly involved in restoration activities" (Liam Heneghan, personal communication, June 2010).

For more information on the latter two programs see Liam Heneghan et al., "Review: Lessons Learned from Chicago Wilderness—Implementing and Sustaining Conservation Management in an Urban Setting," *Diversity* 4, no. 1 (2012): 74–93, doi:10.3390/d4010074. For a lively online discussion of these programs, and much more, see *10 Things Wrong with Environmental Thinking* (blog), http://10thingswrongwithenvironmentalthought.blogspot.com.

64. This quotation, and those that follow, are from Heneghan, author interview, May 2010.
65. "I recognize that the North Branchers were bungling and that the moratorium was a disaster. That is an interesting story, but should not be the whole take on restoration in Chicago" (Heneghan, author interview, May 2010).
66. It would be wrong to give the impression that the science programs will have a clear run. In an e-mail sent shortly after our May 2010 interview, Heneghan added, "One of the folks interested had worked for years with North Branch and was concerned that none of the problems I was discussing (reinvasion, soil changes, loss of arthropod diversity etc.) are on the agenda for rumination in her volunteer group, despite the fact that many of these issues are fairly apparent. The disdain of systematic science may be an attitude on the decline, though, since many volunteer stewards are very interested in having us nominate their site as one of the 100 (for the 100 sites project). In fact, one volunteer recently said that she really hoped the work was not being done by citizen scientists!"
67. Heneghan, author interview, July 2011.

CHAPTER 6

1. For a basic guide to the history and natural history of the Cinque Terre in English, see Mauro Mariotti, *Cinque Terre* (Genoa: Erga edizioni, 1998).
2. For a detailed account of these Mediterranean communities, see Jacques Blondel et al., *The Mediterranean Region: Biodiversity in Space and Time* (Oxford: Oxford University Press, 2010), 118–25.
3. This chapter looks at the record of the Cinque Terre National Park, but not of the important Marine Protected Area that is attached to it. The marine area is a site of significant ecological restoration and, like most marine restoration, this is achieved simply by protection and spontaneous regeneration of biological stocks, not by manipulation of a degraded environment by humans. Very reluctantly, I found I could not accommodate the topic of marine restoration within this book, though it is also briefly referenced in chapter 10.
4. Francisco Moreira, A. Isabel Queiroz, and James Aronson, "Restoration Principles Applied to Cultural Landscapes," *Journal for Nature Conservation* 14 (2006): 223.
5. Ibid., 219.
6. See Miguel N. Bugalho et al., "Mediterranean Cork Oak Savannas Need Human Use to Sustain Biodiversity and Ecosystem Services," *Frontiers in Ecology and the Environment* 9, no. 5 (2011): 278: "More than 75% of all terrestrial ecosystems show such marked evidence of alteration by humans that they are perhaps best perceived as 'human-made systems with natural ecosystems embedded' rather than 'natural systems with human embedded influence.'"
7. Matteo Perrone, author interview, June 2012.
8. Attilio Casavecchia and Enrica Salvatori, *Man's Park 1: History of a Landscape*, (Riomaggiore: Cinque Terre National Park, n.d.), 46.
9. Perrone, author interview, June 2008.
10. Gianfranco Bonanini, author interview, June 2008.
11. Samuele Lercari, author interview, June 2008.
12. See http://www.silviobenedetto.com/sb/ing/index.eng.htm.
13. Perrone, author interview, June 2008.
14. Daniele Moggia, author interview, June 2008.
15. Massimo Evangelisti, author interview, June 2008.
16. Angelo Evangelisti, author interview, June 2008.
17. Bartolomeo Lercari and Lise Bertram, author interviews, June 2008.
18. See especially chapter 12.
19. See note 3, above.
20. Long-term closures of some sections have been in operation since 2010—but because a series of landslips have actually blocked the route, not in order to prevent them.

21. Transition zones between two well-defined ecosystems, which often contain elements of both.
22. Bugalho et al., "Mediterranean Cork Oak Savannas," 278 (emphasis added). For a more detailed treatment of the same subject, see James Aronson et al., *Cork Oak Woodlands on the Edge: Ecology, Adaptive Management and Restoration* (Washington DC: Island Press, 2009).
23. Ricardo Nardelli, e-mail to author, July 2008. Subsequent quotes from Nardelli are also from this e-mail.
24. *Bombina pachypus, Salamandrina tergiditata, Speleomantes ambrosii,* and *Hyla mediterranea.*
25. Perrone, author interviews, June and September 2012.

CHAPTER 7

1. "One of the penalties of an ecological education is that one lives alone in a world of wounds. Much of the damage inflicted on land is quite invisible to laymen. An ecologist must either harden his shell and make believe that the consequences of science are none of his business, or he must be the doctor who sees the marks of death in a community that believes itself well and does not want to be told otherwise" (Aldo Leopold and Luna Leopold, *Round River* [New York: Oxford University Press, 1993], 165).
2. See Oliver Rackham, *Woodlands* (London: Collins, 2006); G. F. Peterken, *Woodland Conservation and Management* (London; New York: Chapman and Hall, 1981); G. F. Peterken, *Natural Woodland: Ecology and Conservation in Northern Temperate Regions* (Cambridge; New York: Cambridge University Press, 1996). Peterken acknowledges in his more recent book that he has radically shifted his view of the significance of "near-natural" forests as reference systems for the restoration of historically managed woodlands in his earlier work.
3. The NWS is an innovative package, initiated by Woodlands of Ireland, a partnership of woodland stakeholders. The Irish Forest Service, a division of the Department of Agriculture, implements and refines the program with the continuing collaboration of Woodlands of Ireland. See http://www.woodlandsofireland.com; "Native Woodland Scheme—Establishment," August 2006 (Irish Forest Service, Department of Agriculture, Fisheries and Food), http://www.agriculture.gov.ie/media/migration/forestry/grantandpremiumschemes/2012/NativeWoodlandEstablishmentScheme.pdf.
4. Deirdre Cunningham, *Brackloon—The Story of an Irish Oak Wood* (Dublin: Programme of Competitive Forestry Research for Development [COFORD], 2006).
5. The state forestry service was historically a division of the Department of Lands, and only developed some conservation expertise when it was reinvented as the Forest and Wildlife Service in 1970. The aim of this earlier

clearance and plantation was the generation of revenue and the provision of employment in a poor rural area.

6. Ireland has a number of state-sponsored companies, colloquially known as "semi-state companies," where the government owns the shares and appoints the boards, but the businesses are run as private commercial enterprises.
7. This was part of its commitment to manage 15 percent of its estate for biodiversity, under the Forest Stewardship Council certification scheme.
8. For background on EU conservation designations, see chapter 13.
9. Cunningham, *Brackloon*, 52.
10. Declan Little, e-mail to author, September 2010.
11. Paudie Blighe, quoted in e-mail from Aileen O'Sullivan, Coillte ecologist, to author, December 1, 2010.
12. No one is suggesting eradicating beech from the entire landscape.
13. Declan Little, "A New Dawn for Native Woodlands" paper presented at the NWS Ireland's Native Woodlands conference, Galway, September 8–11, 2004, *Ireland's Native Woodlands* (Dublin: Woodlands of Ireland, 2005), 179.
14. Ibid.
15. Ibid.
16. Declan Little, author interview, December 2010.
17. Little, "A New Dawn for Native Woodlands," 180.
18. Little, e-mail to author, September 2010.
19. Little, author interview, December 2010.
20. Blighe, quoted in O'Sullivan e-mail, December 2010.
21. Little, e-mail to author, 24 May 20011.
22. LIFE is the EU's funding instrument to support environmental projects. LIFE-Nature funds projects related to the Union's Birds and Habitats Directives. See http://ec.europa.eu/environment/life/about/index.htm.
23. The Burren is the object of a highly regarded LIFE program, with the NPWS. See http://www.burrenlife.com/. The region is a spectacular limestone pavement landscape. It is thought that neolithic overgrazing stripped first forest and then soil from the region's limestone pavement. This created an ecosystem with a combination of microclimates that today sustain a mixture of rare mediterranean and alpine plants unique in the British Isles. In this case, anthropogenic degradation in one millennium appears to have, ironically, produced remarkable biodiversity in another.
24. Sean Quealy, author interview, December 2010.
25. Ibid.
26. Ibid.
27. D. J. Little et al., "Assessment of the Impact of Past Disturbance and Prehistoric *Pinus sylvestris* on Vegetation Dynamics and Soil Development in Uragh Wood, SW Ireland," *The Holocene* 6, no. 1 (1996): 90, 96; Little, e-mail

to author, August 18, 2011. Scots pine was reintroduced widely in Ireland from 1700 onward and is considered a native today.

28. Lusitanian species are those whose "normal" range is in northwest Iberia but that occur in southwest Ireland, for reasons that remain debated. Other examples include the natterjack toad and the Kerry slug.

29. Charcoal does not always indicate human presence, especially since the Scots pine is partly dependant on fire for its propagation. Much charcoal in these woods is the result of natural fires set by lightening. But the valley is also home to many charcoal pits, indicating that the valley was a center for this rural industry, which thrived in the seventeenth and early eighteenth centuries, and of course spurred forest clearance.

30. Comments from Declan Little in this section are from conversations during a field trip to Gleninchaquin, August 31, 2010.

31. Aileen O'Sullivan, Coillte ecologist, and Larry Kelly, Coillte regional manager, e-mails to author, December 1, 2010.

32. Quotations from Peggy and Donal Corkery in this section are from author interviews, August 2010.

33. The Gleninchaquin website can be found at http://www.gleninchaquin.com/.

34. Windthrow is what happens when a big tree is pushed over by a storm. A mass of its roots is torn out of the earth, creating a kind of miniature mound-and-moat effect on the topography.

35. Like Uragh, this name references the Irish word for "yew."

36. Red deer are native to Ireland, but most of the stock has now hybridized with introduced sika.

37. Seamus Heaney, "Planting the Alder," in *District and Circle*, 60 (London: Faber and Faber, 2006).

38. For a full and fascinating treatment of the relationship of between scientists and land managers in restoration, see Robert J. Cabin, *Intelligent Tinkering: Bridging the Gap between Science and Practice* (Washington DC: Island Press, 2011).

39. The reintroduction of the golden eagle in Ireland, far from being a negative for farmers, may actually be a positive, since it preys on these crows. So far, its reintroduction in Donegal has been more successful than that of the sea eagle in Kerry, possibly because the communication with the local community was exceptionally meticulous there. See the Golden Eagle Trust website at http://www.goldeneagle.ie.

40. The annual voluntary clearances have been organized by Groundwork, with international workcamps under Irish leadership. Many Irish environmentalists and ecologists have cut their teeth—and blistered their feet!—working for this group, including Liam Heneghan (see chapter 5). Unfortunately, a disagreement between the group and the NPWS over restoration techniques caused the scheme to be suspended in 2010. See http://www.groundwork.ie.

41. "A Lament for Kilcash," in *An Duanaire, 1600–1900: Poems of the Dispossessed*, selected by Seán Ó Tuama, trans. Thomas Kinsella (Mountrath, Ireland: Dolmen Press, 1981), 328.
42. David Hickie and Mike O'Toole, *Native Trees & Forests of Ireland* (Dublin: Gill & Macmillan, 2002).
43. Kevin Collins, quoted in Paddy Woodworth, "Bringing Back the Forests," *Irish Times*, February 8, 2005. See also The People's Millennium Forest website at http://www.millenniumforests.com and The Woodlands of Ireland at http://www.woodlandsofireland.com.

CHAPTER 8

1. See chapter 3, pp. 25–26 and 33–34.
2. Richard J. Hobbs et al., "Novel Ecosystems: Theoretical and Management Aspects of the New Ecological World Order," *Global Ecology and Biogeography* 15, no. 1 (January 2006): 1–7.
3. Ibid., 2.
4. See Viki A. Cramer, R. J. Hobbs, and Society for Ecological Restoration International, *Old Fields: Dynamics and Restoration of Abandoned Farmland*, (Washington DC: Island Press, 2007); and Viki A. Cramer, Richard J. Hobbs, and Rachel J. Standish, "What's New about Old Fields? Land Abandonment and Ecosystem Assembly," *Trends in Ecology & Evolution* 23, no. 2 (2008): 104–12. doi:10.1016/j.tree.2007.10.005. See also chapter 9, pp. 246–255.
5. Hobbs et al., "Novel Ecosystems," 3.
6. Ibid., 3–4.
7. Ibid., 5.
8. Ibid.
9. James A. Harris et al., "Ecological Restoration and Global Climate Change," *Restoration Ecology* 14, no. 2 (2006): 171.
10. Ibid.
11. Despite this awareness, far too many restoration projects are still attempted without any soil analysis, and often fail as a result. See Liam Heneghan, Susan P. Miller, Sara Baer, Mac A. Callaham, James Montgomery, Mitchell Pavao-Zuckerman, Charles C. Rhoades, and Sarah Richardson, "Integrating Soil Ecological Knowledge into Restoration Management," *Restoration Ecology* 16, no. 4 (2008): 608–17.
12. Jim Harris, author interview, November 2011.
13. James Aronson, e-mail to author, July 4, 2007.
14. Harris et al., "Ecological Restoration," 171.
15. Ibid.
16. Ibid., 175.
17. Ibid. (emphasis added).
18. Ibid.

19. Ibid.
20. James A. Harris, "Introduction: Key Concepts and Research Questions in Restoration Ecology" (paper presented at the ESA/SER Joint Conference, San José, California, August 6–10, 2007).
21. Ibid.
22. Young D. Choi, "Restoration of 'Self-Sustaining' Ecosystems: Challenges and Perspectives" (paper presented at the ESA/SER Joint Conference, San José, California, August 6–10, 2007).
23. Patrica Townsend, "Migrating to Cooler Climates" Patricia A. Townsend, "Migrating to Cooler Climates: Restoring Tropical Forest Connectivity in an Elevational Gradient" (paper presented at the ESA/SER Joint Conference, San José, California, August 6–10, 2007); see also Anna J. Mello, Patricia A. Townsend, and Katie Filardo, "Reforestation and Restoration at the Cloud Forest School in Monteverde, Costa Rica: Learning by Doing," *Ecological Restoration* 28, no. 2 (2010): 148–50.
24. F. Stuart Chapin, "Using Climate Change to Restore Biodiversity," paper presented at the ESA/SER Joint Conference, San José, California, August 6–10, 2007.
25. See also F. Stuart Chapin et al., "Building Resilience and Adaptation to Manage Arctic Change," *Ambio* 35, no. 4 (2006): 198–202.
26. Eric Higgs, "How Will Restorationists Adapt to Climate Change?" (paper presented at the ESA/SER Joint Conference, San José, California, August 6–10, 2007); Eric Higgs, author interview, August 2007.
27. Harris, author interview, August 2007.
28. James Aronson, author interview, August 2007.
29. See chapter 14, p. 414.
30. George Gann, comments at closing lunch, ESA/SER Joint Conference, San José, California, August 6–10, 2007.

CHAPTER 9

1. Stephen Hopper, author interview, November 2009.
2. Steve Easton, Kings Park and Botanic Garden staff member, author interview, September 2009. As well as Hopper and Dixon, other key staff on this project included Kathy Meney, Bob Dixon, Nathan McQuoid, Peter Mooney, and Josh Smith.
3. Kingsley Dixon, author interview, September 2009.
4. John H. Gardner and David T. Bell, "Bauxite Mining Restoration in by Alcoa World Alumina in Western Australia: Social, Political, Historical and Environmental Contexts," *Restoration Ecology* 15, no. 4 (2007): S5.
5. "Bauxite Mining Rehabilitation," Alcoa.com, http://www.alcoa.com/australia/en/info_page/land_management_bau_mine_rehab.asp.
6. *Restoration Ecology* 15, no. 4: S1–S144.
7. Dixon, author interview, September 2009.

8. John Koch, comments on SER jarrah forest field trip, September 2009.
9. John Koch, e-mail to author, September 2009.
10. Richard J. Hobbs, author interview, September 2009.
11. Dixon, author interview, September 2009.
12. John Koch, e-mail to author, September 2009.
13. Hobbs, author interview, September 2009.
14. Hopper, author interview, November 2009.
15. Keith Bradby, author interview, September 2009. Quotations from Bradby, unless otherwise indicated, are from conversations and interviews during our field trip and drive from Albany to Kalgoorlie and back.
16. This phrase has been attributed to, among others, the Italian Marxist Antonio Gramsci.
17. Many farming families in the wheatbelt make huge efforts to maintain social and cultural networks, and travel great distances to do so. Bradby pointed out signs of this resilience in every small town we drove through—quirky sculpture displays on main street, traveling opera shows ("Carmen in a Cowshed"), and simple placards plaintively advertising a little town's social charms ("Varley: Small Place, Great Crowd"). But the loneliness takes a heavy toll. Simon Smale of Greening Australia remembers well the only comment made by a woman leaving one of the farms the Link was about to restore: "I've lived in this place for twenty-five years, and I've hated every single day of it." One of the Link's many ambitions is to restore a greater sense of community, and of relationship to the environment, among the scattered inhabitants of the region.
18. Keith Bradby, e-mail to author, September 2009.
19. This is a subtle distinction between two variations (morphs) of the same species that only a knowledgeable birder would make.
20. Keith Bradby, "Achieving Restoration at the Scale a Hotspot Deserves" (paper presented at the Third SER World Conference, Perth, Western Australia, August 23–27, 2009).
21. Justin Jonson, comments on Hobbs/Harris field trip, September 2009.
22. See chapter 14, pp. 246–255.
23. Hobbs, author interview, September 2009.
24. Robert J. Cabin, *Intelligent Tinkering: Bridging the Gap between Science and Practice* (Washington DC: Island Press, 2011).
25. Dixon, author interview, September 2009.
26. Many names in the region end in "up," which is the Noongar suffix for "place of."
27. Kingsley Dixon, comments made during the kwongan field trip, August 2009.
28. Hopper is a most effective advocate for the view that botanic gardens should make the study of restoration ecology one of their core theoretical missions, and ecological restoration one of their key practices. See Kate Hardwick et al., "The Role of Botanic Gardens in the Science and Practice

of Ecological Restoration," *Conservation Biology* 25, no. 2 (2011), 265–75. See also chapter 14, note 30.

29. I consulted several South African restoration ecologists on the applicability of OCBIL theory to fynbos. They generally agreed with the inclusion of their floristic kingdom as an OCBIL, but were skeptical about any blanket application there of the conclusions Hopper draws from his theory for restoration in southwest Australia. Some fynbos systems yield quickly to IAPs without prior disturbance, and some fynbos systems recover quickly and spontaneously once IAPs are removed (e-mails to author from Pat Holmes, May 24, 2011; Karen Esler, June 9, 2011; Richard Cowling May 6, 2011).
30. Stephen D. Hopper et al., "Globalising Restoration: A Role for Botanic Gardens" (plenary address at SER Conference, Perth, August 23–27, 2009).
31. Hopper, author interview, November 2009. Unless otherwise indicated, all comments from Hopper in this section are from this interview.
32. The richness of roadway verges in native plants is now widely recognized, to the extent that there a number of "mini-national parks" only a few metres wide, but which run for many miles between highways and farmland.
33. He acknowledgesthat there are of course big exceptions to this generalization, with aggressive invasive aliens like rhododendron (*Rhododendron ponticum*) and European buckthorn (*Rhamnus cathartica*) wreaking havoc in native plant communities in Europe and the United States, respectively.
34. Keith Bradby, e-mail to author, May 2010.
35. Hopper remains at odds with Bradby on pollination: "His suggestion that plants in WA [Western Australia] need to move their genes long distances around the landscape via mobile pollinators is true for a few, but certainly not the majority, of plants for which we have any pertinent evidence" (personal communication, March 2012).

CHAPTER 10

1. See Daniel H. Janzen and Winnie Hallwachs, "Conservation History and Future of Área de Conservación Guanacaste (ÁCG) in Northwestern Costa Rica," in *Costa Rican Ecosystems*, ed. Maarten Kappelle (Chicago: University of Chicago Press, forthcoming).
2. Tom Butler and Antonio Vizcaino, *Wildlands Philanthropy: The Great American Tradition* (San Rafael, CA: Earth Aware Editions, 2008), 131.
3. Janzen, informal presentation to guests in Santa Rosa, CA, January 7, 2010.
4. Daniel H. Janzen, author interview, January 2010.
5. Daniel H. Janzen, e-mail to author, May 5, 2011. See Butler and Vizcaino, *Wildlands Philanthropy*, 131. The species estimate began at 235,000, in Daniel Janzen, "Prioritization of Major Groups of Taxa for the All Taxa Biodiversity Inventory (ATBI) of the Guanacaste Conservation Area in north-

western Costa Rica, a Biodiversity Development Project," *ASC Newsletter* 26, no. 4 (1996): 45, 49–56.

6. That is not to say that the old-growth forest is absolutely "undisturbed." The impact of indigenous peoples here was relatively light, but recognizable.
7. Daniel H. Janzen, "Tropical Dry Forest: Área de Conservación Guanacaste, Northwestern Costa Rica," in *Handbook of Ecological Restoration*, vol. 2, *Restoration in Practice*, ed. M. R. Perrow and A. J. Davy (Cambridge: Cambridge University Press, 2002), 570.
8. Ibid.
9. One must qualify this portrait of the tough-minded side of Janzen—ever present though that is—with recognition of his warm and passionate (one is very tempted to say *loving*) appreciation for every detail of the vast range of biodiversity he engages with, and for the complex interactions between all its parts.
10. Daniel H. Janzen and Winnie Hallwachs, e-mails to author, January 1, 2010.
11. Daniel H. Janzen, "Gardenification of Wildland Nature and the Human Footprint," *Science*, 279 (1998): 1312, doi: 10.1126/science.279.5355.
12. William Allen, *Green Phoenix: Restoring the Tropical Forests of Guanacaste, Costa Rica* (New York: Oxford University Press, 2002). This book is essential reading for anyone who wants to grasp the full extent and "audacity"—Allen's word—of what Janzen, Hallwachs, and their Costa Rican colleagues are achieving in the ÁCG, along with enthralling and insightful vignettes of the complex biological dramas performed by the forest's flora and fauna. It was the first book on restoration recommended to me when I conceived the idea for this book. For another useful account, especially on the orange pulp experiment, and of the complex financial mechanisms—"debt for nature swaps"—that channeled international funds to the ÁCG—see Gretchen Daily and Katherine Ellison, "Costa Rica: Paying Mother Nature to Multitask," chapter 8 of *The New Economy of Nature* (Washington DC: Island Press, 2002).
13. Colonel Oliver North built an airstrip that supplied the Contras on Santa Rosa's neighboring Santa Elena peninsula. Today the peninsula is part of the ÁCG, and all obvious signs of this aspect of President Reagan's Central American intervention have vanished beneath restored forest. "There's nothing to see here now, that's the whole point," Janzen wryly told his guests on a field trip to Santa Elena, January 10, 2010.
14. Daniel H. Janzen and Winnie Hallwachs, e-mails to author, 5 May 2011.
15. Daniel H. Janzen and Winnie Hallwachs, e-mails to author, January 2, 2010.
16. For updates on these figures, and more information on this vast caterpillar inventory, see http://janzen.sas.upenn.edu/caterpillars/introductory/preface.htm. For further information on parataxonomists' work and on their use of DNA bar coding in this inventory, see Daniel H. Janzen and

W. Hallwachs, "Joining Inventory by Parataxonomists with DNA Barcoding of a Large Complex Tropical Conserved Wildland in Northwestern Costa Rica," *PLoS ONE* 6, no. 8 (2011), doi:10.1371/journal.pone.0018123.

17. I never heard them use the word "restoration" during two long sessions I spent with them and a group of guests in the ÁCG. It just seems to have dropped out of their vocabulary.
18. "Dry" is of course a relative term—rainfall in dry tropical forest ranges from 10 to 80 inches annually, and in rain forest from 80 to 160, including the further subcategories of wet forest and cloud forest.
19. Daniel H. Janzen, "Costa Rica's Área de Conservación Guanacaste: A Long March to Survival through Non-Damaging Biodiversity and Ecosystem Development" (paper presented at the Norway/UN Conference on the Ecosystem Approach for Sustainable Use of Biological Diversity (Trondheim: Norway Norwegian Directorate for Nature Research and Norwegian Institute for Nature Research, 2000), 122–32.
20. Allen, *Green Phoenix*, 33.
21. Janzen, informal presentation, January 7, 2010.
22. This phrase has its origins in William Shakespeare's dedication to his sonnets, where he gives all the credit for them to their mysterious "onlie begetter . . . Mr. W.H."
23. See Allen, *Green Phoenix*, 65. Janzen has repeatedly given public and private credit to the "huge effort by a very large number of Costa Ricans," and insists that he and Hallwachs are "just part of the team" (Butler and Vizcaino, *Wildlands Philanthropy*, 131). It is quite consistent with his complex nature that he may, in the next sentence, say something that implies that he has thought the whole thing up himself. Regarding Janzen's role, Blanco says, "We have arguments and, yes, conflicts. Daniel has very good ideas, but sometimes they don't fit with reality" (Róger Blanco, author interview, January 2010).
24. Janzen, conversation with guests on field trip to Santa Elena, January 10, 2010.
25. James Aronson, personal communication.
26. Daniel H. Janzen and Winnie Hallwachs, e-mails to author, May 5, 2011.
27. Janzen, "Tropical Dry Forest," 569.
28. Janzen, personal communication, January 5, 2011, which included his own electronic version of "Tropical Dry Forest," prefaced by an abstract not included in the published version, but which he has widely circulated. The quotation here is from part 9 of this abstract.
29. Janzen, "Costa Rica's Área de Conservación Guanacaste," 124.
30. Janzen, author interview, January 2010. Janzen's high valuation of everyday social contact between conservationists and regular citizens is similar to Keith Bradby's (see chapter 9, p. 232).
31. Janzen, informal presentation, January 7, 2010.
32. Janzen, "Costa Rica's Área de Conservación Guanacaste," 123.

33. Ibid.
34. Janzen, Janzen, personal communication, January 5, 2011, abstract, part 9, from "Tropical Dry Forest."
35. Daniel H. Janzen, "Tropical and Biocultural Restoration," *Science* 239 (1988): 243–44.
36. I should add that all three meals were accompanied by delicious fruit juices and, a happy inevitability in Costa Rica, good coffee.
37. Nevertheless, the ÁCG is one of the best-funded parks in the country. Several members of staff commented that they felt very lucky to have the use of jeeps, and access to computers. Janzen and Hallwachs have played a major role here, first by generous personal contributions to the trust that provides vital ancillary income to the park through Costa Rica's Fundación de Parques Nacionales, and then by establishing the independent and international Guanacaste Dry Forest Preservation Foundation, which has been a very significant force behind the ÁCG's purchases of additional land. See http://www.gdfcf.org/.
38. See Janzen, "Tropical Dry Forest," for an account of the most likely explanations.
39. Daniel Janzen, "Use of Gmelina Plantations as a Tool to Restore Ancient Rainforest Pastures to Natural Rainforest and Simultaneously Build a Living Endowment for a Conserved Rainforest" (proposal to the Wege Foundation from the Área de Conservación Guanacaste, March 2, 1999).
40. Felix Carmona, author interview, January 2010.
41. All quotations from Róger Blanco in this section are from an author interview, January 2010.
42. Julio Díaz, author interview, January 2010.
43. This resort has turned much of this small peninsula into a golf course, but its success in otherwise minimizing its environmental footprint, in contrast to the rash of coastal hotels to the south, was praised by several park staff. Red tides from poor waste management by tourism ventures are a bigger threat to the park's marine life than illegal fishing (María Marta Chavarría, ÁCG biologist, author interview, January 2010).
44. Janzen, "Costa Rica's Área de Conservación Guanacaste," 126.
45. Janzen, author interview, January 2010.
46. Ibid.
47. Janzen, "Costa Rica's Área de Conservación Guanacaste," 127.
48. Janzen, author interview, January 2010.
49. Ibid.
50. In fairness, Janzen did not and does not see it as a panacea: "The pulp is a particularly good tool if soil is compacted [by grazing] and beat down by fire and rain; if you had reasonably good soil you would not use it" (author interview, January 2010).
51. Some of Bonilla's comments are truly astounding to anyone who visited the site—which he had. He claimed, for example, that only a handful of

acid-tolerant plants had grown there, and had been again overwhelmed by jaragua by late 2000. "Time reveals the truth," he concluded. Indeed it does, as he would know if he returned to the site today (Alexander Bonilla Duran, presenter, "Ecología en Acción," Radio Monumental, November 23, 2000).

52. Janzen, "Costa Rica's Área de Conservación Guanacaste," 127.
53. "Absolutely Sweet Marie," Bob Dylan, Dwarf Music, 1966.
54. See the online material at http://www.press.uchicago.edu/sites/woodworth/.
55. María Fernández, author interview, January 2010.
56. Janzen and Hallwachs see her work as absolutely central to the success of the ÁCG's new venture in restoring the ocean: "Marine restoration means '**stop assaulting** it and leave it be,'" they told James Aronson in an e-mail of July 12, 2009. "Hardly complex biologically, VERY complex sociologically . . . the cost is not in moving one oyster from one rock to another rock, but rather in convincing (many forms to that verb) society to let the oysters move themselves from one rock to another, and even to have their own rock in the first place.

"Put another way, María Marta Chavarría does vastly more for restoration of the Sector Marino of ÁCG by doing natural history field trips with Cuajiniquil high-school-age kids than all the oyster moving that could be done with a million dollars."

Marine restoration is in its infancy, but is obviously enormously important for the future of the biosphere. Space and time have forced me to largely exclude it here, and it deserves a book in its own right. Bill Ballantine, pioneer of "no-take" reserves in New Zealand and internationally, takes a fairly similar approach to Janzen's in terms of seeing marine restoration as spontaneous recovery through passive preservation rather than active management, see http://www.marine-reserves.org.nz/index.html. The success of this approach has been encouraging, with threatened populations and damaged food chains recovering very rapidly once all fishing ceases. However, new and more activist schools of marine reservation are now emerging. See, for example, two exciting presentations led by Gilles Lecaillon at the SER conference in Mérida, Mexico, August 21–25, 2011: "Fish Restoration by Larval Restocking," and "BioRestore®: An Operational Solution for Accelerating the Regeneration of Marine Coastal Biodiversity in Compliance to French Regulations," 119, http://www.ser.org/docs/default-document-library/re-establishing-the-link-between-nature-and-culture.pdf.

57. María Marta Chavarría, author interview, January 2010.
58. Janzen, "Tropical and Biocultural Restoration"; Rosibel Elizondo Cruz and Róger Blanco Seguro, "Developing the Bioliteracy of School Children for 24 Years: A Fundamental Tool for Ecological Restoration and Conservation in Perpetuity of the Área de Conservación Guanacaste, Costa Rica," *Ecological Restoration* 28, no. 2 (2010): 193–98.

CHAPTER 11

1. See "Te Urewera Mainland Island: DOC's Work: Conservation Achievements," at http://www.doc.govt.nz/conservation/land-and-freshwater/land/mainland-islands-a-z/te-urewera/docs-work/.
2. Alan Tennyson and Paul Martinson, *Extinct Birds of New Zealand* (Wellington, NZ: Te Papa Press, 2006); John Craig et al., "Conservation Issues in New Zealand," *Annual Review of Ecology and Systematics* 31 (2000): 63, gives the slightly higher figure of fifty-four New Zealand bird extinctions for this period. See also T. H. Worthy, "What Was on the Menu—Avian Extinction in New Zealand," *New Zealand Journal of Archaeology* 19 (1999): 125–60.
3. Tommy Tyrberg, "Holocene Avian Extinctions," in *Holocene Extinctions*, ed. Samuel T. Turvey (New York: Oxford University Press, 2009), 63.
4. There is an good short account of this disaster at "Greater Short-Tailed Bat," TerraNature.org, http://terranature.org/batGreaterShort-tailed.htm.
5. David Butler and Don Merton, *The Black Robin* (Oxford: Oxford University Press, 1992), 21.
6. Colin Miskelly, author interview, February 2010.
7. "Turning the Tide," Kakapo Recovery, http://www.kakaporecovery.org.nz/index.php?option=com_content&view=article&id=78&Itemid=169.
8. John Nichols, "History of the Vegetation," in *Botany of the Waikato*, ed. Bruce Clarkson, Merilyn F. Merrett, and Theresa Downs (Hamilton: Waikato Botanical Society, 2002), 26.
9. The exact dates of Maori settlement of New Zealand remain somewhat contentious among scholars.
10. As in Australia, "paddock" is a synonym for "agricultural field."
11. Bruce Clarkson and Joanna C. McQueen, "Ecological Restoration in Hamilton City" (paper presented at the 16th International Conference, Society for Ecological Restoration, Victoria, Canada, August 24–26, 2004).
12. Information about Tui 2000 can be found on the group's website at http://www.envirocentre.org.nz/tui_2000/.
13. Bruce Clarkson, author interview, February 2010.
14. Joanna C. McQueen, ed., *An Ecological Restoration Plan for Maungatautari* (Cambridge, New Zealand: MEIT, 2004), 7.
15. For updates, see http://netlist.co.nz/communities/MaungaTrust/Project_Timeline.cfm.
16. This and subsequent quotes from David Wallace and MEIT staff and visitors are from an author field trip to Maungatautari, February, 2010.
17. There are more fenced enclosures within that perimeter to provide added security for takahe and kiwi.
18. "Key Facts about Brodifacoum," fact sheet, New Zealand Department of Conservation, October 2007, http://www.doc.govt.nz/upload/documents/conservation/threats-and-impacts/animal-pests/northland/brodifacoum-factsheet.pdf.

19. See "Co-Chairs Annual Report," MEIT, August 27, 2012, http://www.maungatrust.org/files/6626/Co%20Chairs%20Annual%20Report.pdf.
20. See especially Colin Miskelly, *Mana Island Ecological Restoration Plan* (Wellington, NZ: Department of Conservation, 1999); Colin Miskelly, "Restoring Kapiti Island," in *Restoring Kapiti: Nature's Second Chance,* ed. Kerry Brown (Dunedin, NZ: University of Otago Press, 2004), 109–15. Unattributed quotations from Miskelly in this section are from an author interview, February 2010.
21. All three species belong to the genus of Austral snipe, *Coenocorypha,* whose members are also found (formerly, only subfossil bones remain) in Fiji and Norfolk Island. They are smaller, more given to heathland habitats, and less given to flight than their relatives in the much more globalized *Gallinago* genus, familiar in north America and Europe.
22. Considered the same species as the previously extinct South Island snipe.
23. Colin Miskelly, "The Identity of the Hakawai," *Notornis* 34, no. 2 (1987): 98
24. Ibid., 95–116.
25. Miskelly and Allan J Baker formally described the subspecies, *C. aucklandica perseverance,* in "Description of a new subspecies of *Coenocorypha* snipe from subantarctic Campbell Island, New Zealand," *Notornis* 56, no. 3 (2010):113–123.
26. *Pakeha* originally seems to have referred to settlers of British origin, and may have been derogatory, though its first meaning was probably simply "foreign" or "exotic." *Pakeha* is increasingly used in contemporary New Zealand as the acceptable term for anyone, or any ideas, of non-Maori origin.
27. Miskelly added that Spencer had died, aged ninety, just a week after this conversation took place.
28. Miskelly, *Mana Island Ecological Restoration Plan*.
29. Miskelly, author interview, February 2010. Unless otherwise noted, subsequent quotes from Miskelly are from this interview.
30. See Alan J Saunders, *A Review of Department of Conservation Mainland Restoration Projects and Recommendations for Further Action* (Wellington, NZ: Department of Conservation, 2000), http://www.DOC.govt.nz/upload/DOCuments/science-and-technical/Mainland.pdf; Craig Gillies et al., "A Review of Conservation Outcomes from Mainland Islands: Progress Report" (workshop presentation, 2010), http://www.sanctuariesnz.org/meetings/DOCuments/Gillies2010.pdf. A useful list of mainland island projects, public and private, with links, can be found at http://www.sanctuariesnz.org/projects.asp.
31. Saunders, *A Review of Department of Conservation Mainland Restoration Projects,* 7
32. Ibid.
33. Ibid., 27 (emphasis added).
34. Clarkson, author interview, February 2010.

35. Ibid. Unless otherwise noted, subsequent quotations from Clarkson are from this interview.
36. Gillies et al., "A Review of Conservation Outcomes from Mainland Islands," slide 3.
37. Greg Moorcroft, author interview, February 2010.
38. Craig et al., "Conservation Issues in New Zealand."
39. Craig is harsh in his criticism of the DOC's stewardship of reserves, especially where the agency is not involved in active restoration, dismissing it as "so-called management . . . more akin to 'wanton neglect'" in another context. See "Restoring Natural Capital Reconnects People to their Natural Heritage: Tiritiri Matangi Island, New Zealand" in *Restoring Natural Capital: Science, Business and Practice*, ed. James Aronson, Suzanne J. Milton, and J. N. Blignaut (Washington DC: Island Press, 2007), 103.
40. Daniel H. Janzen, "Tropical and Biocultural Restoration," *Science* 239 (1988): 243.
41. Craig et al., "Conservation Issues in New Zealand," 65.
42. Ibid.
43. Ibid., 66.
44. Ibid., 68–69.
45. Ibid., 66 (emphasis added).
46. Ibid. See chapters 8 and 14 for further discussion on "novel ecosystems," and on the legitimacy of using invasive aliens to provide ecosystem services and fill ecological niches in restoration projects.
47. See panel 3,"Restoration by Any Means Necessary" (p. 292).
48. See David A. Norton and Craig J. Miller, "Some Issues and Options for the Conservation of Native Biodiversity in Rural New Zealand," *Ecological Management & Restoration* 1, no. 1 (2000): 33; Wren Green and Bruce Clarkson, *The Synthesis Report—Turning the Tide—A Review of the First Five Years of the NZ Biodiversity Strategy* (Wellington, NZ: Department of Conservation, 2006), 1–2, http://www.doc.govt.nz/publications/conservation/a-review-of-the-first-five-years-of-the-nz-biodiversity-strategy/; and Clarkson, author interview, February 2010: "The statutory battle has been won, the big national parks declared. The big issue now is the devastation of our lowland zones, building on very degraded zones, starting from scratch. A lot of people talk about the restoration of existing [biodiverse] sites. We are in trouble if we keep it that narrow. The concept of broadening the restoration concept out beyond existing [biodiverse] resources is recent and has not got much traction. I am interested in that end of it."
49. Colin Miskelly, "Ecological Restoration and Threatened Species Management in New Zealand," *Ecological Management & Restoration* 10, no. 2 (2009): 161–62. This brief note is an excellent summary of island restoration history in New Zealand, and I have drawn on it heavily here.
50. Herbert Guthrie-Smith, *Sorrow and Joys of a New Zealand Naturalist* (Dune-

din, NZ: A. H. & A. W. Reed, 1936), 183–84. Again, I am indebted to Colin Miskelly for this quotation. See also "Boundary Stream Mainland Island," Department of Conservation website, http://www.doc.govt.nz/conservation/land-and-freshwater/land/mainland-islands-a-z/boundary-stream/features/history/.

51. For accounts of the battle that turned this tide of extinctions, and the people who led it, see Rod Morris and Hal Smith, *Wild South: Saving New Zealand's Endangered Birds* (Auckland: TVNZ, in association with Century Hutchinson, 1988); Alison Ballance and Don Merton, *Don Merton: The Man Who Saved the Black Robin* (Auckland, NZ: Reed Books, 2007).
52. The success of traps and poison in turn extended the range of islands available for use as sanctuaries. Previously, any island within swimming distance of a stoat was ruled out for restoration, but now it is reckoned that a line of traps and tunnels ringing an island, plus a line on the opposite mainland shoreline, will suffice to prevent reinfestation by stoats. This has permitted the use of islands, like Maud, much closer to the mainland than would have been previously considered.
53. Miskelly, "Ecological Restoration and Threatened Species," 162.
54. Ibid.

CHAPTER 12

1. Bracken is an invasive plant in many circumstances, but it is not an alien in the Lacandon jungle—indeed it is a very cosmpolitan species, found naturally in many parts of the world.
2. See James D. Nations and Ronald B. Nigh, "The Evolutionary Potential of Lacandon Maya Sustained-Yield Tropical Forest Agriculture," *Journal of Anthropological Research* 36, no. 1 (1980): 2–3.
3. Since the Lacandones are a very small group, with an iconic international reputation as a repository of ancient rain forest lore, it seems to have suited Mexican governments to support them as a kind of showcase of their respect for indigenous culture. Meanwhile, the same governments often trampled on the cultural rights of more numerous and troublesome peoples.
4. See chapter 7, note 23.
5. See chapter 10, p. 259.
6. See also chapter 11, p. 309.
7. "Maori hunting eliminated 26 species (30%) of endemic land birds and 4 (18%) of the endemic sea birds, while ecosystem loss and companion animals eliminated a further 8 land birds" (John Craig, Sandra Anderson, Mick Clout, Bob Creese, Neil Mitchell, John Ogden, Mere Roberts, and Graham Ussher, "Conservation Issues in New Zealand," *Annual Review of Ecology and Systematics* 31 [2000]: 63).
8. While in New Zealand, I spoke to two Maori elders, from different *iwi*,

each of whom had dedicated their lives to defending their people's cultural and political rights about this issue. Neither of them had any difficulty in accepting that Maori were indeed responsible for historical ecological damage. Both expressed regret, and some frustration, that younger Maori today often present a romantic and unscientific view of their relationship to the environment before the arrival of European settlers.

9. A biblical text, the Book of Deuteronomy (22:6), enjoins us not to take the mother bird if we harvest eggs from a nest; but that endorsement of sustainable hunting was in the context of, and subordinate to, the prior and repeated insistence, according to some readings, in the Book of Genesis (1:26–28) that our relationship with the Earth is one of dominion, not power-sharing.
10. Stewart A. W. Diemont and Jay F. Martin, "Lacandon Maya Ecosystem Management: Sustainable Design for Subsistence and Environmental Restoration," *Ecological Applications* 19, no. 1 (2009): 262.
11. David Douterlungne, speaking in *Chujúm: una alterniva tradicional de maneja agroforestal en la Selva Lacondona* (directed by Alberto Nulman Magidin and Ana Luisa Montes de Oca, Ad Astra Producciones, Mexico City, 2005), DVD.
12. Samuel Levy Tacher, author interview, January 2010.
13. Samuel Levy Tacher and John Duncan Golicher, "How Predictive Is Traditional Ecological Knowledge? The Case of the Lacandon Maya Fallow Enrichment System," *Interciencia* 29, no. 9 (2004): 496.
14. Levy Tacher, author interview, January 2010.
15. Nations and Nigh, "Evolutionary Potential," 20.
16. Several observers have speculated that much—or even all—of the remaining primary forest in the region may itself at some stage have passed through the *milpa* system, and there is little doubt that its composition has been, to a greater or lesser degree, conditioned by millennia of Mayan agroforestry. So "old-growth forest" is a better term than "primary forest" here, as in so many places in the world. See Anabel Ford and Ronald Nigh, "Origins of the Maya Forest Garden: Maya Resource Management," *Journal of Ethnobiology* 29, no. 2 (2009): 213–36; David. G. Campbell, John Guittar, and Karen S. Lowell, "Maya Home Gardens as Biodiversity Hotspots" (paper presented at the VII Congreso Internacional de Mayistas, Merida, Mexico, 2007); and David G. Campbell, John Guittar, and Karen S. Lowell, "Are Colonial Pastures the Ancestors of the Contemporary Maya Forest?," *Journal of Ethnobiology* 28, no. 2 (2008):278–89, doi: 10.2993/0278–0771–28.2.278.
17. For an account of the problems with this project, see Caroline Fraser, *Rewilding the World: Dispatches from the Conservation Revolution* (New York: Picador, 2010), 64–70.
18. All these practices suggest that the agriculture of the classic Maya period, which supported their moumental systems, may often have been closer to a mosaic of cultivation and forest, and indeed of cultivation-within-

forest, than the massive clearances that have often been inferred from their great engineering and architectural achievements, and especially their engineering of water courses. The evidence is sparse, however, and there may have been both such mosaics and big cleared areas. Commenting on their achievement in his period, Nations and Nigh point out that "the Maya achieved what has yet to be attained in the twentieth century: the creation of a diverse, long-term, stable food production system in the tropical forest biome" (Nations and Nigh, "Evolutionary Potenial," 1–2). For other accounts of the extraordinary productivity of Maya forest gardens and their complex relationship with rain forest ecology, see Ford and Nigh, "Origins of the Maya Forest Garden"; Campbell et al., "Maya Home Gardens"; and Campbell et al., "Are Colonial Pastures the Ancestors."

19. D. Douterlungne et al., "Applying Indigenous Knowledge to the Restoration of Degraded Tropical Rain Forest Clearings Dominated by Bracken Fern," *Restoration Ecology* 18, no. 3 (2010): 322.
20. Levy Tacher, author interview, January 2011.
21. Douterlungne et al., "Applying Indigenous Knowledge," 322.
22. Levy Tacher, "Restauración de la conectividad del paisaje a partir del conocimiento ecológico tradicional maya en los poblados de Nueva Palestina y Plan de Ayutla, selva Lacandona, REBIMA, Chiapas" (unpublished teaching presentation shown to author, Lacanhá, January 2011).
23. The Lacandon and the scientific methods do not always coincide. One of Levy Tacher's students, Francisco Román Dañobeytia, told me an instructive story on a later visit to the jungle: "It is not that Don Manuel knows everything. *We construct this knowledge together.* And sometimes he is wrong. He is very apprehensive about harming the plants we are propagating in the nursery. We found that when transplanting cuttings, it is often beneficial to prune the roots. He would not do it. We did tests with him, and proved we were right. But he is still very reluctant to cut off the roots of living plants" (author interview, August 2011; emphasis added).
24. Douterlungne et al., "Applying Indigenous Knowledge," 322.

CHAPTER 13

1. One should not underestimate the environmental impact of this practice, however. It is reckoned that, by the time industrial exploitation began in the 1940s, "nearly half of the . . . large Midland raised bogs recorded in 1814 had been cut away by hand" (F. H. A. Aalen, Kevin Whelan, and Matthew Stout, *Atlas of the Irish Rural Landscape* [Cork: Cork University Press, 1997], 111).
2. Catherine Farrell, e-mail to author, October 10, 2010.
3. Patrick Kavanagh, "The One," in *The Complete Poems* (Newbridge, Ireland: Goldsmith Press, 1972), 291.
4. Technically, a bog that has been exploited industrially is "cutaway"; one

exploited by hand is "cutover." But Kavanagh was almost certainly referring to the latter, as industrial exploitation was rarely feasible among the humpy drumlins of his native County Monaghan in southern Ulster.

5. F. Renou-Wilson et al., *BOGLAND—Sustainable Management of Peatlands in Ireland* (STRIVE Report No. 75, prepared for the Environmental Protection Agency, Johnstown Castle, County Wexford, Ireland, 2011), 11.
6. National Parks and Wildlife Service, *The Status of EU Protected Habitats and Species in Ireland* (Department of the Environment, Heritage, and Local Government, 2008), 47–49, http://www.npws.ie/publications/eu conservationstatus/NPWS_2007_Conservation_Status_Report.pdf.
7. Michael Viney, *Ireland, a Smithsonian Natural History* (Belfast: Blackstaff Press, 2003), 102.
8. It had unusual internal drainage systems known as "soaks"; it subsequently became one of the most intensively studied bogs in the world.
9. While Ireland should undoubtedly be grateful to the Dutch for their helpful intervention on so many levels, this episode recalls a familiar environmental relationship between the developed and developing world: the metropolis, having consumed its own natural capital, puts pressure on the periphery to preserve biodiversity—a case of "Don't do as I did then, do as I say now."
10. Garrett Hardin, "The Tragedy of the Commons," *Science* 162, no. 3859 (1968): 1243–48. In this very influential essay, Hardin argues that if every farmer has ten cows on the commons, and one more cow will degrade it, the value of that extra cow to the farmer who puts it out will always exceed the immediate damage that degradation will do to him in particular.
11. Coillte had already been very busy on *intact* bogs, having planted 6,400 hectares of formerly intact raised bogs and 211,000 hectares of formerly intact blanket bog by the end of the 1980s. See Catherine Farrell, "Review of Peatland Restoration in Ireland," *Peatlands International*, no. 1 (2009): 19.
12. The creation of commercially productive grasslands and tillage fields was also attempted, but it failed for various reasons. There were also innovative and much more successful efforts to explore recreational uses for the bogs, incorporating outdoor pursuits like hiking and fishing, sculpture trails, archaeological exhibitions and wildlife trails. See http://www.loughbooraparklands.com/parklands.
13. As so often in such cases, the biodiversity issue is arguable. Some national rarities—crossbills, for example—did find suitable habitats in the new forests.
14. One factor that influenced Bord na Móna to consider restoration here was pressure from the angling community. Despite its claims to have set up an effective system of silt traps on its bogs, internationally renowned salmonid-breeding rivers like the Owenmore and Oweninny were inundated up with peat silt, which was destroying fish stocks. Rivers and stocks have recovered well since restoration began.

15. Mitsch used this phrase in his presentation at the SER First International Conference in Zaragoza in 2005 (see chapter 3, p. 27).
16. Catherine Farrell, author interview, September 2010.
17. Catherine Farrell and G. J. Doyle, "Rehabilitation of Industrial Cutaway Atlantic Blanket Bog in County Mayo," *Wetlands Ecology and Management* 11 (2003): 34.
18. Catherine Farrell, e-mail to author, October 2010.
19. Farrell, author interview, September 2010. All other quotations in this section, from this site, are from the same interview.
20. Curiously, this rush, *Schoenus nigricans*, is not characteristic of Atlantic blanket bog elsewhere, not even in broadly similar conditions in Scotland. It is usually considered a classic fenland species. Farrell believes its status on Irish blanket bog may be due to a combination of strong marine-derived elements in the ecosystem, of burning regimes, and of climate.
21. Farrell and Doyle, "Rehabilitation of Industrial Cutaway Atlantic Blanket Bog," 21.
22. Gerry McNally, author interview, October 2010.
23. Catherine Farrell, author interview, November 2010.
24. Ibid.
25. National Parks and Wildlife Service, *The Status of EU Protected Habitats and Species*; backing documents, article 17 forms, maps; vol. 3, Active Raised Bog, Future Prospects, 13, section 7.2, http://www.npws.ie/publications/euconservationstatus/NPWS_2007_Cons_Ass_Backing_V3.pdf. See also Jim Ryan, "Raised Bog Restoration in Ireland" (paper presented at Coillte LIFE Project—Restoring Raised Bog in Ireland End of Project Conference, Carrick-on-Shannon, Co. Leitrim, Ireland, May 8, 2008), http://www.raisedbogrestoration.ie/life04/downloads/raised-bog-conference-pres-jim-ryan.pdf. For IPCC data, see "Raised Bog Monitoring," Irish Peatland Conservation Council website, http://www.ipcc.ie/peatland-action-plan/over-exploitation-of-peatlands-for-peat/.
26. *Ireland: 4th National Report to the Convention on Biological Diversity* (Dublin: Department of the Environment, Heritage and Local Government, 2010), sec. 1.2.4, p. 13, http://www.cbd.int/doc/world/ie/ie-nr-04-en.pdf.
27. Turf Cutters and Contractors Association, "Submission to the Working Group on Cessation of Turf Cutting on Certain Raised Bog Special Areas of Conservation and Natural Heritage Areas" (2009), http://dl.dropbox.com/u/14708387/WG_Submission_Mini-A5.pdf.
28. From *druim*, a ridge.
29. Judit Keleman, e-mail to author, November 23, 2010.
30. Unpublished NPWS survey, attached to e-mail from Judit Keleman, November 23, 2010.
31. This problem dogs numerous bog restorations.
32. All quotations from Killyconny Bog field trip, author interviews, October 2010.

33. Keleman e-mail, November 23, 2010.
34. Ibid.
35. See chapter 8, p. 207.
36. All unattributed quotations from Jim Ryan here are from Killyconny Bog field trip, October 2010.
37. Catherine O'Connell, comment made during Killyconny Bog field trip, October 2010.
38. Ryan, "Raised Bog Restoration in Ireland."
39. Farrell, author interview, November 2010.
40. National Parks and Wildlife Service, *The Status of EU Protected Habitats and Species*, backing documents, article 17 forms, maps; vol. 3, Active Raised Bog, Future Prospects, 14, section 8, http://www.npws.ie/publications/eu conservationstatus/NPWS_2007_Cons_Ass_Backing_V3.pdf
41. Ibid., section 1.
42. Farrell, author interview, November 2010.
43. See note 5, above.
44. Florence Renou-Wilson, e-mail to author, December 20, 2010.
45. M. G. C. Schouten, "Peatland Research and Peatland Conservation in Ireland: Review and Prospects,"in *After Wise Use: The Future of Peatlands: 13th International Peat Congress*, June 8–13, (Tullamore, Ireland: International peat Society, 2008), 660–61 (original emphasis).
46. Ibid. I only deal here with wetland creation in the context of bog afteruse. But remarkable work has been done in this field in Ireland by Rory Harrington, see Mark Everard, Rory Harrington, and Robert J. McInnes, "Facilitating Implementation of Landscape-Scale Water Management: The Integrated Constructed Wetland Concept," *Ecosystem Services* 2 (2012): 27—37.
47. All remaining quotations in this chapter were comments at the launch of Bord na Móna's Biodiversity Action Plan in Tullamore on November 10, 2010, and the subsequent field trip to Ballycon bog.

CHAPTER 14

1. See especially Jared M. Diamond, *Collapse: How Societies Choose to Fail or Succeed* (Camberwell, Victoria: Penguin, 2007). For a critical view of Diamond's very influential perspective, see Patricia Ann McAnany and Norman Yoffee, *Questioning Collapse: Human Resilience, Ecological Vulnerability, and the Aftermath of Empire* (Cambridge: Cambridge University Press, 2010).
2. Henry David Thoreau's speech at Concord was published after his death as the essay "Walking," available at http://thoreau.eserver.org/walking.html.
3. I am indebted to Simon Schama's *Landscape and Memory* (London: Fontana Press, 1996) for a stimulating discussion of Thoreau's relationship to the wild. Schama's key quotation from Thoreau makes an excellent case for seeing the human in wildness, and wildness in the human: "It is vain to dream of a wildness distant from ourselves. There is none such. It is the

bog in our brain and bowels, the primitive vigor of Nature in us, that inspires that dream."

4. Thoreau, "Walking."
5. A particularly impressive example of environmental education, campaigning for the restoration of direct childhood experience of the natural world, is the No Child Left Inside Coalition in the US: http://www.cbf.org/page.aspx?pid=687.
6. James Aronson et al., *Restoring Natural Capital: Science, Business, and Practice* (Washington DC: Island Press, 2007), ix. The Global Footprint Network estimates that we are already using the equivalent of 1.5 planets to provide resources and absorb waste. That is, it takes the Earth 18 months to regenerate what we consume/use in a year (http://www.footprintnetwork.org/en/index.php/GFN/page/world_footprint/).
7. Ecological restoration is also a major plank in the Global Strategy for Plant Conservation; it features as a leading element in the Aichi Targets (2011–2020) of the Convention on Biological Diversity, including the very ambitious goal of restoring 15 percent of degraded ecosystems across the world over the next ten years (http://www.cbd.int/doc/strategic-plan/2011-2020/Aichi-Targets-EN.pdf). In response to this target, a Global Restoration Council, under International Union for the Conservation of Nature and World Resources Institute auspices, was set up in September 2011, with a brief to restore 150 million hectares of degraded land by 2020 (http://www.wri.org/press/2011/09/release-leaders-announce-global-effort-restore-150-million-hectares-deforested-land). Many restorationists on the ground are understandably skeptical about such ballpark aspirations. Here is one—necessarily anonymous—response from the field to this announcement: "Well, we all know how far the biodiversity targets were missed at the last UN biodiversity conference! Frankly, this target of 150 million hectares by 2020 is a figure plucked out of the hat by some politician. They could equally easily have said 50 million or 500 million hectares."
8. In Cuba, Jesús Matos Mederos and Domingo Ballate Denis produced an *ABC de la Restauración Ecológica* (Santa Clara: Editorial Feijóo, 2006) that draws on the primer and on the distinctive experience of Cuban restorationists.
9. Miguel Calmon et al., "Emerging Threats and Opportunities for Large-Scale Ecological Restoration in the Atlantic Forest of Brazil," *Restoration Ecology* 19, no. 2 (2011): 157.
10. C. T. S. Nair and R. Rutt, "Creating Forestry Jobs to Boost the Economy and Build a Green Future," *UNASYLVA—FAO* no. 233 (2009): 3–10.
11. For a clear account of the debate in Brazil, but with much wider implications, see James Aronson et al., "What Role should Government Regulation Play in Ecological Restoration? Ongoing Debate in São Paulo State,

Brazil," *Restoration Ecology* 19, no. 6 (2011): 690–95. doi:10.1111/j.1526-100X.2011.00815.x.

12. L. Roberts, R. Stone, and A. Sugden, "The Rise of Restoration Ecology," *Science* 325, no. 5940 (2009): 555.
13. J. M. Rey Benayas et al., "Enhancement of Biodiversity and Ecosystem Services by Ecological Restoration: A Meta-Analysis," *Science* 325, no. 5944 (2009): 1121–24.
14. A. D. Bradshaw, "Restoration: An Acid Test for Ecology," in *Restoration Ecology: A Synthetic Approach to Ecological Research*, ed. William R. Jordan, M. E. Gilpin, and H. J. D. Aber, 23–31 (Cambridge: Cambridge University Press, 1987).
15. Bruce Pavlik, e-mail to author, November 8, 2011. Here is how he described it to his colleagues in Kew: "Our developing collaborative project will focus on a species rich ecosystem that has been degraded or destroyed by human activity. Typically, restoration efforts begin years after degradation, long after it is possible to know the original diversity and key synergisms that make an ecosystem functional and self-sustaining. We are, therefore, proposing a novel [not novel ecosystem!] approach: careful disassembly of the intact ecosystem prior to mining, deforestation or other impending disturbance. Once inventoried of microbes, plants and animals, we can test for critical interrelations that facilitate reassembly to ultimately improve the recreated ecosystem to meet human, commercial and biodiversity needs. This 'disassembly-reassembly' project will produce useful science and economic benefits over a 30-year period. We are currently engaging with extractive industries to find the necessary funding, project location and support facilities" (Bruce Pavlik, "Building Kew's Programme in Restoration Ecology," *Samara: The Newsletter of the Partners of the Millennium Seed Bank Partnership* (May 2011), http://www.kew.org/ucm/groups/public/documents/document/kppcont_037869.pdf.
16. See William Jordan and George Lubick, *Making Nature Whole: A History of Ecological Restoration* (Washington DC: Island Press, 2011).
17. Meanwhile, unsurprisingly, many practitioners of ecological restoration just get on with doing the best job they can, largely oblivious of its theoretical ramifications.
18. William R. Jordan, e-mail to author, September 27, 2011.
19. For more information on the New Academy, see http://csh.depaul.edu/centers-and-institutes/inc/about/partners/Pages/default.aspx.
20. William R. Jordan, *The Sunflower Forest: Ecological Restoration and the New Communion with Nature* (Berkeley: University of California Press, 2003).
21. Jordan, e-mail to author, April 12, 2010. Unless specified otherwise, the quotations that follow are from this e-mail, the culmination of a series of face-to-face and e-mail conversations about *The Sunflower Forest*.
22. Jordan, *The Sunflower Forest*, 22, 125, 76.
23. Jordan and Lubick, *Making Nature Whole*.

24. Jordan, e-mail to author, September 28, 2011.
25. Ibid.
26. Rudolf De Groot et al., "Integrating the Ecological and Economic Dimensions in Biodiversity and Ecosystem Service Valuation," in *The Economics of Ecosystems and Biodiversity: Ecological and Economic Foundations (TEEB DO)*, ed. Pushpam Kumar (London: Earthscan, 2010), 19.
27. Jordan, *The Sunflower Forest*, 12.
28. Ibid., 114.
29. Ibid., 88.
30. Ibid., 114.
31. Ibid., 50 (emphasis and letters in brackets added).
32. See p. 120.
33. Jordan, *The Sunflower Forest*, 114.
34. Ibid.
35. See chapter 5, p. 120.
36. James Aronson et al., "Restoration and Rehabilitation of Degraded Ecosystems in Arid and Semi-Arid Lands. I. A View from the South," *Restoration Ecology* 1, no. 1 (1993): 8–17; James Aronson et al., "Restoration and Rehabilitation of Degraded Ecosystems in Arid and Semi-Arid Lands. II. Case Studies in Southern Tunisia, Central Chile and Northern Cameroon," *Restoration Ecology* 1, no. 3 (1993): 168–87.
37. Aronson et al., "Restoration and Rehabilitation II," 185.
38. Van Andel and Aronson, *Restoration Ecology*, 228.
39. Aronson et al., "Restoration and Rehabilitation I," 15.
40. Andre F. Clewell and James Aronson, *Ecological Restoration: Principles, Values and Structure of an Emerging Profession* (Washington DC: Island Press, 2007), 110.
41. Ibid., 88.
42. Ibid., 14–15.
43. S. T. A. Pickett and V. Thomas Parker, "Avoiding the Old Pitfalls: Opportunities in a New Discipline," *Restoration Ecology* 2, no. 2 (1994): 75–79.
44. James Aronson, S. Dhillion, and E. Floc'h, "On the Need to Select an Ecosystem of Reference, however Imperfect: A Reply to Pickett and Parker," *Restoration Ecology* 3, no. 1 (1995): 1–3 (emphasis added).
45. Ibid., 1.
46. James Aronson, e-mail to author, November 21, 2011.
47. Ibid., 2
48. James Aronson, author interview, March 2011.
49. Ibid. Or, as Don Falk, a former executive director of SER, once put it, "Restoration uses the past not as a goal but as a reference point for the future. . . . It is not to turn back the evolutionary clock, but to set it ticking again" (Donald Falk, "Discovering the Future, Creating the Past: Some Reflections on Restoration," *Restoration & Management Notes* 8, no. 2 (1990): 71–72.

50. Aronson, Dhillion, and Le Floc'h, "On the Need to Select an Ecosystem of Reference," 2.
51. Eric Higgs gives an interesting participant's account of the process and thinking behind the development of the SER primer in chapter 3 of *Nature by Design: People, Natural Process, and Ecological Restoration*, 107–10 (Cambridge: MIT Press, 2003).
52. Clewell and Aronson, *Ecological Restoration*.
53. Ibid., 46–47, 96.
54. James Aronson Sue Milton, and James Blignaut, eds., *Restoring Natural Capital: Science, Business and Practice* (Washington DC: Island Press, 2007) Clewell (and, I should point out, this author) were among some forty contributing writers. See chapter 3, pp. 35–48, for a discussion of restoration of natural capital.
55. James A. Harris et al., "Ecological Restoration and Global Climate Change," *Restoration Ecology* 14, no. 2 (2006): 170–76; Richard J. Hobbs et al., "Novel Ecosystems: Theoretical and Management Aspects of the New Ecological World Order," *Global Ecology and Biogeography* 15, no. 1, (2006): 1–7 (see also chapter 8, pp. 201–203).
56. Richard Hobbs, "Novel Ecosystems: Looking Forward Rather than Back" (paper presented in *Novel Ecosystems: When and How Do We Intervene in the New Ecological World Order?*, special session at the SER 4th World Conference on Ecological Restoration, Mérida, Mexico, August 22, 2011).
57. Young D. Choi, "Shooting at a Moving Target: Restoration Ecology in a Changing World" (paper presented at the "Moving Targets and Shifting Sands" session at the SER 4th World Conference on Ecological Restoration, Mérida, Mexico, August 22, 2011).
58. Eric Higgs, "Novel Ecosystems: When and How Do We Intervene in the New Ecological World Order," Mérida, 2011. Higgs's transition to the novel ecosystem trend has obviously been a uncomfortable one. He has warned specifically that restoration "must not abandon history entirely. If it does, we will be giving in too much to the capricious nature of contemporary judgment" (Higgs, *Nature by Design*, 131).
59. Hobbs et al., "Novel Ecosystems," 1–7 (emphasis added).
60. President Bush used the phrase "New World Order" to describe the global political landscape after the fall of communism, supposedly ripe for universal democracy, in UN and State of the Union addresses in the early 1990s.
61. Richard J. Hobbs, "From Our Southern Correspondent," *Bulletin of the British Ecological Society* 43 no. 1 (2012): 40–41.
62. Richard Hobbs, author interview, September 2009.
63. Ibid.; and John M. Koch and Richard J. Hobbs, "Synthesis: Is Alcoa Successfully Restoring a Jarrah Forest Ecosystem After Bauxite Mining in Western Australia?" *Restoration Ecology* 15, no. 4 (2007): S137–44. For Hobbs's comments on Justin Jonson's work, see chapter 9, p. 236.

64. Stephen D. Hopper, "OCBIL Theory: Towards an Integrated Understanding of the Evolution, Ecology and Conservation of Biodiversity on Old, Climatically Buffered, Infertile Landscapes," *Plant and Soil* 322 (2009): 49–86 (see chapter 9, p. 248). Hopper was referring to Viki A. Cramer, Richard J. Hobbs, and Rachel J. Standish, "What's New about Old Fields?" *Trends in Ecology and Evolution* 23, no. 2 (2008): 104–12. Standish and Hobbs replied to Hopper in "Restoration of OCBILs in South-Western Australia: Response to Hopper," *Plant and Soil* 330, no. 1–2 (2010): 15–18, DOI: 10.1007/s11104–009–0182-z.
65. Standish and Hobbs, "Response to Hopper," 18.
66. Hobbs, e-mail to author, October 6, 2012.
67. Richard J. Hobbs, Eric Higgs and James A. Harris, "Novel Ecosystems: Implications for Conservation and Restoration," *Trends in Ecology and Evolution* 24, no. 11 (2009): 599 (emphasis added).
68. Ibid., 601.
69. They concede that the distinction they make between hybrid and novel ecosystems is "somewhat arbitrary" (Hobbs, Higgs, and Harris, "Novel Ecosystems: Implications," 601).
70. Ibid., 604 (emphasis added).
71. See chapter 9, pp. 234–237.
72. James Aronson, e-mail to author, July 10, 2010.
73. James Aronson, Carolina Murcia, Daniel Simberloff, Gustavo Kattan, Kingsley Dixon, and David Moreno-Mateos, "The Future We Want—Deciding the Fate of Global Ecosystems" (unpublished manuscript, June 2013), Microsoft Word.
74. Hobbs, Higgs and Harris, "Novel Ecosystems: Implications," 604.
75. Ibid.
76. Stephen T. Jackson and Richard J. Hobbs, "Ecological Restoration in the Light of Ecological History," *Science* 325, no. 5940 (2009): 567, 568.
77. Ibid., 568.
78. Richard J. Hobbs, Eric S. Higgs, and Carol M. Hall, ed., *Novel Ecosystems: Intervening in the New Ecological World Order* (Hoboken, NJ: Wiley-Blackwell, 2013), 358.
79. Richard J. Hobbs et al., "Intervention Ecology: Applying Ecological Science in the Twenty-First Century," *Bioscience* 61, no. 6 (2011): 442–50.
80. Ibid., 442.
81. Ibid., 443.
82. See chapter 3, p. 30. I put this point to Hobbs, who responded: "Sure, some leading literature very much follows this idea. Whether current practice does is another question. Look at most policy and legislation, and dynamism is largely absent" (Hobbs, e-mail to author, October 6, 2012). But this is a different problem, and the answer is for restorationists to communicate better with managers and policy makers, not to suggest that the discipline itself still hankers after "previous static states."

83. Hobbs et al., "Intervention Ecology," 444.
84. Ibid., 448.
85. Ibid., quoting Paul Wapner, *Living through the End of Nature: The Future of American Environmentalism*. (Cambridge, MA: MIT Press, 2010).
86. See Jackson and Hobbs, "Ecological Restoration," 568.
87. Kingsley Dixon, author interview, August 2011.
88. Hobbs et al., *Novel Ecosystems*.
89. Ibid., 58 (emphasis added).
90. Ibid., 60.
91. Ibid., 358.
92. Ibid., 355.
93. Ibid., 16–29.
94. Van Andel and Aronson, *Restoration Ecology*, 240.
95. Of course, such a neat theoretical division of sites into three scenarios will rarely match realities on the ground precisely; painful questions of triage, as well as unexpected outcomes, will always be part of the territory of restoration. But these issues can be minimized by a clear conceptual understanding of what the options are.

Bibliography

"After Wise use: The Future of Peatlands." Tullamore, Ireland: International Peat Society, 2008.

Aalen, F. H. A., Kevin Whelan, and Matthew Stout. *Atlas of the Irish Rural Landscape*. Cork: Cork University Press, 1997.

Adler, J. "Daniel H. Janzen." *Smithsonian* 36, no. 8 (2005): 88–89.

Allen, William. *Green Phoenix: Restoring the Tropical Forests of Guanacaste, Costa Rica*. New York: Oxford University Press, 2001.

Aronson, James, James N. Blignaut, Sue J. Milton, and Andre F. Clewell. "Natural Capital: The Limiting Factor." *Ecological Engineering* 28, no. 1 (2006): 1–5.

Aronson, James, Pedro H. S. Brancalion, Giselda Durigan, Ricardo R. Rodrigues, Vera L. Engel, Marcelo Tabarelli, José M. D. Torezan, et al. "What Role should Government Regulation Play in Ecological Restoration? Ongoing Debate in São Paulo State, Brazil." *Restoration Ecology* 19, no. 6 (2011): 690–95. doi:10.1111/j.1526–100X.2011.00815.x.

Aronson, James, Andre F. Clewell, James N. Blignaut, and Sue J. Milton. "Ecological Restoration: A New Frontier for Nature Conservation and Economics." *Journal for Nature Conservation* 14, no. 3–4 (2006): 135–39.

Aronson, James, S. Dhillion, and E. Floc'h. "On the Need to Select an Ecosystem of Reference, however Imperfect: A Reply to Pickett and Parker." *Restoration Ecology* 3, no. 1 (1995): 1–3.

Aronson, James, and Edouard Floc'h. "Hierarchies and Landscape History: Dialoguing with Hobbs and Norton." *Restoration Ecology* 4, no. 4 (1996): 327–33.

———. "Vital Landscape Attributes: Missing Tools for Restoration Ecology." *Restoration Ecology* 4, no. 4 (1996): 377–87.

Aronson, James, C. Floret, E. Floc'h, C. Ovalle, and R. Pontanier. "Restoration and Rehabilitation of Degraded Ecosystems in Arid and Semi-Arid Lands. I. A View from the South." *Restoration Ecology* 1, no. 1 (1993): 8–17.

———. "Restoration and Rehabilitation of Degraded Ecosystems in Arid and Semi-Arid Lands. II. Case Studies in Southern Tunisia, Central Chile and Northern Cameroon." *Restoration Ecology* 1, no. 3 (1993): 168–87.

Aronson, James, Sue Milton, and James Blignaut. "Conceiving the Science, Business, and Practice of Restoring Natural Capital." *Ecological Restoration* 24, no. 1 (2006): 22–24.

Aronson, James, Sue J. Milton, James N. Blignaut, and Andre F. Clewell. "Nature Conservation as if People Mattered." *Journal for Nature Conservation* 14, no. 3–4 (2006): 260–63.

Aronson, James, Suzanne J. Milton, and J. N. Blignaut; Society for Ecological Restoration International. *Restoring Natural Capital: Science, Business, and Practice*. Washington DC: Island Press; Tucson, AZ: Society for Ecological Restoration International, 2007.

Aronson, James, João S. Pereira, and Juli G. Pausas; Society for Ecological Restoration International. *Cork Oak Woodlands on the Edge: Ecology, Adaptive Management, and Restoration*. Washington DC: Island Press, 2009.

Arrow, Kenneth, Partha Dasgupta, Lawrence Goulder, Gretchen Daily, et al. "Are We Consuming Too Much?" *Journal of Economic Perspectives* 18, no. 3 (2004): 147–73.

Ballance, Alison, and Don Merton. *Don Merton: The Man Who Saved the Black Robin*. Auckland, NZ: Reed Books, 2007.

Batker, D., I. de la Torre, R. Costanza, P. Swedeen, J. Day, R. Boumans, and K. Bagstad. "Gaining Ground: Wetlands, Hurricanes, and the Economy: The Value of Restoring the Mississippi River Delta." *Environmental Law Reporter News and Analysis* 40, no. 11 (2010): 11106–10.

Berry, Wendell. *A Timbered Choir: The Sabbath Poems, 1979–1997*. Washington DC: Counterpoint, 1998.

Beschta, Robert L., and William J. Ripple. "Large Predators and Trophic Cascades in Terrestrial Ecosystems of the Western United States." *Biological Conservation* 142, no. 11 (2009): 2401–14. doi:10.1016/j.biocon.2009.06.015.

Blignaut, James, Christo Marais, and Jane Turpie, "Determining a Charge for the Clearing of Invasive Alien Plant Species (IAPs) to Augment Water Supply in South Africa," *Water SA* 33, no. 1 (2007): 27–34.

Blignaut, James, and Christina Moolman. "Quantifying the Potential of Restored Natural Capital to Alleviate Poverty and Help Conserve Nature: A Case Study from South Africa." *Journal for Nature Conservation* 14, no. 3–4 (2006): 237–48.

Blondel, Jacques, James Aronson, Jean-Yves Bodiou, and Gilles Boeuf. *The Mediterranean Region: Biological Diversity in Space and Time*. Oxford: Oxford University Press, 2010.

Brown, Kerry. *Restoring Kapiti: Nature's Second Chance*. Dunedin, NZ: University of Otago Press, 2004.

Bugalho M. N., M. C. Caldeira, J. S. Pereira, J. Aronson, and J. G. Pausas. "Mediterranean Cork Oak Savannas Require Human use to Sustain Biodiversity and Ecosystem Services." *Frontiers in Ecology and the Environment* 9, no. 5 (2011): 278–86.

Burns, M., M. Audouin, and A. Weaver. "Advancing Sustainability Science in South Africa." *South African Journal of Science* 102, no. 9/10 (2006): 379–84.

Butler, David, and Don Merton. *The Black Robin*. Oxford: Oxford University Press, 1992.

Butler, Tom, and Antonio Vizcaíno. *Wildlands Philanthropy: The Great American Tradition*. San Rafael, CA: Earth Aware, 2008.

Cabin, Robert J. *Intelligent Tinkering: Bridging the Gap between Science and Practice*. Washington DC: Island Press, 2011.

Calmon M., S. C. da Silva, P. H. S. Brancalion, A. Paese, J. Aronson, P. Castro, and R. R. Rodrigues. "Emerging Threats and Opportunities for Large-Scale Ecological Restoration in the Atlantic Forest of Brazil." *Restoration Ecology* 19, no. 2 (2011): 154–58.

Campbell, David G., John Guittar, and Karen S. Lowell. "Are Colonial Pastures the Ancestors of the Contemporary Maya Forest?" *Journal of Ethnobiology* 28, no. 2 (2008): 278–89.

———. "Maya Home Gardens as Biodiversity Hotspots." Paper presented at the VII Congreso Internacional de Mayistas, Merida, Mexico, 2007.

Caro Baroja, Julio. *The Basques*. Reno: Center for Basque Studies, University of Nevada, Reno, 2009.

Carson, Rachel. *Silent Spring*. London: Penguin, 1999.

Casavecchia, Attilio, and Enrica Salvatori. *Man's Park 1: History of a Landscape*. Riomaggiore: Cinque Terre National Park, n.d.

Chapin, F. Stuart. "Using Climate Change to Restore Biodiversity." Paper presented at the ESA/SER Joint Conference, San José, CA, August 6–10, 2007.

Chapin F. S. III, M. Hoel, S. R. Carpenter, J. Lubchenco, B. Walker, T. V. Callaghan, C. Folke, et al. "Building Resilience and Adaptation to Manage Arctic Change." *Ambio* 35, no. 4 (2006): 198–202.

Clarke, Donal. *Brown Gold: A History of Bord Na Móna and the Irish Peat Industry*. Dublin: Gill & Macmillan, 2010.

Clarkson, Bruce D., Merilyn F. Merrett, and Theresa Downs. *Botany of the Waikato*. Hamilton, NZ: Waikato Botanical Society, 2002.

Clewell, Andre F., and James Aronson; Society for Ecological Restoration International. *Ecological Restoration: Principles, Values, and Structure of an Emerging Profession*. Washington DC: Island Press, 2007.

Clewell, Andre F., James Aronson, and Keith Winterhalder; Society for Ecological Restoration. *The SER Primer on Ecological Restoration*. Tucson, AZ: Society for Ecological Restoration Science & Policy Working Group, 2004.

Costanza, Robert, Ralph d'Arge, Rudolf de Groot, Stephen Farber, Monica Grasso, Bruce Hannon, Karin Limburg, et al. "The Value of the World's Ecosystem Services and Natural Capital." *Nature* 387 (1997): 253–60.

Cox, Barry. "The Biogeographic Regions Reconsidered." *Journal of Biogeography* 28, no. 4 (2001): 511–23. doi:10.1046/j.1365-2699.2001.00566.x.

Craig, John, Sandra Anderson, Mick Clout, Bob Creese, Neil Mitchell, John Ogden, Mere Roberts, and Graham Ussher. "Conservation Issues in New Zealand." *Annual Review of Ecology and Systematics* 31 (2000): 61–78.

Cramer, Viki A., and Richard J. Hobbs; Society for Ecological Restoration International. *Old Fields: Dynamics and Restoration of Abandoned Farmland*. Washington DC: Island Press, 2007.

Cramer, Viki A., Richard J. Hobbs, and Rachel J. Standish. "What's New about Old Fields? Land Abandonment and Ecosystem Assembly." *Trends in Ecology & Evolution* 23, no. 2 (2008): 104–12. doi:10.1016/j.tree.2007.10.005.

Cronon, William. *Changes in the Land: Indians, Colonists, and the Ecology of New England*. 20th Anniversary ed. New York: Hill and Wang, 2003.

Cruz, Rosibel Elizondo and Róger Blanco Segura. "Developing the Bioliteracy of School Children for 24 Years: A Fundamental Tool for Ecological Restoration and Conservation in Perpetuity of the Área De Conservación Guanacaste, Costa Rica." *Ecological Restoration* 28, no. 2 (2010): 193–98.

Cullis, J. D. S., A. H. M. Gorgens, and C. Marais. "A Strategic Study of the Impact of Invasive Alien Plants in the High Rainfall Catchments and Riparian Zones of South Africa on Total Surface Water Yield." *Water SA* 33, no. 1 (2007): 35–42.

Cunningham, Deirdre. *Brackloon: The Story of an Irish Oak Wood*. Sandyford: Coford, 2005.

Daily G. C., J. Goldstein, H. A. Mooney, L. Pejchar, S. Polasky, P. M. Kareiva, T. H. Ricketts, J. Salzman, and R. Shallenberger. "Ecosystem Services in Decision Making: Time to Deliver." *Frontiers in Ecology and the Environment* 7, no. 1 (2009): 21–28.

Daily, Gretchen C. *Nature's Services: Societal Dependence on Natural Ecosystems*. Washington DC: Island Press, 1997.

Daily, Gretchen C., and Katherine Ellison. *The New Economy of Nature: The Quest to Make Conservation Profitable*. Washington DC: Island Press, 2002.

De Groot, Rudolf, et al. "Integrating the Ecological and Economic Dimensions in Biodiversity and Ecosystem Service Valuation." In *The Economics of Ecosystems and Biodiversity: Ecological and Economic Foundations (TEEB DO)*, edited by Pushpam Kumar, 3–36. London: Earthscan, 2010.

De Groot, R. S., M. A. Wilson, and R. M. Boumans. "A Typology for the Classification, Description and Valuation of Ecosystem Functions, Goods and Services." *Ecological Economics* 41, no. 3 (2002): 393–408.

De Lange, W. J., and B. W. van Wilgen. "An Economic Assessment of the Contribution of Biological Control to the Management of Invasive Alien Plants and to the Protection of Ecosystem Services in South Africa." *Biological Invasions* no. 12 (2010): 4113–24.

Diemont, S. A. W., and J. F. Martin. "Lacandon Maya Ecosystem Management: Sustainable Design for Subsistence and Environmental Restoration." *Ecological Applications* 19, no. 1 (2009): 254–66.

Donlan, J. "Re-Wilding North America." *Nature* 436, no. 7053 (2005): 913–14.

Douterlungne, D., S. J. Levy-Tacher, D. J. Golicher, and F. R. Danobeytia. "Applying Indigenous Knowledge to the Restoration of Degraded Tropical Rain Forest Clearings Dominated by Bracken Fern." *Restoration Ecology* 18, no. 3 (2010): 322–29.

Du Preez, M., S. Tessendorf, and S. G. Hosking. "Application of the Contingent Valuation Method to Estimate the Willingness-to-Pay for Restoring Indigenous Vegetation in Underberg, Kwazulu-Natal, South Africa." *South African Journal of Economic and Management Sciences* 13, no. 2 (2010): 135–57.

Ehrlich, Paul R., Peter M. Kareiva, and Gretchen C. Daily. "Securing Natural Capital and Expanding Equity to Rescale Civilization." *Nature* no. 468 (2012): 68–73.

Eliot, T. S. *Collected Poems 1909–1962*. London: Faber and Faber, 1963.

Elliot, Robert. *Faking Nature: The Ethics of Environmental Restoration*. London: Routledge, 1997.

Esler, Karen J., Patricia M. Holmes, David M. Richardson and Edward T.F. Witkowski, eds., "Riparian Vegetation Management in Landscapes Invaded by Alien Plants: Insights from South Africa." Special issue, *South African Journal of Botany* 74 (2008).

Eugenio, Figueroa B. "New Linkages for Protected Areas: Making them Worth Conserving and Restoring." *Journal for Nature Conservation* 14, no. 3–4 (2006): 225–32.

Everard, Mark, Rory Harrington, and Robert J. McInnes. "Facilitating Implementation of Landscape-Scale Water Management: The Integrated Constructed Wetland Concept." *Ecosystem Services* 2 (2012): 27–37. doi:10.1016/j.ecoser.2012.08.001.

Falk, Donald A. "Discovering the Future, Creating the Past: Some Reflections on Restoration." *Restoration & Management Notes* 8, no. 2 (1990): 71–72.

Falk, Donald A., Margaret A. Palmer, and Joy B. Zedler; Society for Ecological Restoration International. *Foundations of Restoration Ecology*. Washington DC: Island Press, 2006.

Farrell, Catherine A."Peatland Utilisation and Research in Ireland 2006." Proceedings of a symposium, October 10, 2006. 68pp. Tullamore, Ireland: Irish Peat Society, 2006.

———. "Review of Peatland Restoration in Ireland." *Peatlands International* 1 (2009): 18–21.

Farrell, Catherine. A., and G. J. Doyle. "Rehabilitation of Industrial Cutaway Atlantic Blanket Bog in County Mayo, North-West Ireland." *Wetlands Ecology and Management* 11, no. 1–2 (2003): 1–2.

Farrell, Catherine A., and University College Dublin. Department of Botany. "An Ecological Study of Intact and Industrial Cutaway Atlantic Blanket Bog at Bellacorick, North-West Mayo." PhD diss., University College Dublin, 2001.

Ford, Anabel, and Ronald Nigh. "Origins of the Maya Forest Garden: Maya Resource Management." *Journal of Ethnobiology* 29, no. 2 (2009): 213–36.

Fraser, Caroline. *Rewilding the World: Dispatches from the Conservation Revolution*. New York: Picador, 2010.

Friederici, Peter. *Nature's Restoration: People and Places on the Front Lines of Conservation*. Washington DC: Island Press, 2006.

Gardner, J. H., and D. T. Bell. "Bauxite Mining Restoration by Alcoa World Alumina Australia in Western Australia: Social, Political, Historical, and Environmental Contexts." *Restoration Ecology* 15 (2007): S3–S10.

Gobster, Paul H. "The Chicago Wilderness and its Critics." *Ecological Restoration* 15, no. 1 (1997): 32–37. doi:10.3368/er.15.1.32.

Gobster, Paul H., and R. Bruce Hull. *Restoring Nature: Perspectives from the Social Sciences and Humanities*. Washington DC: Island Press, 2000.

Good, Ronald. *The Geography of the Flowering Plants*. 3rd ed. New York: J. Wiley, 1964.

Green, Wren, and Bruce Clarkson. *Turning the Tide? A Review of the First Five Years of the New Zealand Biodiversity Strategy: The Synthesis Report*. Wellington, NZ: Dept. of Conservation, 2005.

Guthrie-Smith, H. *Sorrows and Joys of a New Zealand Naturalist*. Dunedin, NZ: A. H. and A. W. Reed, 1936.

Haffajee, Ferial. "DOSSIER—South Africa: Water for Everyone." *Unesco Courier* 52, no. 2 (1999): 26.

Hardin, Garrett. "The Tragedy of the Commons." *Science* 162, no. 3859 (1968): 1243–48.

Hardwick, Kate A., Peggy Fiedler, Lyndon C. Lee, Bruce Pavlik, Richard J. Hobbs, James Aronson, Martin Bidartondo, et al. "The Role of Botanic Gardens in the Science and Practice of Ecological Restoration." *Conservation Biology* 25, no. 2 (2011), 265–75.

Harris, James A., Richard J. Hobbs, Eric Higgs, and James Aronson. "Ecological Restoration and Global Climate Change." *Restoration Ecology* 14, no. 2 (2006): 170–76.

Heaney, Seamus. *Selected Poems, 1965–1975*. London: Faber & Faber, 1980.

Heneghan, Liam, Susan P. Miller, Sara Baer, Mac A. Callaham, James Montgomery, Mitchell Pavao-Zuckerman, Charles C. Rhoades, and Sarah Richardson. "Integrating Soil Ecological Knowledge into Restoration Management." *Restoration Ecology* 16, no. 4 (2008): 608–17.

Heneghan, Liam, Christopher Mulvaney, Kristen Ross, Lauren Umek, Cristy Watkins, Lynne M. Westphal, and David H. Wise. "Review: Lessons Learned from Chicago Wilderness—Implementing and Sustaining Conservation Management in an Urban Setting." *Diversity* 4, no. 1 (2012): 74–93.

Hickie, David and Mike O'Toole. *Native Trees & Forests of Ireland*. Dublin: Gill & Macmillan, 2002.

Higgs, Eric. *Nature by Design: People, Natural Process, and Ecological Restoration*. Cambridge, MA: MIT Press, 2003.

———. "The Two-Culture Problem: Ecological Restoration and the Integration of Knowledge." *Restoration Ecology* 13, no. 1 (2005): 159–64.

Hilderbrand, Robert H., Adam C. Watts, and April M. Randle. "The Myths of Restoration Ecology." *Ecology and Society* 10, no. 1 (2005). http://www.ecologyand society.org/vol10/iss1/art19/

Hobbs, Richard J. "The Future of Restoration Ecology: Challenges and Opportunities." *Restoration Ecology* 13, no. 2 (2005): 239–41. doi:10.1111/j.1526–100X.2005.00030.x.

———. "From Our Southern Correspondent." *Bulletin of the British Ecological Society* 43 no. 1 (2012): 40–41.

———. "Overcoming Barriers to Effective Public Communication of Ecology." *Frontiers in Ecology and the Environment* 4, no. 9 (November 2006): 496–97.

———. "Setting Effective and Realistic Restoration Goals: Key Directions for Research." *Restoration Ecology* 15, no. 2 (2007): 354–57. doi:10.1111/j.1526–100X.2007.00225.x.

———. "The Working for Water Programme in South Africa: The Science Behind the Success." *Diversity and Distributions* 10, no. 5 (2004): 501–3. doi:10.1111/j.1366–9516.2004.00115.x.

Hobbs, Richard J., Salvatore Arico, James Aronson, Jill S. Baron, Peter Bridgewater, Viki A. Cramer, Paul R. Epstein, et al. "Novel Ecosystems: Theoretical and Management Aspects of the New Ecological World Order." *Global Ecology and Biogeography* 15, no. 1 (January 2006): 1–7. doi:10.1111/j.1466–822X.2006.00212.x.

Hobbs, Richard J., Lauren M. Hallett, Paul R. Ehrlich, and Harold A. Mooney. "Intervention Ecology: Applying Ecological Science in the Twenty-First Century." *Bioscience* 61, no. 6 (2011): 442–50.

Hobbs, Richard. J. and J. A. Harris. "Restoration Ecology: Repairing the Earth's Ecosystems in the New Millennium." *Restoration Ecology* 9, no. 2 (2001): 239–46.

Hobbs, Richard J., Eric S. Higgs, and Carol M. Hall, eds. *Novel Ecosystems: Intervening in the New Ecological World Order*. (Hoboken, NJ: Wiley-Blackwell, 2013).

Hobbs, Richard. J., E. Higgs, and J. A. Harris. "Novel Ecosystems: Implications for Conservation and Restoration." *Trends in Ecology and Evolution* 24, no. 11 (2009): 599–605.

Hobbs, Richard J. and David A. Norton. "Towards a Conceptual Framework for Restoration Ecology." *Restoration Ecology* 4, no. 2 (1996): 93–110. doi:10.1111/1526–100X.ep11617675.

Hopper, Stephen D. "OCBIL Theory: Towards an Integrated Understanding of the Evolution, Ecology and Conservation of Biodiversity on Old, Climatically Buffered, Infertile Landscapes." *Plant and Soil* 322, no. 1–2 (2009): 1–2.

Hopper, Stephen D., Kate Hardwick, Peggy Fiedler, Lyndon Lee, Bruce Pavlik, and Richard Hobbs. "Globalising Restoration: A Role for Botanic Gardens." Plenary address at SER Conference, Perth, August 23–27, 2009.

Jackson, Stephen T. and Richard J. Hobbs. "Ecological Restoration in the Light of Ecological History." *Science* 325, no. 5940 (2009): 567.

Janzen, Daniel H. "Coevolution of Mutualism between Ants and Acacias in Central America." *Evolution* 20, no. 3 (1966): 249–75.

———. "Costa Rica's Area de Conservación Guanacaste: A Long March to Survival through Non-Damaging Biodiversity and Ecosystem Development." Paper presented at the Norway/UN Conference on the Ecosystem Approach for

Sustainable Use of Biological Diversity (Trondheim: Norway Norwegian Directorate for Nature Research and Norwegian Institute for Nature Research, 2000), 122–32.

———. "Gardenification of Wildland Nature and the Human Footprint." *Science* 279, no. 5355 (1998): 1312–13.

———. "Tropical Ecological and Biocultural Restoration." *Science* 239, no. 4837 (1988): 243–44.

Janzen, Daniel H. and W. Hallwachs. "Conservation History and Future of Area De Conservación Guanacaste (ACG) in Northwestern Costa Rica." In *Costa Rican Ecosystems*, edited by Maarten Kappelle. Chicago: University of Chicago Press, forthcoming.

———. "Joining Inventory by Parataxonomists with DNA Barcoding of a Large Complex Tropical Conserved Wildland in Northwestern Costa Rica." *PLoS ONE* 6, no. 8 (2011): e18123. doi:10.1371/journal.pone.0018123.

Janzen, Daniel H., W. Hallwachs, T. Dapkey, P. Blandin, J. M. Burns, J. P. W. Hall, D. J. Harvey, et al. "Integration of DNA Barcoding into an Ongoing Inventory of Complex Tropical Biodiversity." *Molecular Ecology Resources* 9, Suppl. 1 (2009): 1–26.

Janzen Daniel H., and P. S. Martin. "Neotropical Anachronisms: The Fruits the Gomphotheres Ate." *Science* 215, no. 4528 (1982): 19–27.

Jordan, William R. *The Sunflower Forest: Ecological Restoration and the New Communion with Nature*. Berkeley: University of California Press, 2003.

Jordan, William R., Michael E. Gilpin, and John D. Aber. *Restoration Ecology: A Synthetic Approach to Ecological Research*. Cambridge: Cambridge University Press, 1987.

Jordan, William R., and George M. Lubick. *Making Nature Whole: A History of Ecological Restoration*. Washington DC: Island Press, 2011.

Katz, Eric. "Restoration and Redesign: The Ethical Significance of Human Intervention in Nature." *Restoration & Management Notes* 9, no. 2 (1991): 90–96. doi:10.3368/er.9.2.90.

Kavanagh, Patrick, and Peter Kavanagh. *The Complete Poems*. New York: Peter Kavanagh Hand Press, 1972.

King, Michael. *The Penguin History of New Zealand*. Auckland, NZ: Penguin Books, 2003.

Koch, J. M., and R. J. Hobbs. "Synthesis: Is Alcoa Successfully Restoring a Jarrah Forest Ecosystem After Bauxite Mining in Western Australia?" *Restoration Ecology* 15, (2007): S137–S144.

Koenig R. "Unleashing an Army to Repair Alien-Ravaged Ecosystems." *Science* 325, no. 5940 (2009): 562–63.

Kumar, Pushpam, ed. *The Economics of Ecosystems and Biodiversity: Ecological and Economic Foundations (TEEB DO)*. London: Earthscan, 2010.

Lamb, A. J., ed. *Jonkershoek Forestry Research Centre, Pamphlet 384*. Pretoria, South Africa: Forestry Branch of the Dept. of Environmental Affairs, 1985.

Le Maitre, David C., B. W. van Wilgen, R. A. Chapman, and D. McKelly. "Invasive Plants and Water Resources in the Western Cape Province, South Africa: Modelling the Consequences of a Lack of Management." *Journal of Applied Ecology* 33 (1996): 161.

Leopold, Aldo, and Luna B. Leopold. *Round River: From the Journals of Aldo Leopold*. Oxford: Oxford University Press, 1993.

Leopold, Aldo, and Charles Walsh Schwartz. *A Sand County Almanac, and Sketches here and there*. London: Oxford University Press, 1968.

Little, D. J., F. J. G. Mitchell, S. von Engelbrechten, and E. P. Farrell. "Assessment of the Impact of Past Disturbance and Prehistoric Pinus Sylvestris on Vegetation Dynamics and Soil Development in Uragh Wood, SW Ireland." *Holocene* 6, no. 1 (1996): 90–99.

Macdonald, I. A. W. "Recent Research on Allen Plant Invasions and their Management in South Africa: A Review of the Inaugural Research Symposium of the Working for Water Programme." *South African Journal of Science* 100 (2004): 21–26.

Marais C., and A.M. Wannenburgh. "Restoration of Water Resources (Natural Capital) through the Clearing of Invasive Alien Plants from Riparian Areas in South Africa—Costs and Water Benefits." *South African Journal of Botany* 74, no. 3 (2008): 526–37.

Marais, C., B. W. van Wilgen, and D. Stevens. "The Clearing of Invasive Alien Plants in South Africa: A Preliminary Assessment of Costs and Progress." *South African Journal of Science* 100, (2004): 97–102.

Mariotti, Mauro. *Cinque Terre*. Genoa: Erga, 1998.

Matos Mederos, Jesús and Domingo Ballate Denis. *ABC De La Restauración Ecológica*. Santa Clara, Cuba: Editorial Feijóo, 2006.

Matthiessen, Peter. *Lost Man's River*. New York: Vintage Books, 1998.

Matthiessen, Peter, and Robert Bateman. *The Birds of Heaven: Travels with Cranes*. New York: North Point Press, 2001.

McQueen, Joanna C., ed. *An Ecological Restoration Plan for Maungatautari*. Cambridge, NZ: MEIT, 2004.

Meine, Curt. *Correction Lines: Essays on Land, Leopold, and Conservation*. Washington DC: Island Press, 2004.

Mello, Anna J., Patricia A. Townsend, and Katie Filardo. "Reforestation and Restoration at the Cloud Forest School in Monteverde, Costa Rica: Learning by Doing." *Ecological Restoration* 28, no. 2 (2010): 148–50.

Millennium Ecosystem Assessment. *Living Beyond Our Means: Natural Assets and Human Well-Being. Statement of the MA Board*. 2005. http://www.maweb.org/en/BoardStatement.aspx.

Mills, Anthony J., and Richard M. Cowling. "Rate of Carbon Sequestration at Two Thicket Restoration Sites in the Eastern Cape, South Africa." *Restoration Ecology* 14, no. 1 (2006): 38–49.

Miskelly, Colin M. "Ecological Restoration and Threatened Species Management in New Zealand." *Ecological Management & Restoration* 10, no. 2 (2009): 161–62.

———. "The Identity of the Hakawai." *Notornis* 34, no. 2 (1987): 95–116.

Miskelly, Colin M., and A. J. Baker. "Description of a New Subspecies of Coenocorypha Snipe from Subantarctic Campbell Island, New Zealand." *Notornis* 56, no. 3 (2010): 113–23.

Miskelly, Colin M., and Helen Gummer; New Zealand. Department of Conservation. *Third and Final Transfer of Fairy Prion (Titiwainui) Chicks from Takapourewa to Mana Island, January 2004*. Wellington, NZ: New Zealand Dept. of Conservation, 2004.

Miskelly, Colin M., Graeme A. Taylor, Helen Gummer, and Rex Williams. "Translocations of Eight Species of Burrow-Nesting Seabirds (Genera Pterodroma, Pelecanoides, Pachyptila and Puffinus: Family Procellariidae)." *Biological Conservation* 142, no. 10 (2009): 1965–80.

Miskelly, Colin M.; New Zealand Department of Conservation. *Mana Island Ecological Restoration Plan*. Wellington, NZ: New Zealand Dept. of Conservation, 1999.

Moreira, Francisco, A. Isabel Queiroz, and James Aronson. "Restoration Principles Applied to Cultural Landscapes." *Journal for Nature Conservation* 14, no. 3–4 (2006): 217–24.

Morris, Rod, and Hal Smith. *Wild South: Saving New Zealand's Endangered Birds*. Auckland, NZ: TVNZ, in association with Century Hutchinson, 1988.

Nair, C. T. S., and R. Rutt. "Creating Forestry Jobs to Boost the Economy and Build a Green Future." *UNASYLVA-FAO* no. 233 (2009): 3–10.

Nations, James D., and Ronald B. Nigh. "The Evolutionary Potential of Lacandon Maya Sustained-Yield Tropical Forest Agriculture." *Journal of Anthropological Research* 36, no. 1 (1980): 1–30.

National Parks and Wildlife Service. *The Status of EU Protected Habitats and Species in Ireland*. Short Reports. Department of the Environment, Heritage, and Local Government, 2008, 47–49. http://www.npws.ie/publications/eu conservationstatus/NPWS_2007_Conservation_Status_Report.pdf.

New Zealand Parliament. "Treaty of Waitangi Settlements Process." August 23, 2006. http://www.parliament.nz/en-NZ/ParlSupport/ResearchPapers/4/9/2/492fd68a5b8340fd931db848731d3eed.htm.

Norton, David A., and Craig J. Miller. "Some Issues and Options for the Conservation of Native Biodiversity in Rural New Zealand." *Ecological Management & Restoration* 1, no. 1 (2000): 26–34. doi:10.1046/j.1442-8903.2000.00005.x.

Ó Tuama, Seán, and Thomas Kinsella. *An Duanaire, 1600–1900: Poems of the Dispossessed*. Portlaoise, Ireland: Dolmen Press, 1981.

Packard, Steve. "Chronicles of Restoration." *Ecological Restoration* 6, no. 1 (1988): 13–22. doi:10.3368/er.6.1.13.

———. "No End to Nature." *Ecological Restoration* 8, no. 2 (1990): 72. doi:10.3368/er.8.2.72.

———. "Restoring Oak Ecosystems." *Ecological Restoration* 11, no. 1 (1993): 5–16. doi:10.3368/er.11.1.5.

———. "Successional Restoration." *Ecological Restoration* 12, no. 1 (1994): 32–39. doi:10.3368/er.12.1.32.

Perrow, Martin R., and A. J. Davy, eds. *Handbook of Ecological Restoration*. 2 vols. Cambridge: Cambridge University Press, 2002.

Peterken, G. F. *Natural Woodland: Ecology and Conservation in Northern Temperate Regions*. Cambridge: Cambridge University Press, 1996.

———. *Woodland Conservation and Management*. London: Chapman and Hall, 1981.

Pickett, S. T. A., and V. Thomas Parker. "Avoiding the Old Pitfalls: Opportunities in a New Discipline." *Restoration Ecology* 2, no. 2 (1994): 75–79.

Rackham, Oliver. *Woodlands*. London: Collins, 2006.

Rey Benayas, J. M., A. C. Newton, A. Diaz, and J. M. Bullock. "Enhancement of Biodiversity and Ecosystem Services by Ecological Restoration: A Meta-Analysis. *Science* 325, no. 5944 (2009): 1121–24.

Renou-Wilson, Florence, T. Bolger, C. Bullock, F. Convery, J.P. Curry, S. Ward, D. Wilson, and C. Müller. *BOGLAND—Sustainable Management of Peatlands in Ireland*. STRIVE Report No. 75, prepared for the Environmental Protection Agency, Johnstown Castle, County Wexford, Ireland, 2011.

Roberts, L., R. Stone, and A. Sugden. "The Rise of Restoration Ecology." *Science* 325, no. 5940 (2009), 555.

Roche, Shauna, Kingsley W. Dixon, and John S. Pate. "For Everything a Season: Smoke-Induced Seed Germination and Seedling Recruitment in a Western Australian Banksia Woodland." *Austral Ecology* 23, no. 2 (1998): 111–20.

Rokich, Deanna P., Kingsley W. Dixon, K. Sivasithamparam, and Kathy A. Meney. "Smoke, Mulch, and Seed Broadcasting Effects on Woodland Restoration in Western Australia." *Restoration Ecology* 10, no. 2 (2002): 185–94.

Román Dañobeytia, Francisco. *Raíces Mayas para la restauración de selvas*. DVD. Instituto Nacional de Ecología, 2011.

Ross, Laurel M. "The Chicago Wilderness and its Critics." *Ecological Restoration* 15, no. 1 (1997): 17–24. doi:10.3368/er.15.1.17.

———. "Illinois' Volunteer Corps." *Restoration & Management Notes* 12, no. 1 (1994): 57–59. doi:10.3368/er.12.1.57.

Saunders, Alan J.; New Zealand Department of Conservation. *A Review of Department of Conservation Mainland Restoration Projects and Recommendations for further Action*. Wellington, NZ: New Zealand Dept. of Conservation, 2000.

Schama, Simon. *Landscape and Memory*. London: Fontana Press, 1996.

Schumacher, E. F. *Small Is Beautiful; Economics as if People Mattered*. New York: Harper & Row, 1973.

Seastedt, Timothy R., Richard J. Hobbs, and Katharine N. Suding. "Management of Novel Ecosystems: Are Novel Approaches Required?" *Frontiers in Ecology and the Environment* (2008): 6. doi: 10.1890/070046.

Standish, Rachel J., and Richard J. Hobbs. "Restoration of OCBILs in South-Western Australia: Response to Hopper." *Plant and Soil* 330, no. 1–2 (2010): 15–18.

Stevenson, Michael J. "Problems with Natural Capital: A Response to Clewell." *Restoration Ecology* 8, no. 3 (2000): 211–13.

Stevens, William K. *Miracle under the Oaks: The Revival of Nature in America*. New York: Pocket Books, 1995.

Suding, Katharine N., and Richard J. Hobbs. "Threshold Models in Restoration and Conservation: A Developing Framework." *Trends in Ecology & Evolution* 24, no. 5 (2009): 271–79. doi:10.1016/j.tree.2008.11.012.

Tacher, S. I. L., and J. D. Golicher. "How Predictive Is Traditional Ecological Knowledge? The Case of the Lacandon Maya Fallow Enrichment System." [In English.] *INTERCIENCIA-CARACAS* 29 (2004): 496–503.

Tennyson, Alan J. D., and Paul Martinson. *Extinct Birds of New Zealand*. Wellington, NZ: Te Papa Press, 2006.

Thoreau, Henry David, and Carl Bode. *The Portable Thoreau: With an Introduction*. Harmondsworth, UK: Penguin Books, 1981.

Towns, D. R., Charles H. Daughtery, and I. A. E. Atkinson. "Ecological Restoration of New Zealand Islands." Papers presented at the Conference on Ecological Restoration of New Zealand Islands, University of Auckland, November 20–24, 1989, Auckland, NZ: New Zealand Dept. of Conservation, 1990.

Turner, R. Eugene, Ann M. Redmond, and Joy B. Zedler. "Count it by Acre or Function—Mitigation Adds Up to Net Loss of Wetlands." *National Wetlands Newsletter* 23, no. 6 (2001): 5–16.

Turpie, Jane K., Barry J. Heydenrych, and Stephen J. Lamberth. "Economic Value of Terrestrial and Marine Biodiversity in the Cape Floristic Region: Implications for Defining Effective and Socially Optimal Conservation Strategies." *Biological Conservation* no. 112 (2003): 233.

Turpie Jane K., C. Marais, and J. N. Blignaut. "The Working for Water Programme: Evolution of a Payment for Ecosystem Services Mechanism that Addresses both Poverty and Ecosystem Service Delivery in South Africa." *Ecological Economics* 65, no. 4 (2008): 788–98.

Turvey, Sam. *Holocene Extinctions*. Oxford: Oxford University Press, 2009.

Van Andel, J. and James Aronson. *Restoration Ecology: The New Frontier*. Malden, MA: Blackwell, 2006.

van Wilgen, B. W. "Working for Water—Scientific Challenges in the Field of Invasive Alien Plant Management." *South African Journal of Science*. 100, no. 1 (2004): 19.

van Wilgen, B. W., M. P. de Wit, H. J. Anderson, D. C. Le Maitre, I. M. Kotze, S. Ndala, B. Brown, and M. B. Rapholo. "Costs and Benefits of Biological Control of Invasive Alien Plants: Case Studies from South Africa." *South African Journal of Science* 100 (2004): 113–22.

Viney, Michael. *Ireland: A Smithsonian Natural History*. Belfast: Blackstaff Press, 2003.

Wapner, Paul Kevin. *Living through the End of Nature: The Future of American Environmentalism*. Cambridge, MA: MIT Press, 2010.

"Welcome to the Anthropocene." *Economist*, May 26, 2011.

Whisenant, S., A. D. Bradshaw, J. L. Craig, W. Jordan, S. L. Pimm, D. S. Saunders and M. B. Usher. *Repairing Damaged Wildlands: A Process-Orientated, Landscape-Scale Approach*. Cambridge: Cambridge University Press, 1999.

Whitman, Walt. *Leaves of Grass and Other Writings: Authoritative Texts, Other Poetry and Prose, Criticism*. New York: Norton, 2002.

Wilson, David. "Climate Change, Carbon and Irish Peatlands." In *Peatland Utilisation and Research in Ireland*, ed. Catherine Farrell, 56–60 (Gorey: Irish Peat Society, 2006).

Wilson, David, Jukka Alm, Jukka Laine, Kenneth A. Byrne, Edward P. Farrell, and Eeva-Stiina Tuittila. "Rewetting of Cutaway Peatlands: Are we Re-Creating Hot Spots of Methane Emissions?" *Restoration Ecology* 17, no. 6 (2009): 796–806.

Wolf, Gary. "Barcode of Life—A Team of Insect Researchers has Invented a Device to Identify Every Living Creature. so Why do Other Biologists Hate the Idea?" *Wired*. 16, no. 10 (2008): 200.

Woodworth, Paddy. "What Price Ecological Restoration?" *Scientist*. 20, no. 4 (2006): 38.

———. "Working for Water in South Africa: Saving the World on a Single Budget?" *World Policy Journal* 23, no. 2 (2006): 31–44.

Worthy, T. H. "What was on the Menu—Avian Extinction in New Zealand." *New Zealand Journal of Archaeology* 19 (1999): 125–60.

Index

acacia, 50, 52, 65, 235, 249
acahual, 327, 340–41, 344, 347–48
Afghanistan, 22
African National Congress (ANC), 54–55, 57, 65–66
agave, 164, 216
Albany, 226–27
Alcoa World Alumina, 218–23, 416, 428
alder, 173, 178, 180, 182, 184–85, 188–89
Alexander River, 35
Allen, Bill, 261
Amazon River, 87, 114, 406
anthropocene era, 32, 110, 159, 300, 386, 396, 426
Archibald, George, 3, 14, 16, 22
Arctic Council, 211
Área de Conservación Guanacaste (ÁCG), 256–57, 261–67, 269, 271–72, 275–81, 283–86, 391
Aronson, James, 21, 29, 35–36, 38, 42, 45, 47, 133, 196, 201, 211, 267, 384–85, 395, 398, 400, 406–10, 412–13, 420, 423, 429
ash, 98, 106, 170–72, 176–77, 347
Asmal, Kadar, 49, 56–59, 66, 70
ATLANTIC (Alliance to Let Nature Take Its Course), 113
Atlantic Rainforest Restoration Pact, 392
Attenborough, David, 274
Auckland, 295–96, 306, 321
Australia, 82–83, 160, 164, 200, 214–55, 268, 288, 300, 308, 401, 416
  southwest Australia, 200, 222, 224–26, 246, 248, 250, 416
  Western Australia, 214, 217–18, 220–22, 227, 232–33, 251, 416, 427
Australian Wilderness Society, 243
Avonbeg River, 437
Azerbaijan, 22

Baker, Allan J., 314
Balaban, Jane, 124, 126
Balaban, John, 124, 126
bald eagle, 13, 23
Ballinacor Estate, 437
Ballycon bog, 381, 383
Ballygannon Woods (Ballygannon), 194–95, 438
Ballykine Castle, 176
Balsa (*Ochroma pyramidale*, *chujúm*), 341, 344–45, 347–48
Baraboo, 3, 16
Baraboo Hills, 15
Baroja, Julio Caro, 327
Barzen, Jeb, 1, 3–4, 8–9
Basque conflict, 87, 429
Basque Country (Basque people), 26, 327
Baviaanskloof (Baboon Canyon), 60, 62, 64, 76

Beara, 180
beech, 73, 171, 175–77, 192, 310–11, 401
Beilfuss, Rich 1, 9–10
Bellacorick, 351–53, 359, 361, 364–66
Bellamy, David, 357
Benedetti, Silvio, 145
Berg River, 75
Berlusconi, Silvio, 150
Berre Lagoon, 411
Berry, Wendell, 166
Bertram, Lise, 154
Betz, Bob, 99
Big South Cape Island, 293, 312–14, 324
biocontrol, 82–84
biodiversity, 5, 30–31, 33, 50, 53, 55, 101, 109, 110, 125, 129, 131, 133, 136, 138–43, 158, 160–161, 163, 164, 165, 169, 171, 175, 178, 182, 183, 184, 189, 192, 200, 209, 211, 218, 219, 225, 228, 232, 233, 234, 235, 243, 245, 250, 252, 257, 259, 260, 270, 271, 273, 280, 283, 300, 302, 310, 320, 322, 324, 325, 344, 353, 354, 358, 359, 361, 364, 366, 374–75, 382, 383, 388, 389, 390, 393, 396, 403, 408, 421, 427, 429, 430, 434
biodiversity hotspot, 226, 236, 247
biosphere reserve, 232, 328
birch, 169–71, 173, 176, 180–83, 185, 188, 194, 373–75
Birman, Bathsheba, 87, 114
black robin, 294
Blake, William, 36
Blanco, Róger, 269, 277–84
blanket bogs, 174, 182, 351–58, 360–61, 363–64, 367, 379–80
Blighe, Paudie, 166, 171, 174
Blignaut, James, 21, 36, 47, 86
Blix, Petra, 97, 123–28, 131
Blom, Gertrude, 330
Blouberg, 58
Bluff Knoll/Bular Meila, 232
Bohr, Niels, 33
Boland, Jackie, 123–24, 126
Bonanini, Franco, 140–41, 144–51, 153–54, 157–58, 161–62, 164
Bonilla, Alexander, 283
Bord na Móna, 351–52, 357–60, 367, 373, 375, 378, 380–83
Botha, P. W., 56
Boundary Stream Mainland Island (BSMI), 292
Bournay, Emmanuelle, 41
Bowers, Keith, 21, 35
bracken, 181, 183, 186, 327–28, 344, 348
Brackloon Woods, 166, 168–70, 172–76, 179–80, 184, 187, 193
Bradby, Keith, 225, 227, 229–30, 232–34, 237–44, 246, 252–54, 390
Bradshaw, Anthony D., 100, 393
Brazil, 392
Bremner, Kobey, 289
Brewer River, 228
Brown, Phil, 303
Buena Vista (Georgia), 18
Bugalho, Miguel N., 133, 160
Bunker Hill Forest Preserve, 116–17, 121, 123, 125
Burren, 175, 335
Bush, George H. W., 415
Bush Heritage Australia, 234, 236

Cacao, 273
Cahorra Bassa dam (Mozambique), 9–10
Caldeira, Maria C., 133
California, 63, 73, 142
Calumet River, 90, 128
Campbell Island, 314, 318
Canada, 20, 23, 354, 377, 391–92
Caoine Cill Cháis, 195
Cape Leeuwin, 226
Cape Naturaliste, 226
Cape to Cape Catchments Group, 226
Cape Town, 54–55, 63, 71
Cape Town University, 73
carbon, 30–31, 61, 75, 105, 122, 206, 208, 212, 228, 235, 375, 377–78
Carmona, Felix, 271–76
carrying capacity, 46, 61, 335–36, 357, 434, 438
Castellanos Chankin, Don Manuel, 327, 331–34, 337, 340, 343–50

Castlearchdale, County Fermanagh, 192
Castro, Rónald, 271–72
cats, 227, 245, 289–92, 294, 300, 312–13
Catskill-Delaware watershed restoration, 37
Centre d'Ecologie Fonctionnelle et Evolutive (Montpelier), 407
Ch'ol people, 329, 331, 342–43, 348, 350
Chankin, Adolfo, 346
Chapin, F. Stuart "Terry," 210
charismatic megafauna, 22
Chassahowitzka National Wildlife Refuge, 19
Chatham Islands, 294, 312–13, 315–17
Chavarría, María Marta, 285–86
Chesapeake Bay Restoration Program, 392
Chiapas, 328–29, 349
Chicago, 73, 87–132, 156, 171, 217, 299, 399, 405
Chicago Wilderness, 87, 112, 125–28, 130–31
Chikwaukee Prairie, 128
China, 22
Choi, Young D., 207, 373, 414
Cinque Terre, 133–65, 391
Citrus County (Florida), 19
Civilian Conservation Corps, 128
Clara, Vale of, 194–95, 438
Clara Bog, 337, 355, 357
Clarkson, Bruce, 287, 297–301, 303–5, 311, 320–21, 391
classic ecosystem, 37, 395, 401, 404
Clewell, Andre F., 29, 384–85, 399, 412, 423
climax ecosystem, 385
Clinton, Bill, 66
Clonbur Woods, 175–76, 178–79
CNRS (Centre national de la recherche scientifique), 407
Coffey, Raymond, 113, 115, 123
Coillte, 166, 169–71, 175–79, 182, 192, 358, 365, 369, 373–76, 381
Cois Abhainn, 437
Collins, Kevin, 193–94
Colquhoun, Ian, 222
Comprehensive Everglades Restoration Plan (CERP), 66, 392
Conradie, Beatrice, 73
conservation biology, 266, 408, 423
Conservation Reserve Program (Montana), 25
consumerism, 37, 387, 395
Cook County (Illinois), 89, 91, 94, 96, 112–15, 117–18
Corcovado National Park, 267
Corkery, Donal, 166, 182–84, 186–91, 194
Corkery, Peggy, 182–83
Corniglia, 136, 138, 156
Corniolo, 151–52, 155–57
Costa Rica, 209, 256, 263–65, 267–68, 270–71, 281, 286
Costanza, Robert, 21, 28, 33, 36–37, 46
Coulter, Ann, 333
Council for Scientific and Industrial Research (South Africa), 71
Cowles, Henry, 90, 92
Cowling, Richard, 49, 56, 66, 68, 74, 81
Craig, John, 292, 321–22, 324, 391
Crann, 192
Cross, John, 194
Cuajiniquil, 285
Cuba, 391–92
culling, 114, 126, 405
Cumeenaloughan, 179
Cummenadillure, 188
Cunningham, Deirdre, 169–71, 187
Curtis prairie, Madison arboretum, 24–25
Czech Republic, 429

D'Arcy, Gabriel, 382
Daly, Herman E., 45
Dana, Dick, 15–16
Dana, Jane, 15
Dandalup Dam, 224
Dane County (Wisconsin), 8
Danobeytia, Francisco Roman, 341
Darling Range, 218, 220
Darling Scarp, 219
Davis, Janny, 315

deer, 7, 22, 89, 114, 126, 181, 187–88, 194, 289–94, 304, 355, 405
Deer Grove (Cook County, Illinois), 89
Del Oro fruit juice company, 280–83
Delaney, Michael, 373–74
Delta Program, 9, 22
DePaul University, 88, 129
Deutsche Bank, 44
Diamond, Jared M., 287, 291
Díaz, Julio, 278
Dikana, Bulelwa, 51
Dish Mountain, 179, 184
Dixon, Kingsley, 214, 216–17, 220–23, 239, 246–48, 426–29, 433
dogs, 113, 128, 141, 269–70, 290, 292, 310, 328, 336
Doñana, 137
Donegal, County, 437
Dos Ríos (Costa Rica), 276
Douglas fir, 172, 175
Droste, Bernd von, 232
Ducks Unlimited, 317
Duff, Joe, 12–13, 16, 18–20
DuPage County (Illinois), 89, 113–14

Eades, Eugene, 241–2
Easter Island, 385, 387, 435
Eastern Cape Province (South Africa), 60, 78
Ecocoffins: Reducing the Burden of Bereavement, 50
ecological engineering, 27
ecological restoration
  accountability, 164, 336
  adaptive management, 24, 30, 74, 86, 250, 325, 341, 366, 410
  agreed terminology (definitions), 409
  as "acid test" (for ecological theory), 100, 393
  "biocultural restoration," 269, 285–86
  catastrophe, 45, 66, 139
  citizens, 8, 44, 88, 95–96, 102–3, 112, 114, 130, 215–17, 242, 289, 291–92, 298–300, 321, 428
  climate change, 1, 26, 33–35, 40, 67, 82, 105, 110, 128, 137, 163–64, 190–91, 196–98, 201, 203–4, 206–7, 209–13, 230, 263, 273, 276, 376, 378, 382, 384, 396, 416, 420, 425–26
  communication, 30, 32–33, 121, 154, 156, 438
  community/communities (ecological), 7, 202, 236, 395, 396, 397, 409, 424
  community/communities (human), 10, 33, 38–40, 57, 61, 93, 102–3, 113, 121–22, 129, 130–31, 150, 231–32, 242, 268, 298–300, 302, 309, 216, 353, 358, 368, 382–83, 404–5, 411, 413, 435
  complexity (ecological), 21, 29, 31, 33, 35, 41–42, 103, 138, 198, 203–4, 212, 217–18, 248, 252, 366, 378, 396, 416, 429
  conflict, 51, 69, 89, 96, 122, 131–32, 382, 405, 414, 429
  conservation, 1, 3, 4, 6, 7, 9, 10, 17, 22, 23, 32, 34, 42, 45, 47, 54, 58, 72, 75, 81, 90, 95, 101, 109, 112, 128, 131, 132, 133, 161, 162, 169, 172, 175, 191, 194, 198, 201, 202, 210, 226, 231, 232, 237, 241, 244, 250, 251, 252, 253, 254, 256, 259, 261–66, 267, 268, 269, 270, 273, 276, 283, 284, 287–326, 338, 356, 357, 360, 365, 367, 369, 376, 380, 386, 388, 389–90, 391, 392–93, 396, 403, 408, 417, 420, 421, 429, 430, 432, 434
  corridor/s, 37, 209–10, 233, 245–46, 250–54, 275–76, 343, 364, 390
  counterintuitive strategies, 22, 105, 137, 359, 361, 379
  cultural landscapes, 111, 133, 137, 159–60, 164, 197, 413
  defeatism, 397, 419
  degradation, 78, 247
  democracy, 96, 108, 116, 133, 155, 164
  "ecocentric restoration trend," 395–406, 432, 436
  ecosystems of reference, 34, 412

education, 58, 80, 91, 115, 120, 175, 177, 245, 260, 269, 299
emerging ecosystems, 364, 415
engagement with nature, 406, 435–37
evolution, 1, 31, 114, 120, 203, 208, 225, 233, 424
failure, 6, 20, 25, 67, 74, 85, 198, 223, 228, 280, 318–19, 398, 426
farmers and farming, 10, 25, 182–91, 228–36, 241, 249, 251, 254, 296, 301–2, 306–7, 309–10, 328–29, 339–42, 345, 357, 408, 438
fire, 4, 35–36, 50, 61, 63, 67, 72, 75, 78, 81–82, 84, 91–92, 99–100, 103–4, 106–7, 111, 113, 115, 122, 124–25, 185, 265, 267, 269, 271, 274, 276–79, 282, 296–97, 348, 352, 405, 430
fragmentation (and scale), 24, 107, 125, 128, 155–56, 252, 273
fundamentalism, 123
gardening, 5, 106–7, 408
global change, 416, 421, 423–25, 432
global strategy, 391–92
habitat creation, 30
historical ecosystems, 34, 96, 201, 204, 208, 323, 397, 402, 415–19, 425
historical range of variation, 30–31, 197, 396, 411, 418, 422, 424–26, 432
historical reference systems, 6, 8, 30, 92, 139, 179, 197, 199–200, 204, 207, 211–12, 258–59, 317, 364, 374, 379, 389, 395–6, 398, 410–14, 416, 420, 422, 424, 430, 433
historical trajectory, 29–30
hubris, 102, 196, 203, 397, 414–15, 421, 424, 431, 437
human absence, 268, 402
human presence, 133, 160, 233, 260, 268, 402
human relationship with the natural world, 2, 8, 38, 138, 337, 349, 386, 388, 434
human values, 107, 413
humility, 252, 421, 431
hunting, 134, 148, 173, 263, 267, 269–70, 276, 289, 291, 293, 304, 306, 324, 336–37, 343, 388, 390, 405
hybrid ecosystems, 199, 418, 431
"intelligent tinkering," 7
invasive alien plants (IAPs), 24–25, 52–58, 60–61, 64, 67–72, 74–76, 79–80, 82, 84–85, 110, 198, 200, 208, 405–6
investment, 75, 221, 419, 426–27
Iowa, 9
irreversible changes, 39, 199, 213, 415, 418, 426–32
legislation, 392
"mainland islands," 288, 292, 295, 297, 301–2, 312, 317, 319–20, 324–25
mainstreaming nature, 66, 81, 392
mammal pests, 289, 307
metaphor, 7, 30, 38–39, 111, 197, 408
microfauna and microflora, 35, 202, 370
monitoring, 24, 30, 74–76, 78, 85, 125, 162, 164, 168–69, 173, 177, 237, 311, 328, 344, 374, 393
moral hazard, 397, 423, 432
native invasive plants, 106
natural succession, 6, 31, 90, 92, 109, 139, 174, 246, 248–49, 259, 264, 267, 300, 340
no-analog ecosystems, 32, 34, 198, 211, 397, 436
nostalgia, 110–11, 201, 384, 397, 435
novel ecosystems, 26, 34, 110, 196, 199–201, 212, 236, 323, 384, 396–97, 409, 413–15, 417–22, 424–32, 436
"novel ecosystems trend," 396, 413–33
pain, 354, 397, 405
paradigm shifts, 260, 387
paradox, 35, 83, 214–17, 256, 259–60
philosophy, 122, 211, 401–6
poisons, 229, 287, 291–93, 300, 303–4, 306–7, 325–26, 348, 374, 438
preservation, 7, 91–92, 182, 251, 266, 317, 322–25, 383
private land, 58, 78–79, 150–56, 235, 321–24, 367–70, 375

ecological restoration (*continued*)
productive land, 390–91
public land, 57, 95, 102–5, 108, 120, 122
reallocation, 409
reconnection, 203, 246, 250, 253
rehabilitation, 30, 85, 220, 351, 358–66, 380–83, 409
relative immaturity as a discipline, 319
replicated experiments, 98, 101, 130, 252, 283, 323, 338
research, 381
ritual, 120, 309, 336, 405
"RNC (Restoration of Natural Capital) trend," 396, 403, 406–13, 423, 436
"shame," 120, 404–5
social capital, 37, 42, 48, 57, 69–70, 86
social restoration, 232, 408
species reintroduction, 21–22, 190, 323–24, 438
spiritual dimension, 38–39, 41, 43, 102, 309, 403, 436
spontaneous restoration, 25, 77, 138–39, 248, 283
stakeholders, 244, 411, 413
stewardship, 183, 310, 321, 337
"straw man" arguments, 421, 423–24
"studied disregard for human interests," 395, 399, 401
surrogate (homologue) species, 315, 318–19
target/s, 8, 21–22, 26, 30–31, 33, 69, 91, 95, 151, 173–74, 196–213, 317–20, 395–433
thresholds, 31–32, 39, 86, 213, 408, 418, 420, 431–32, 187, 199, 435
traditional agricultural practices, 137–38, 146–47, 152, 328, 355
traditional ecological knowledge (TEK), 156, 242, 334–35, 337–38, 343, 350
translocation, 202–3, 293–94, 312–13, 316–17, 319, 334
"tragedy of the commons," 357, 367
trapping, 248, 290–91, 320
triage, 125, 189, 209, 354
"unnatural" methods, 22–23, 325–26
unpredictability, 1, 5, 33, 106–7, 204, 424
urban restoration, 87–132, 214–17, 297–301
volunteers, 77, 88–89, 95, 102–3, 113, 116–18, 124, 126–27, 130–31, 209, 302, 428
wilderness, 2, 7–8, 14, 87, 92, 112, 125–28, 130–31, 226, 229, 243, 385–88, 405–6, 433
"wildland garden," 256, 260
"zoo impression," 303–4
Ecological Restoration, 24, 137, 384, 385, 399
Ecological Restoration: Principles, Values and Structure of an Emerging Profession, 412–13
Economics of Ecosystems and Biodiversity, The (TEEB), 44, 392, 403
Ecosur, 329, 337
ecosystem goods and services, 31, 37–39, 42, 46, 51, 200, 284, 386, 393, 395, 411, 432, 436
ecotourism, 260, 268, 270, 333–34, 342, 383
Edenderry, 381
Edgebrook (Chicago), 114
Egan, Dave, 24, 26, 137
Eliot, T. S., 384
Emerson, Ralph Waldo, 404
End of Nature, 112
Erlich, Paul R., 423
Erris (Bangor), 353, 365
Erris Peninsula, 352
Esler, Karen, 85
Estonia, 354
eucalypts, 50–52, 73, 83, 215, 218, 226–28, 235–36, 239, 243, 249, 299
European buckthorn, 73, 88, 94, 98
European Commission, 376
European Union (EU), 186, 202, 357

European Union (EU), Habitats Directive 175, 181–82, 358–59, 376
European Union (EU), LIFE-Nature program 175, 177, 373
European Union (EU), Special Area of Conservation (SAC) 169, 181, 194, 358–59, 367–69
Evangelisti, Massimo, 133, 151, 153–54, 156
extinctions, 22–23, 54, 137, 180, 190, 210, 216, 244, 259–60, 288, 290–91, 293–94, 310, 312, 314, 318, 324–25, 335–36, 350–51, 356, 375, 381, 383, 386, 432

Farley, Joshua, 46
Farrell, Catherine, 351, 359–68, 374–76, 379, 381
Faulkner, Erin, 97
fen, 126, 175, 177, 354–56, 365
Fermilab, 93
Fernández, María, 285–86
Finland, 354
Fiordland, 294, 319
Fitz-Stirling, 230
Fitzgerald River, 227, 232–34, 237, 239, 241, 245
Fitzgerald River National Park, 227, 232, 253
Florence, 162
Florida, 18–20, 66
flux paradigm, 410
Food and Agriculture Organization, 392
Ford, John, 175
forests, 22, 37, 43, 52, 90, 91, 92, 93, 105, 137, 139, 168–69, 180–81, 191, 192–93, 208, 212, 225, 227, 232, 245, 267, 273, 304, 305, 323, 358, 385, 388, 401, 434
Forest Preserve District (FPD) (Cook County, Illinois), 89, 91–92, 94, 100, 103, 113–19, 124, 126
Four Seasons Resort (Peninsula Papagayo), 279
Frelimo, 10
Fremantle, 236, 415
Frost, Robert, 144
fynbos, 53–56, 64, 69, 71–72, 76–78, 85–86, 200, 208, 421
Fynbos Forum, 53–56, 64, 71

Galapagos Islands, 232
Galway, Co., 175
Gann, George, 212
Garden Route (South Africa), 52, 65
Gardiner, John, 222
Garland, Rex, 307
garlic mustard, 94, 116, 124
Genoa, 134, 141–42, 147, 157
George (South Africa), 73
Georgia, USA, 16–18
Gingin, 249
Glendalough, 166–68
Gleninchaquin, 179, 191, 193
gmelina, 273–75
Gnarlbine Rock, 244–45
Gobster, Paul, 107, 121–23, 129
Golfo dei Poeti, 136
Gómez Sánchez, Antonio (Toño), 327–28, 333–34, 344
Gondwana Link, 225–28, 230–34, 237–39, 241, 243–44, 246–47, 250–55, 391, 416, 428
Gondwanaland, 214, 225, 246
Good, Ronald, 53
Gouza, Jennifer, 78
Grand Rapids, Michigan, 274
Great Lakes Restoration Initiative, 392
great spotted woodpecker, 195, 438
Great Western Woodlands, 227, 234, 243, 254
Greater Cape Floristic Region, 226, 421
Greater Cape Region, 248
Green Phoenix, 261
Green, Gemma, 303
Greene prairie, Madison Arboretum, 24
Greening Australia, 225
greenwashing, 220
grevillea, 225, 233, 235
Grevillea, 233
Groot River, 60

Guam, 291
Guatemala, 265, 331
Guinness family, 176
Guthrie-Smith, Herbert, 324
Gutiérrez, Milena 271–75, 281
Guvano, 139

hach winik, 327, 330–31, 341–42, 346–47
hakawai, 311, 313–16, 326
Hall, Carol M., 431
Hallett, Lauren M., 199, 414, 423, 432
Hallwachs, Winnie, 256, 259–63, 266–70, 282–85, 390
Hamilton, 87, 295, 297, 299–301, 306, 320–21
Hamilton, Alexander, 87, 114
Harm's Wood Forest Preserve, 97
Harris, James A., 29, 196, 201–2, 207, 211–12, 234–35, 413–14, 417, 419, 430
Hawaii, 37, 291
Heaney, Seamus, 189, 351, 369
Hendricks, Lindiwe, 59
Heneghan, Liam, 87–89, 101, 129–32, 400
Henry, Richard, 324
herbicides, 65, 76, 106–7, 117, 122, 124, 127, 230, 348, 374, 405
Hickie, David, 192
Higgs, Eric S., 29, 196, 211–12, 414, 417, 425, 431
hihi, 301, 304
Hiwassee State Wildlife Refuge, 16
Hobbs, Richard, 29, 33–34, 196, 198–201, 205, 207, 210–12, 221, 223, 225, 234, 236, 239, 323, 384, 396–98, 413–17, 419–31
holly, 169–70, 183, 185
Holmes, Patricia, 49, 68–70, 76–77, 85
Hopkins, Aareka, 335–37
Hopper, Stephen, 216, 223, 246, 248–53, 416, 421
Horizontes Forest Experimental Station, 270–72
Hosking, Stephen, 72
Hukanai Gully, 299
human population, 54, 138, 227, 325, 385, 387, 390
Hurricane Katrina, 26–27
Hurricane Mitch, 282
Hussein, Saddam, 35
Huxmann, Jeff, 12, 15

Illinois, 9, 16, 91, 93, 99–101, 109, 128–30, 398
India, 22
Indian Ocean tsunami, 26
Indiana, 92
Indiana Dunes, 90, 128
indigenous species, 77, 208, 306, 427
Inkatha Freedom Party, 65
Institute for Nature and Culture, 88, 132
Intergovernmental Panel on Climate Change (IPCC), 198, 382
International Crane Foundation (ICF), 1, 3–4, 6, 8–9, 13, 16, 22, 105
International Peat Society (IPS), 369
International Union for the Conservation of Nature (IUCN), 391
Iowa City, 3
Iran, 22
Ireland, 73, 167–68, 171, 173–76, 180–81, 189–92, 194–95, 311, 335, 352, 354–58, 361–62, 365–68, 374–75, 377–78, 381, 383, 391
Irish Ecological Monitoring Network (IEEM), 168
Irish Peat Conservation Council (IPCC), 367–68, 374
Island Press, 36
islands, 232, 274–75, 287–326, 385, 387, 390, 435
Israel, 35, 406
Italian Institute for Environmental Protection and Research, 133, 161

Jackson, Stephen T., 414, 422
Janzen, Daniel, 256–70, 272–75, 277, 279–86, 322, 335, 390
Japanese knotweed, 176
jarrah forest, 218–24, 232, 428
Jasper Ridge Biological Preserve, 205–7, 417

Jefferson, Thomas, 106
Jensen, Jens, 90–91
Jezile, Nosipho, 82
Jonkershoek Forestry Research Station, 52
Jonson, Justin, 234–37, 251, 416, 428
Jordan, William R. (Bill), 47, 88, 96, 120, 123, 211, 384, 395, 398–406, 408, 412, 423
Journal for Nature Conservation, 137
Joyce, James, 179
Joyi, Bandile, 66

kahikatea, 300
kaka, 302, 304
kakapo, 292, 294, 324
Kalgoorlie, 227, 230, 243
Kansas, 229
Karoo, 46
karri, 226–27, 232, 245
Kasrils, Ronnie, 59, 64
Katz, Eric, 109, 399
Kavanagh, Patrick, 187, 353, 359
Kazakhstan, 22
Keesing, Amanda, 238
Keleman, Judit, 370–72
Kenworthy, Garry, 17
Kerry, Co., 166, 180, 183–84, 190, 194, 437–38
Kew Royal Botanic Gardens, 223, 246, 249, 392–93
Killarney National Park, 180–81, 191–92, 194
Killyconny bog, 369, 373–74, 379
Kilroy, Michael, 352
Kings Park and Botanic Garden, 214, 416, 426–28
kiwi, 290, 292, 301–3, 324
Kleer, Kris de, 430
Kleinbooi, Cedric "Sappie," 62
Klerk, F. W. de, 65
Knysna, 52
Koch, John, 220–23
Kogelberg Reserve, 65
kokako, 290, 301
Kolotiy, Alexander, 4
Korling, Torkel, 93–94
Kruger National Park, 406

La Spezia, 142, 145–46
Lacandon people, 327–32, 334, 337–46, 348–50, 389, 391
Lacanhá Chansayab, 327, 330–31, 334, 339
Lacanhá river, 334
"land ethic," 7
Lake County (Illinois), 90, 100
Lake Michigan, 90, 128
Lambeck, Robert, 225
Languedoc, 407
Leopold, Aldo, 1, 6–8, 16, 168, 215, 218, 326, 386
Lercari, Bartolomeo, 133, 154–56
Lercari, Samueli, 145
Liffey Head Bog, 355
Liguria, 139, 162
Ligurian Riviera, 146
Limbaugh, Rush, 333
limestone, 175, 177–78, 216
limiting factor, 45
Lishman, Bill, 13
Little, Declan, 168, 171, 173–74, 179, 181, 184, 187, 189–92, 194
lodgepole pine, 185, 358, 365, 369, 373
Logan, Betty, 244
Lough Inchaquin, 180, 183, 186
Lough Mask, 175
Louisiana, 26–27
Lower Zambezi Valley, 9, 22
Lowveld Escarpment, 52
Lukas, Abbey-gail, 62, 64
Lullymore Heritage Centre, 383

macchia, 134, 136, 139, 149, 151–52, 154, 161–62
Macdonald, Ian, 71, 74
Madagascar, 37, 287
Madison Arboretum, 1, 8, 24–25, 399–400
Maitre, David Le, 71, 75
Malagasy, 385
male fern, 184

Mana Island, 316–18
Manarola, 142, 145–46, 148, 151, 156–57
Mandela, Nelson, 53–56, 65–66
Mangaiti Gully, 299
Maori, 291, 295, 298, 301–3, 305–11, 313–16, 319, 323, 335–36, 388, 391
Marais, Christo, 55, 57, 63, 65–66, 68–70, 73–75, 77, 80–81
Marcos, Subcomandante, 329
Margaret River, 226–27, 245
Marsh Arabs, 35
Martinez, Denis, 29
Massachusetts, 110–11
Masters, Linda, 98
Mathiessen, Peter, 3
Mayan people, 327, 329–31, 339, 343–44, 349–50
Mayo, Co., 168, 175, 352–53, 355, 360, 362, 373
McHenry County (Illinois), 90
Mciteka, Mncedisi, 65–66
McKibben, Bill, 112
McNally, Gerry, 367
Mead, S.B., 100
Mediterranean (Basin), 133–36, 140, 147, 160–61, 163
mediterranean (climate biome type), 206–7, 232
Mendelson, Jon, 109–10, 112–13, 399
Mérida (Mexico), 212, 414, 426
Merton, Don, 293–94
Mesoamerican Biological Corridor, 343
Mexico, 265, 343
mice, 305–6
Middle Fork Forest Preserve, 100
Midewin National Tallgrass Prairie, 90
Millennium Ecosystem Assessment (MA), 37–41, 44
milpa, 340–42, 345–47, 350, 389, 391
Milton, Sue, 36, 46
Miskelly, Colin, 292–93, 311–16, 318, 320
Mississippi, 13, 20, 28
Mississippi Flyway, 19
Missouri Botanical Garden, 36
Mitsch, Bill, 27–28, 361
moa, 310, 318, 388
Moggia, Daniele, 150, 158
Mongolia, 22
Monterey, 299
Monterosso, 135, 141, 145–46, 148, 150, 156
Montes Azules Natural Park, 328
Monteverde (Costa Rica), 209
Mooney, Harold A., 206, 423
Moorcroft, Greg, 289, 292, 320
Moreira, Francisco, 137
Morley, Asher, 289
mosaic, 98, 140, 155, 160, 252, 362, 410, 434
Mount Katahdin, 386
Mount Maungatautari, 287, 301–2, 304–5, 307, 309–10, 321, 326
Mozambique, 9–10
multiple stable states, 31
Mumbai, 137
Muraviovka Park, 4
Murcia, Carolina, 29
Murphy, Tom, 129
Museum of Natural History—Mozambique, 9
Mussolini, Benito, 144

Nardelli, Riccardo, 133, 161–64
National Audubon Society, 95
National Conference on Ecosystem Restoration, 392
National Geographic, 357
National Heritage Areas (NHAs), 359, 368
National Parks and Wildlife Service (NPWS), 181, 187, 191, 193–94, 355–57, 359, 368–71, 375, 378–80
Native Americans, 109–11, 146, 308, 335, 337, 388
Native Woodlands Scheme (NWS), 168–69, 172–73, 182, 189, 193
Natura 2000 Network, 202
nature, 2, 5, 7, 16, 38, 40, 41, 42, 43, 45, 47, 66, 92–93, 102, 103, 109, 110, 112, 114, 119, 120, 123, 127, 135, 137, 138, 139,

158, 167, 168, 174, 203, 211–12, 218, 231, 236, 260, 263, 283, 300, 313, 325, 330, 338, 386, 388, 391, 395, 397, 398, 400, 402–3, 405–6, 410, 411–12, 419, 435, 437
Nature Conservation Foundation India, 392
Nebraska, 16
Necedah Wildlife Refuge, 13–14, 20–21, 23
Negev Desert, 406
Nelson Mandela Metropolitan University, 72
Nelson, Carol, 53, 72, 123–28, 131, 414
Nevius, Joe, 119
New Academy for Nature and Culture, 399, 404
New England, 92, 144
New Orleans, 26–27
New York, 16, 37, 137, 148
New Zealand, 37, 232, 287–326, 335–36, 391–92
New Zealand pigeon, 336
Newby, Ken, 241
Nijmegen University, 357
No Child Left Inside Campaign, 128
Noongar people, 216, 232–33, 239, 241–42
Norminton, Gregory, 3
North Branch Restoration Project (NBRP), 87–88, 90, 93–98, 101, 103–9, 112–14, 116–28, 156, 299, 405
North Carolina, 110
North Park, Chicago, 123
Northbrook Village (Cook County, Illinois), 97
Noubous, Tersia, 62
Novel Ecosystems: Intervening in the New Ecological World Order, 431–32
Nowanup Farm, 241, 245
Noyce, Philip, 243
Nueva Palestina, 328–31, 334, 339, 349

O'Connell, Catherine, 374
O'Toole, Mike, 192
oak, 94, 98–101, 104, 109, 113–14, 118, 124, 133–34, 136, 139–40, 161, 166, 168–74, 177, 180–82, 184–85, 187–88, 194, 203, 205, 249, 437
Obama, Barack, 128
OCBIL (old, climatically buffered, infertile landscape), 246, 248, 250–51
Ohio State University, 27
Operation Migration, 12–14, 16, 19–20
Operation Vuselela (Renewal), 64–65
"Orange Pulp Restoration Project," 279–84
Otamatuna, 290
Outeniqua forest, 52
overgrazing, 60–62, 76, 296, 328, 335, 358, 367
Owenee River, 170
Oweninny, 353, 362, 378–79

Packard, Steve, 87, 93–101, 103–10, 112–14, 116, 118, 120–22, 124, 127, 399
Pakeha, 316
Pakistan, 22
Palenque, 330, 343
Palestine, 35
Pallinup River, 228
Palo Alto, 196
Palos Fen (Illinois), 126
Palos Heights (Illinois), 90
Pan Africanist Congress (PAC), 65–66
Panama, 343
Panama Canal, 265
Paniagua, Enrique, 334, 345
Pantepui Highlands, 248
Parcela del Príncipe, 277
Parker, V. Thomas, 410
Parque Nacional Santa Rosa, 257
Parks Canada, 391
Pausas, Juli G., 133
Pavlik, Bruce, 393–94
peat bogs, 295, 298, 354, 377
Peniup, 234–37, 251–52, 416, 419, 428
Pennsylvania, 110
People's Millennium Forest, 192, 194
peregrine falcon, 163, 173

Pereira, João S., 133
Perkins, Dwight, 90
Perrone, Matteo, 133, 144, 147, 149, 157–58, 164–65
Perth (Western Australia), 214–18, 220–21, 224, 227
Peterken, George, 168
Pew Charitable Trust, 234
Phytophthora cinnamomi (dieback), 219, 232, 250, 253
Piazza Marconi, 146, 154
Pickett, S. T. A., 410
Pitt Island, 313
Pocosol, 278–79
polar bear, 208
Pollardstown Fen, 355
Port Elizabeth, 63, 72
Portofino, 146
possum, 287–90, 301–6, 325, 405
poverty, 1, 10, 18, 37, 48, 56, 63, 69–70, 78–80, 156, 183, 308, 352, 396
Powell, Mike, 60–64, 70, 72, 82
Prach, Karel, 430
prairie, 1–7, 9, 11, 15, 23–25, 90–91, 93–95, 97–101, 103–7, 110–11, 113–14, 116, 123–24, 126–29, 351–52, 354, 385, 388, 400, 404–5
prehensile-tailed porcupine, 263
Preston, Guy, 49–52, 55–57, 64, 68, 70, 72, 74–75, 77, 79–81, 83–85
Prince Bernhard Nature Fund, 357
protea, 54–55
Provence, 73
Pukaha Mount Bruce Wildlife Centre, 292
Purdue University, 207
Putauhinu Island, 314–15, 326
Pyrenees, 26

Quealy, Sean, 166, 176–79
Queensland, 142
Quiros, Julio Díaz, 269–70

Rackham, Oliver, 168, 174, 191
rain forest, 52, 54, 137, 203, 247, 257, 264–65, 267, 272–76, 279–81, 288, 327–31, 333, 335, 337, 339, 342–43, 345, 347, 349, 374
rainbow lorikeet, 215, 217
raised bog, 351–52, 354–59, 367–70, 372–76, 378–81
Rapallo, 146
rats, 46, 287, 290, 292–93, 301, 305–6, 310, 312, 314–15, 318, 324–25, 336
rata, 301–2, 304–5, 326
Raven, Peter, 387
Ravensthorpe, 230, 242
Ray, Heather, 14–15, 17–19
Rekacewicz, Philippe, 41
Renamo, 10
Renou-Wilson, Florence, 380
resilience, 109, 135, 210, 264, 345, 425
restiad bog, 298
Restoration & Management Notes, 109
restoration ecology, 1, 21, 42, 197, 212, 216, 225, 252, 319, 393–94, 397, 407–8, 410, 415–17, 422–23, 428
Restoration Ecology, 47, 219, 385, 408
Restoration of Natural Capital (RNC), 8, 21, 36–37, 39, 42, 46–48, 50, 69–70, 80, 198, 267, 388, 396–97, 400, 403–4, 406, 413, 423, 432, 436
Restoration of Natural Capital: Science, Business and Practice, 413
Rhode Island, 257
rhododendron, 170, 190–92, 366, 437
Rieger, John, 29
Rincón de la Vieja, 272–73
Rincón-Cacao, 276
Riomaggiore, 136, 141, 145–46, 148, 151–52, 156–57, 159
Robinson, Heath, 352
rock wren, 318–19
Rodriguez, Karen, 120
Ross, Laurel, 102, 108, 112, 115–21, 127–29
Russia, 4, 22
Ryan, Gerry, 383
Ryan, Jim, 351, 369, 373–74, 376

saddleback, 301, 324
salination, 230
San Bernardino, 138–39, 142

San Cristóbal de las Casas, 272, 329, 338
San Cristóbal Station, 272
San Francisco, 207
San Francisco South Bay Salt Pond Restoration Project, 392
San José (California), 198, 205, 208, 212, 395
San José (Costa Rica), 276
Sand County Almanac, 6–8
Sand Ridge Nature Preserve, 104
Sandberg, Carl, 89
Santa Elena (Costa Rica), 267, 278, 285
Santa Rosa (Costa Rica), 257, 261–62, 264–68, 271–72, 277
Sauganash, 114, 117
Saunders, Alan J., 319
savanna, 23, 93–94, 98–102, 104–6, 111, 113–14, 116, 121, 123–24, 126, 133, 205, 343, 365–66, 388, 434
Schouten, Matthijs, 357, 380–81
Schreiner, Barbara, 74
Sciaccetrà, 150
Science, 260, 384, 393, 422
Scientist, The, 83
Scots pine, 172, 177, 180, 182–83, 188
Seeley, A. J., 299
Sentiero Azzurro, 143, 146, 148, 158
SER International Primer on Ecological Restoration (SER primer), 29–33, 198, 207, 391–92, 412, 424
Shark Bay, 231–32
Shell (oil company), 241
Shelley, Percy Bysshe, 136
Sitka spruce, 172, 358
Smale, Simon, 241
Smith, Jim, 374
Snares Island, 312, 314–15
snipe, 293, 311–18, 324, 326
Society for Ecological Restoration (SER), 21, 26, 29, 32–35, 88, 120, 198, 204–8, 212, 220–21, 247, 297, 311, 335, 391–92, 395, 400, 409, 412, 414, 419–20, 424–25, 429–30
socio-ecosystems, 47
soil, 4, 25, 30, 35, 38, 39, 53, 55, 61, 67, 72, 77–78, 88, 129–30, 139, 143, 144, 151, 160, 164, 172, 173, 174, 175, 177, 180–81, 183, 184, 185, 187, 188, 189, 192, 202, 208, 216, 218, 220, 222, 228, 234, 235–36, 247, 250, 251, 267, 278, 281, 282, 298, 328, 339, 340, 341, 344, 345, 347, 348, 356, 358, 372
Somme Prairie, 111, 129
Somme Prairie Grove, 95, 97–98, 101, 106–7, 110, 116, 127
Somme Woods, 97–98
Sonjica, Buyelwa, 59
South Africa, 36–37, 46–54, 56–59, 63–67, 70–73, 78–85, 197, 200, 208, 226, 232, 248, 320
Southwest Australian Floristic Region, 248
spekboom, 60, 62, 76
Spencer, Rongo, 314–15
Sphagnum moss, 190, 298, 354–55, 359, 361–64, 371, 373–74, 379–80
St. Louis (Missouri), 21, 45–46, 48, 86, 400, 406, 413
Standish, Rachel, 416
Stanford University, 196, 205–6, 417, 423
Stanford University Linear Accelerator Center (SLAC), 205
Stapleton, Ray, 383
Stevens, William K., 104, 118
Stevenson, Michael J., 21, 47
Stewart Island (New Zealand), 294, 312, 314–15, 324
Stirling Range (Western Australia), 227, 230, 232–34, 237–39, 241, 243, 245, 250, 253–54
stitchbird, 301, 304
Stone, Richard, 384
Subtropical Thicket Restoration Project, 61, 64, 76
Sugden, Andrew, 384
Sukhdev, Pavan, 44
Sunflower Forest, 88, 400–406
sustainability (sustainable development), 29–30, 37, 82, 86, 230, 329, 337, 384, 388, 393, 411
Swan River (Western Australia), 215–17
Sweet Freedom Farm, 15–16, 23

Table Mountain, 65
Tacher, Samuel Levy, 327–34, 337–39, 342–49
Tairi, Ally, 305, 307, 309
takahe, 292, 294, 302–3, 318
tall goldenrod, 106–7
tapu (taboo), 336, 388, 391
Tauroa, Lance, 309
Te Urewera, 288–92, 295, 320, 326
Tennessee, 16
Texas, 20
The Nature Conservancy (TNC), 102, 112, 234, 234
Thoreau, Henry David, 111, 386, 402
Tilaran Mountains, 209
tipping points, 31, 39, 41, 45, 386
Tiritiri Matangi Island, 37
Torre Guadiola, 159
Townsend, Patricia, 196, 209
trapdoor spider, 214, 216
Trees for Life, 124
Trinity College, Dublin, 56
tropical dry forest, 257, 261, 265, 267–68, 283
trout, 301, 358
Tui 2000, 297
Tulija, 343
Turkmenistan, 22
Tuz Chi, Lázaro Hilario, 327
Tzeltal people, 328–31, 339, 342–43, 348, 350

Ugalde, Álvaro, 267
Umkhonto we Sizwe (Spear of the Nation), 65
UNESCO, 54, 60, 148, 153–54, 232, 328
United Nations (UN), 37, 148, 392
University College Dublin, 378
University of Pennsylvania, 262
University of Waikato, 287, 297
University of Western Australia, 225, 416
University of Wisconsin–Madison, 7, 24
Uragh Woods, 180–82, 184, 186
Urban Wildlife Coalition, 87, 114
Uzbekistan, 22

Van Andel, Jelte, 385
Van Wilgen, Brian, 55, 68, 71–72
Venezuela, 248
Vermont, 209
Vernazza, 133, 135–36, 138, 141–42, 145–46, 150–51, 153–54, 156, 161, 165
Vestal Grove, 98
Vietnam, 93
Viney, Michael, 356
Volastra, 145–46, 158
Volunteer Stewardship Network, 102

Waikato, 295, 298
Waikato District Council, 292
Waikato River, 295, 299
Waitangi, Treaty of, 307–8
Waiwhakareke National Heritage Centre, 297–98
Walden Pond, 386
Wallace, David, 287, 301–2, 305, 307
Walpole-Nornalup National Park, 227
Wargowsky, Larry, 23
Warren River (Western Australia), 227–28
Warren, Norman, 280
Washington Post, 25
Waukegan Road, 98
Wege Foundation, 273
weka, 312–13
Wellington (New Zealand), 311, 316
Western Cape (South Africa), 52, 58, 69
Western Mining Corporation (Western Australia), 216
Westphal, Lynne, 129
Westport (County Mayo), 168
weta, 289, 301
wetlands, 3, 9, 14, 22, 27, 35, 38–39, 45, 61, 81, 84, 89, 137, 201, 203, 212, 214–15, 217, 249, 260, 300, 361, 381–83, 385, 388
Whitman, Walt, 384, 403
whooping crane, 3, 13, 15–16, 18, 22–23, 12–14, 21

Whooping Crane Eastern Partnership, 13, 15–16, 22–23
Wicht, C. L., 52
Wicklow, Co., 167, 194, 355, 437
wild boar, 134
willow, 63, 170–71, 178, 180, 185, 188, 296, 298–99, 358
Wilson, David, 378
Winterhalder, Keith, 29
Wisconsin, 1, 3–4, 7–9, 14, 18, 24, 128
Wise, David, 129
Wood, Grant, 398
woodland puccoon, 97
woodlands, 52, 92, 144, 162, 167, 168, 170, 173, 174, 189, 191, 193, 194, 195, 225, 243, 245, 374, 391, 392, 438
Woodlands of Ireland, 173, 192
Working for Water (WfW), 37, 48–86, 88, 156, 320, 391
Working for Wetlands, 61, 81, 84
Working for Woodlands, 61, 84
Working on Fire, 61, 81, 84
World Environmental Center, 220
"world of wounds," 168, 215, 218, 326
World Policy Journal, 83

Xolocotzi, Efraín Hernández, 338

Yale School of Forestry and Environmental Studies, 21
yew, 172, 175, 177–80, 192–93
YODFEL (young, often-disturbed, fertile landscape), 246, 248, 250
Yosemite National Park, 22, 137, 406
Yucatan peninsula, 343

Zambezi, 1, 9
Zapatistas (Ejército de Liberación Nacional Zapatista; EZLN), 329, 331
Zaragoza, 21, 23, 25–27, 33, 35–37, 39, 43, 45–47, 198, 210
Zedler, Joy, 25–26, 33, 198, 399